District/Central Solar Hot Water Systems Design Guide

ASHRAE
1791 Tullie Circle, NE
Atlanta, GA 30329
www.ashrae.org

Reviewed and approved by ASHRAE Technical Committee (TC) 6.7, Solar Energy Utilization

ISBN 978-1-936504-52-7

Library of Congress Control Number: 2013939920

Contents

List of Figures and Tables

Figure — Page

Figure **Page**

Figure **Page**

Figure		Page

1. Introduction

Solar thermal systems are commonly used for the domestic water heating, space heating, (industrial) process heating, and even for cooling of goods and buildings. The Energy Independence and Security Act (EISA) 2007 SEC. 523 requires that "if lifecycle cost-effective, as compared to other reasonably available technologies, not less than 30% of the hot water demand for each new Federal building or Federal building undergoing a major renovation be met through the installation and use of solar hot water heaters." In the United States, different types of solar water heating systems are available and primarily used for standalone buildings. Different design guidelines are available from the National Renewable Energy Laboratory (NREL) and American Society of Heating, Refrigerating, and Air-Conditioning Engineers (ASHRAE) for small size systems. These systems are relatively complex and application of solar based heating in the United States is often limited by economical considerations when compared to traditional heating systems and local cost of fossil fuels.

In recent years, numerous innovations in solar thermal technologies have resulted in cost effective large scale systems including integrated solar supported heating networks. Examples of such systems installed in Denmark, Germany, Austria and other countries have proven that such systems are reliable and may be more economical compared to small scale systems. Such systems may be cost effective for clusters of buildings containing e.g., Army barracks, dining facilities, gyms, child development centers and swimming pools. Similar opportunities exist on large hospital campuses, family housing complexes, etc.

The District/Central Solar Water Heating Systems – Design Guide is the first attempt to develop recommendations on optimal and reliable configurations of solar water heating systems in different climates along with design specifications, planning principles, and guidelines for such systems that serve building clusters with significant domestic hot water (DHW) needs (e.g., barracks, dining facilities, Child Development Center [CDC], Gyms) that operate in combination with district/central heating systems. Note that, throughout the industry, the terms "district heating system" and "central heating system" are commonly used interchangeably. For the purposes of this document, the term "Central Solar Hot Water System" is used to denote systems that serve clusters of buildings from a large centralized solar thermal field(s) (in contrast with small size solar thermal systems that serve standalone buildings). The guidelines are complemented by numerous case studies of successfully implemented solar supported thermal networks along with results of exemplary simulations of different system options based on real world scenarios. This document also discusses the benefits and disadvantages of large scale centralized versus decentralized solar thermal systems.

The Guide was developed by a group of government, institutional, and private-sector parties funded by the US Army Installations Management Command (IMCOM), US Army Corps of Engineers (USACE) and the US Department of Energy Federal Energy Management Program (DOE FEMP), and reviewed and approved by American Society of Heating, Refrigerating, and Air-Conditioning Engineers (ASHRAE) Technical Committee (TC) 6.7. The work on the Guide was managed and executed by the Energy Branch of Engineer Research and Development Center Construction Engineering Research Laboratory (ERDC-CERL). The project manager and principle investigator was Dr Alexander Zhivov. Major contributors to the Guide were: Alfred Woody (Ventilation and Energy Applications, USA), Andy Walker (USDOE National Renewable Energy Laboratory), Reiner Croy (ZfS – Rationelle Energietechnik GmbH, Germany), Dr. Stephan Richter (GEF Ingenieur AG, Germany), Dr. Rolf Meißner and Detlev Seidler (Ritter XL Solar GmbH, Germany), Wolfgang Striewe and Stefan Fortuin (Dept. Thermal Systems and Buildings, Fraunhofer Institute for Solar Energy Systems ISE, Germany), Dieter Neth (Senergy Consulting, Germany), Franz Mauthner and Dr. Werner Weiss (AEE – Institute for Sustainable Technologies, Austria), Harald Blazek (S.O.L.I.D. Gesellschaft für Solarinstallation und Design m.b.H, Austria), and Anders Otte Jørgensen and Rene Rubak (ARCON Solar, Denmark).

2. Solar Energy

2.1 Solar radiation intensity

The efficiency and performance of solar water heating systems depend on a site's solar energy resource. Solar resource is measured by the solar radiation intensity of an area, but cloud cover and latitude must also be factored into the system purchase decision. The amount of solar energy available for heating water varies by geographical location. Figures 2.1 and 2.2 show that tilted flat plate collectors located in the United States at an angle equal to the latitude receive the most sunlight in the desert Southwest and the least sunlight in the Pacific Northwest and states in the Northeast. Table 2.1 lists the maximum daily solar radiation and average daily solar radiation for selected US locations. Detailed monthly solar radiation information for a multitude of locations in the United States is available from the National Renewable Energy Laboratory at: www.nrel.gov/gis/solar_map_development.html

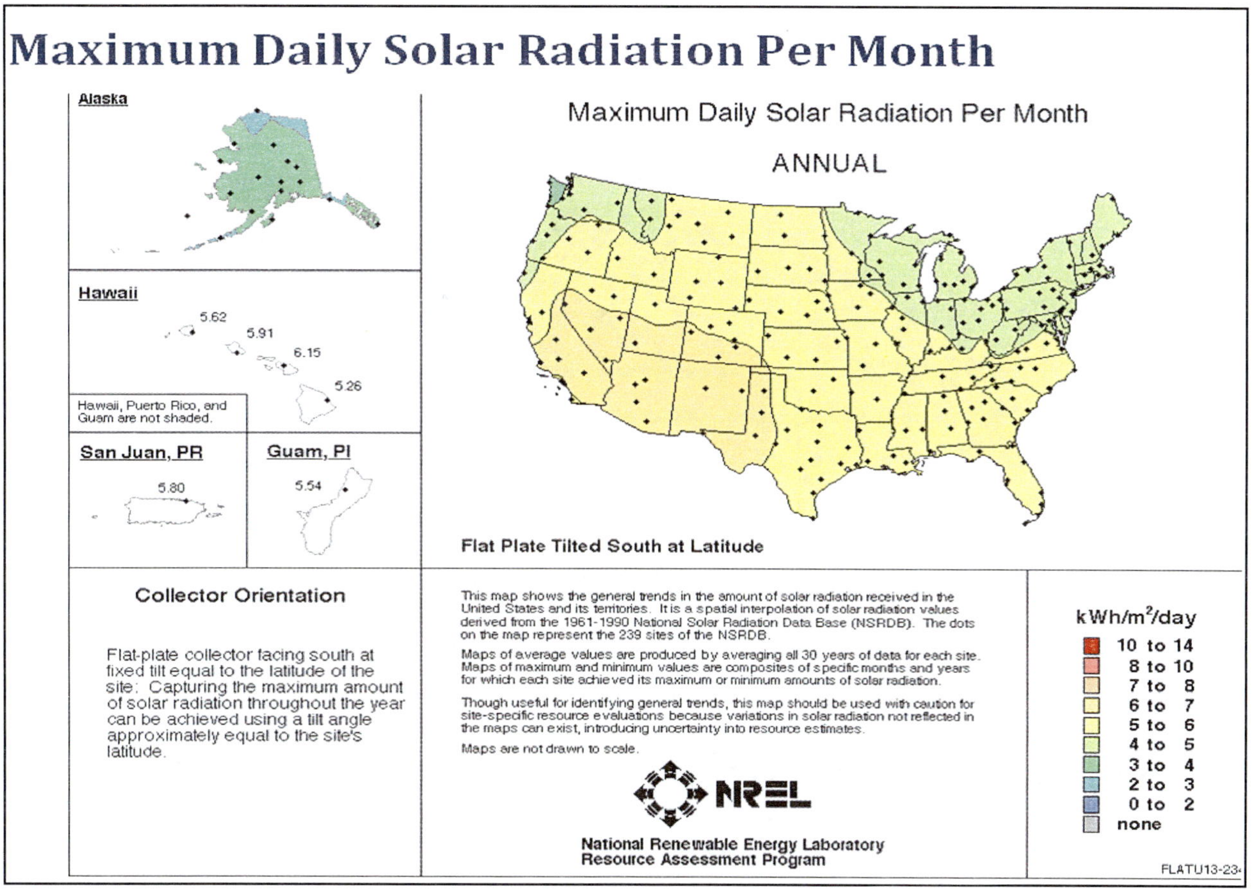

Figure 2.1. Map showing the maximum daily solar resource available for tilted flat plate solar collectors in the United States.

Average Daily Solar Radiation Per Month

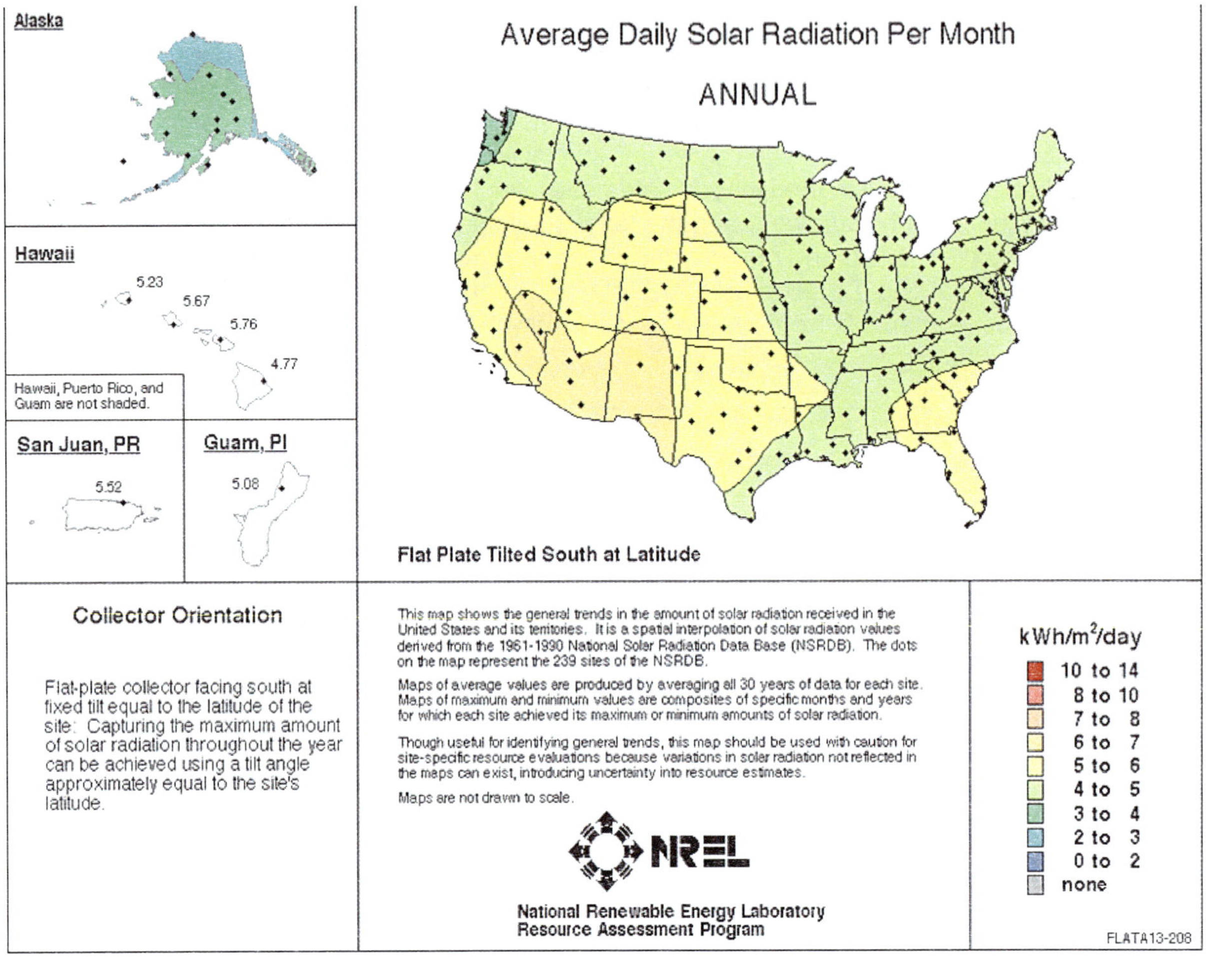

Figure 2.2. Map showing the average daily solar resource for tilted flat plate collectors in the United States.

Table 2.1. Daily solar radiation values for specific locations, I [kWh/m²/day].

Location	I Max	I Ave
Anchorage, AK	4.6	3.0
Austin, TX	6.3	5.3
Boston, MA	5.6	4.6
Chicago, IL	5.7	4.4
Denver, CO	6.1	5.5
Fargo, ND	6.5	4.6
Honolulu, HI	6.5	5.5
Jacksonville, FL	6.1	4.9
Knoxville, TN	5.6	4.7
Sacramento, CA	7.2	5.5
San Diego, CA	6.5	5.7
Seattle, WA	5.7	3.7

The energy delivery of solar water heating systems depends on the environmental conditions of sunlight and temperature. The output of a solar energy system is proportional to the intensity of the sunlight, but its efficiency also depends on temperature. Efficiency decreases as the solar collector gets hotter and more heat is lost to the ambient environment.

2.2 Cloud cover

The power of solar radiation entering the atmosphere, or the "solar constant," is 1367 W/m^2 (50,169 Btu/hr/sq ft). Within the atmosphere, this power is reduced by absorption, scattering, and reflection effects to about 1000 W/m^2 (36,700 Btu/hr/sq ft) on the earth's surface if there is a clear sky (Figure 2.3). The solar radiation that reaches the earth's surface is further reduced by clouds, which reflect part of the radiation back into space, and absorb another part. In addition, a part of the radiation is dispersed into diffuse radiation by multiple reflections. Diffuse irradiation on the earth's surface consists of the irradiation coming from angles different than the solar incidence angle (i.e., the actual sun position). Therefore, cloud cover reduces the total amount of irradiation (global radiation), and (due to scattering), it also changes the relation between beam and diffuse radiation. As cloud cover changes with the seasons, these effects are also seasonal dependent. The diffuse fraction of the total annual global radiation can be higher than 50%, depending on the location.

Even on a clear day a significant amount (>10%) of solar radiation is scattered due to the effects of molecules in the atmosphere. This causes the sky at daytime to appear blue and colors the sun red at sunset, when the distant light waves that enter the atmosphere are longer.

2.3 Site latitude

The latitude of the site will affect the solar radiation collected, so it is important to tilt panels based on the latitude of the installation site. The United States lies closer to the middle latitudes, which means it receives more solar energy in the summer when the sun's path is most perpendicular to locations below. The sun slants far south during the shorter days of the winter months when the sun follows a southern path in the sky. For this reason, solar collectors should face true south in the northern hemisphere, i.e.:

Collector tilt angle = latitude to maximize annual gain

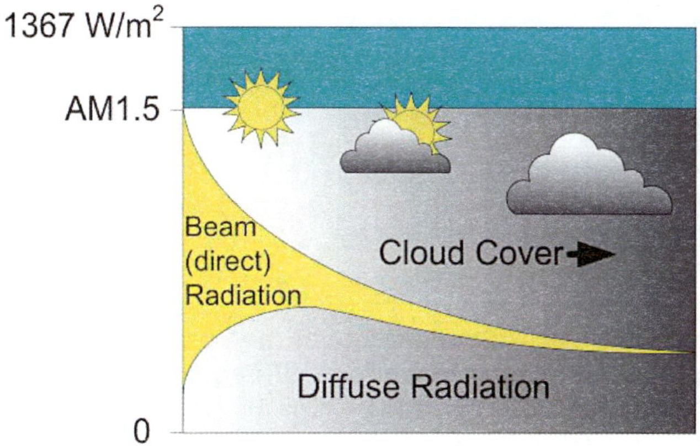

Figure 2.3. Left: Solar Constant and proportions of beam and diffuse solar radiation in relation to cloud cover; Right: Photo of the effects of scattering at sunset taken at 1640 ft (500 m) height.

2.3.1 Season of year

The height of the sun's travel on the horizon varies by season of the year. In the summer, it is more directly overhead; in the winter, it is much lower. Figure 2.4 shows the different paths* of the sun over the sky for Washington, DC. The maximum incidence angle during (solar) noon for the location in this example is 75 degrees on the 21st of June (the summer solstice). The minimum incidence angle occurs on the 21st of December (the winter solstice). And the average incidence angle occurs on the 21st of March and the 21st of September (the autumnal and vernal equinoxes). For locations further north, these curves will be flatter, and for locations further south, steeper.

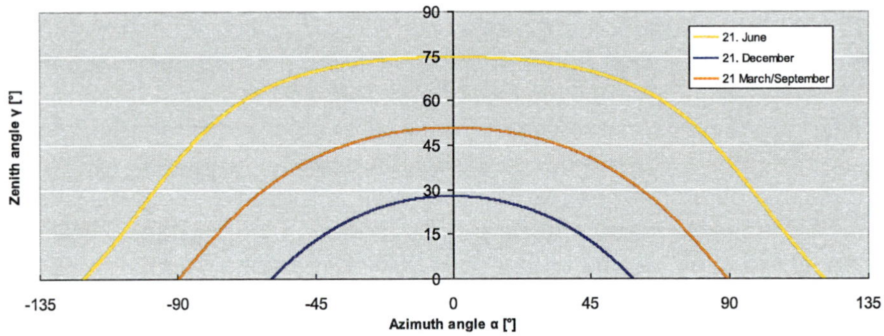

Figure 2.4. Path of the sun for latitude = 38° 53" (Washington, DC) relative to a horizontal plane.

Thus, the amount of solar energy that can be collected varies throughout the year. Monthly available solar radiation values are used to determine the size of the solar collector system required to satisfy the heating demand. Figure 2.5 shows the variation of beam and diffuse radiation over the year for three locations. (The values apply to a collector surface tilted 40 degrees south facing.)

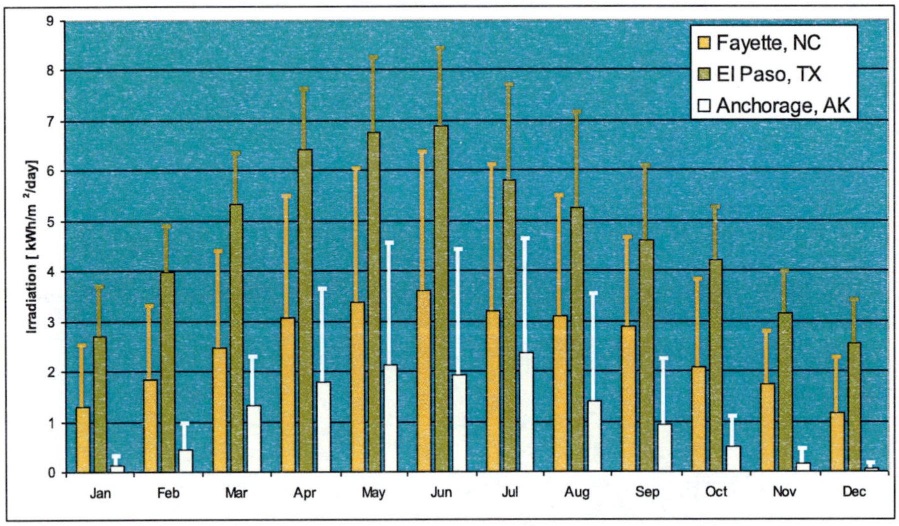

Figure 2.5. Variation of daily month average beam and diffuse radiation over the year for three locations. The beam (bar) and diffuse (stick) daily radiation are shown for a 40-degree tilted, south facing collector surface.

* The relative movement of the Sun around a position on Earth consists of two movements. The daily movement consists of a circular movement around a tilted axis. The annual movement consists of a sinusoidal variation of the tilted axis over the year. In addition, the latitude of the location of the solar thermal system adds an offset to the tilted axis, due to the curvature of the Earth's surface. The latitude ('offset') is equal to the average tilt angle of the sun above the horizon.

The sum of beam and diffuse radiation falling on a horizontal surface is usually called "global solar radiation." The annual total in the figure above, 40-degree tilted south facing, for Fayette NC is ~507,504 Btu/y*sq ft (~1600 kWh/y*m^2); and for El Paso, TX, 697,818 Btu/y*sq ft (2200 kWh/y*m^2); and for Anchorage, AK, 272,783 Btu/y*sq ft (860 kWh/y*m^2). This results in an annual daily average of 1,396, 1,935, and 761 Btu/sq ft*day (4.4, 6.1, and 2.4 kWh/m$^{2)}$*day respectively. The fraction of diffuse radiation is 44%in Fayette, 21% in El Paso, and 53% in Anchorage. Day and seasonal variation can be very high; a nice summer day may provide as much as 2,030.02 Btu/sq ft*day (6.4 kWh/m^2)*day in Fayette and 2664 and 1459 Btu/sq ft*day (8.4 and 4.6 kWh/m^2 *day) in El Paso and Anchorage and a cloudy winter day may have as little as 63.44 Btu/sq ft*day (0.2 kWh/m^2*day) on average in the month of December in Anchorage (Figure 2.6).

2.3.2 Solar collector tilt

Solar systems should be designed to match the heating demands with the solar energy intensity that varies throughout the year. In the northern hemisphere, a solar thermal system will receive less solar radiation in the winter than in summer. To improve the seasonal solar energy collection, the solar collector can be tilted so that it would be more perpendicular to the sun's path when the heating demand is greatest. Tilting of collectors (from horizontal) should be done to maximize the radiation collection skewed for usage. Thus a system that is providing heat for the winter should have a tilt angle equal to the site's latitude, plus up to 15 degrees; a system for year round heating should have a tilt angle equal to the site's latitude. Table 2.2 lists collector tilt angles at selected locations for maximizing winter solar energy collection.

Maximum radiation can be collected when the collector is faced due South in the Northern Hemisphere. If the system must be East or West facing, using a Westerly azimuth is more optimal than an Easterly one since the collector can take advantage of the warmer part of the day.

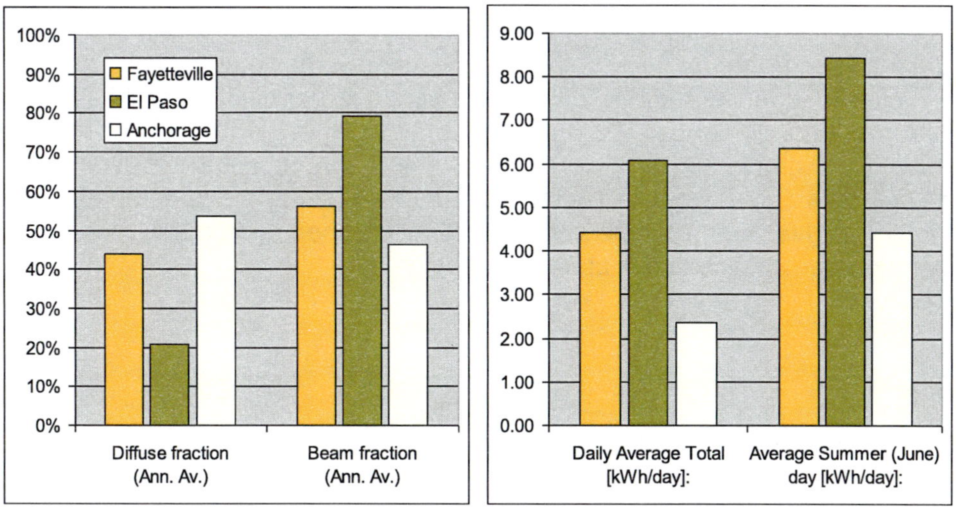

Figure 2.6. Annual average beam and diffuse fractions (left) and daily June average total radiation for three locations (right).

Table 2.2. Tilt angles corrected for maximum winter season energy collection.

Latitude	Angle at +15 degrees
25 degrees (Key West, Florida)	40 degrees
30 degrees (Houston, Texas)	45 degrees
35 degrees (Albuquerque, New Mexico)	50 degrees
40 degrees (Denver, Colorado)	55 degrees
45 degrees (Minneapolis, Minnesota)	60 degrees

2.4 Orientation to path of sun

Two angles describe the orientation of the collector:

- The azimuth* angle α, also called "compass orientation": The angle in a horizontal surface between the collector and the due south direction. Due south, towards the equator, is by definition an orientation of 0°.
- The tilt angle β ("sky ward orientation"): The angle between the collector and the horizontal surface.†

Costs for supporting structures can be saved by mounting collectors flush with an existing, suitable surface or structure. Tools (brackets etc.) are usually available from the supplier of the collector, which will suit various common surface types like tiled or sheet metal roofs, brick or wood walls etc. The fittings should provide sufficient strength to endure extreme weather conditions like wind and snow loads. It is important that the mounting and structure comply with local standards and regulations.

If mounted in line with an existing structure (e.g., a wall surface) the orientation may be less than optimal, in many cases. The effects of these compromises are often less than expected. This is shown in Figure 2.7. A deviation from the maximum yield tilt and orientation can be compensated for by a larger collector area, if space and costs permit.

A rule of thumb for optimum exposure to solar radiation is that a collector should face the equator (orientation of 0 degrees, due south) with a tilt angle of 0.7 times the Latitude of the location (but at least 10 degrees, or the minimum working angle of the collector). This applies to heating domestic water systems.

The optimum collector tilt angle for the usage of solar energy during winter months is higher because the average sunlight incidence angle is lower. Choosing a larger tilt angle provides a yield that is geared towards winter heat demands. The optimum collector tilt angle for the winter months is higher than the latitude, for example 50 degrees for a site at 45 degrees latitude. Deviations up to 60 degrees from these optimum angles generally will lead to a loss of solar radiation of less than 15% compared to optimum angles. Note that the annual yield also depends on many other factors such as the heat demand characteristics or the storage capacity.

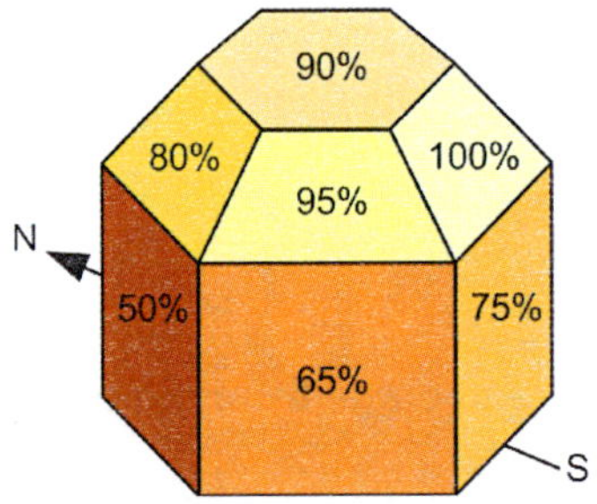

Figure 2.7. Influence of tilt and orientation on the percent of total solar radiation received annually. In this example the maximum annual radiation on a 45-degree tilted surface facing south at Latitude = 50 degrees is indicated by 100%.

---

* Azimuth: The angular deviation of the collector surface with respect to the direction of due south. Deviation to the west is positive and deviation to the right is negative.

† "Tilt" = angle between horizontal and tilted plane. "Zenith angle" which is the angle between the vertical and the tilted plane, thus "Tilt" = 90°-"'Zenith angle'")

Choosing a smaller tilt angle gears the yield towards the summer. Usually a summer bias increases the potential overall annual yield of a collector, but only if this extra yield can be used. Solar thermal systems are thus often sized by matching the collector yield during the summer period with the hot water heating need (called economical collector sizing), any increase in collector area will increase the fraction covered in winter, but will also increase the stagnation periods and duration during summer. Refer to Section 3.4.2 (p 35) for an explanation of stagnation.

2.5 Shading

Shading causes less radiation to reach the collector surface. The positioning decision of the collector may be influenced by anything that can cause shade on the collector surface such as mountain ranges on the horizon, nearby or tall buildings, nearby trees (in particular when carrying leaves during winter), and nearby roof construction. Local fog conditions can also cause a loss in sunlight. Shading should be avoided during the peak sun hours of the day, 9 a.m.–3 p.m.

The loss of incoming radiation due to shading must be taken into account during simulations for calculating the prospective yield and usually involves mapping the obstacles on the horizon and sky in the face of the collector orientation (Figure 2.8).

Causes of shade such as fallen leaves and snow depend on location and climatic conditions and should also be considered.

2.6 Collector placement within a building cluster

The solar collectors can be placed on buildings or on the ground. In some cases they could be placed in an elevated position over parking lots providing shade below as an additional benefit. Placement on a building is normally on the roof where a sloped or flat roof exists. Placement on a sloped roof normally creates a collector tilt similar to that of the roof. Also, the orientation may not be directly to the south. In these cases, heating energy obtained from the collector must be derated from that of an optimum placement. Such a placement normally gives a better appearance than that of tilted collectors on a flat roof. Building integrated collectors are assimilated into the original construction of the roof or can be placed slightly above the pitched roof. In either case, the collectors will use the structure framework of the roof as the main support.

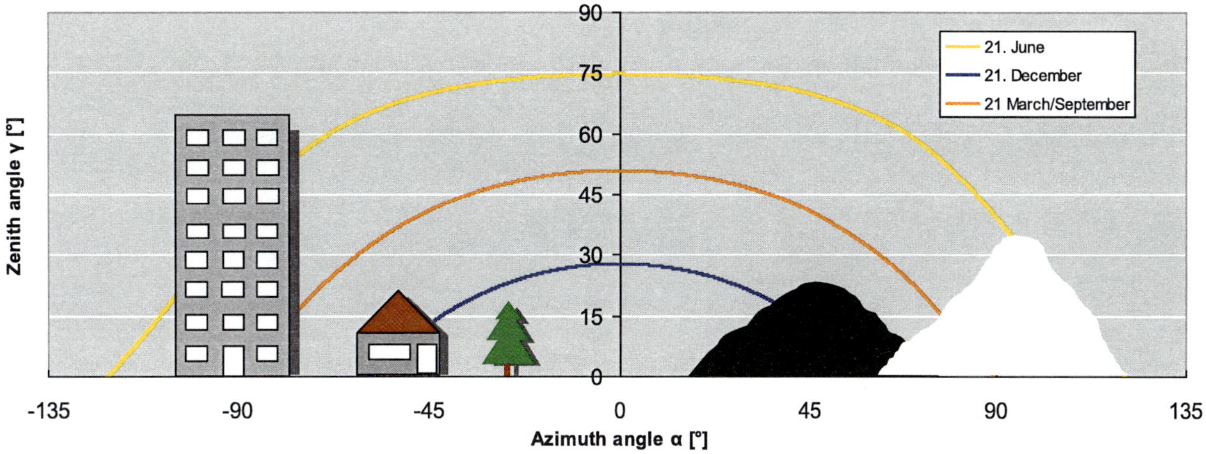

Figure 2.8. Influence of structures on shading of incoming solar radiation. Height and distance both need to be taken into account.

The most important considerations for collector placement is to encourage integration in existing infrastructure, and to avoid shading at periods when solar radiation is plentiful and heating is needed. For large central systems the option of creating a large collector field is usually chosen. This system can be integrated with carport roofing or placed on the flat roof of a large building.

Figures 2.9 and 2.10 show examples of installed collector fields. Integration into new building structures may be more aesthetic and may provide for savings in roof cover material otherwise used. Specially constructed flat plate collectors can provide a closed, insulated surface, which may serve as a roof cover (Figure 2.11).

Figure 2.9. Placement of flat plate air collectors on a flat roof.

Figure 2.10. Flat plate collectors on mounting construction.

Figure 2.11. Two examples of aesthetic placements of collectors. At the right the collectors are integrated into the roof cover together with PV collectors at either side.

Collectors placed on the ground or on flat roofs need a supplemental support to generate the collector tilt to the sun. The rows of collectors must be separated by a short distance so that one row does not shade the row behind it. The distance to assure no shading on the winter solstice can be calculated by:

Separation distance = collector height X tangent of the angle (90 - latitude- 23.5)

Since the energy collected in the winter is a small percentage of the total, then the spacing between rows can be slightly reduced with only minor loss in performance. Figure 2.12 shows how the winter midday sun-angle is (usually) used to determine the angle.

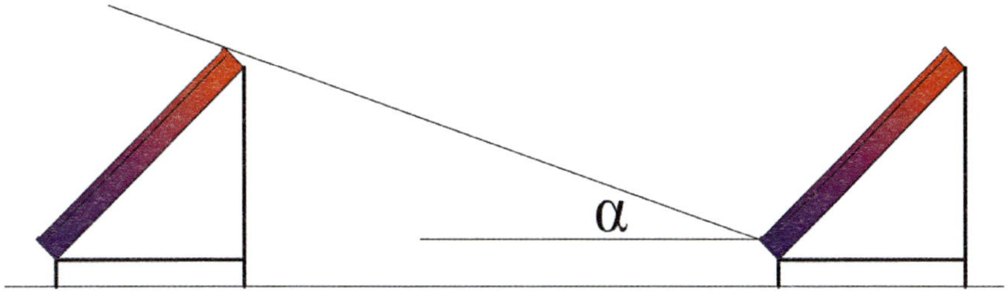

Figure 2.12. Flat plate collectors with mounting construction.

3. Solar Hot Water Thermal System

The main components added to a conventional heating system when solar thermal energy is used are:

- collector field with collector field piping and support structure
- heat transfer fluid (water or water glycol mixture)
- a storage tank system
- pump for solar loop and other pumps for other loops
- heat exchangers to transfer heat from one loop to another
- expansion and safety devices for each closed loop
- a controller with temperature sensors in collector field and storage tank and that turns the pump on and off.

Since a solar thermal system does not usually act as the main heat source, an auxiliary (back-up) heater is necessary to cover periods of high energy demand or too little solar radiation (usually in winter). Figure 3.1 shows a schematic of a solar domestic hot water system.

3.1 Collector performance indices

Various performance characteristics are used to assess and compare solar thermal systems. The most important ones are the "solar fraction" (***SF***), the "specific solar energy yield" (***SE***), and the "solar system efficiency" (***SN***). The following sections discuss these indices.

To ensure that the reference quantities and energy flows can be clearly assigned, the essential parameters are entered in accordance with their definition and their "position within the system" (Figure 3.2). When selecting the proper components of a large solar hot water thermal system it is best to use analytical modeling tools such as f-Chart or TRNSYS. These computer programs can estimate annual energy collected, storage and heat exchanger effects, system heat losses, etc. Refer to Section 3.3.2 (p 25) for information regarding these modeling tools.

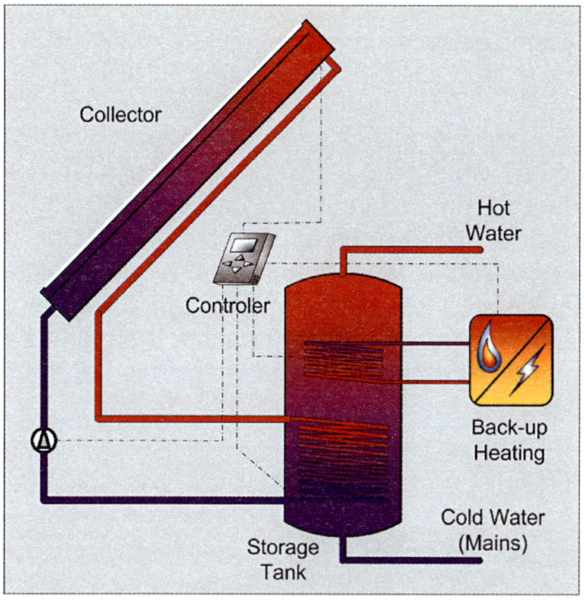

Figure 3.1. A schematic example of a solar domestic hot water system, showing the characteristic components.

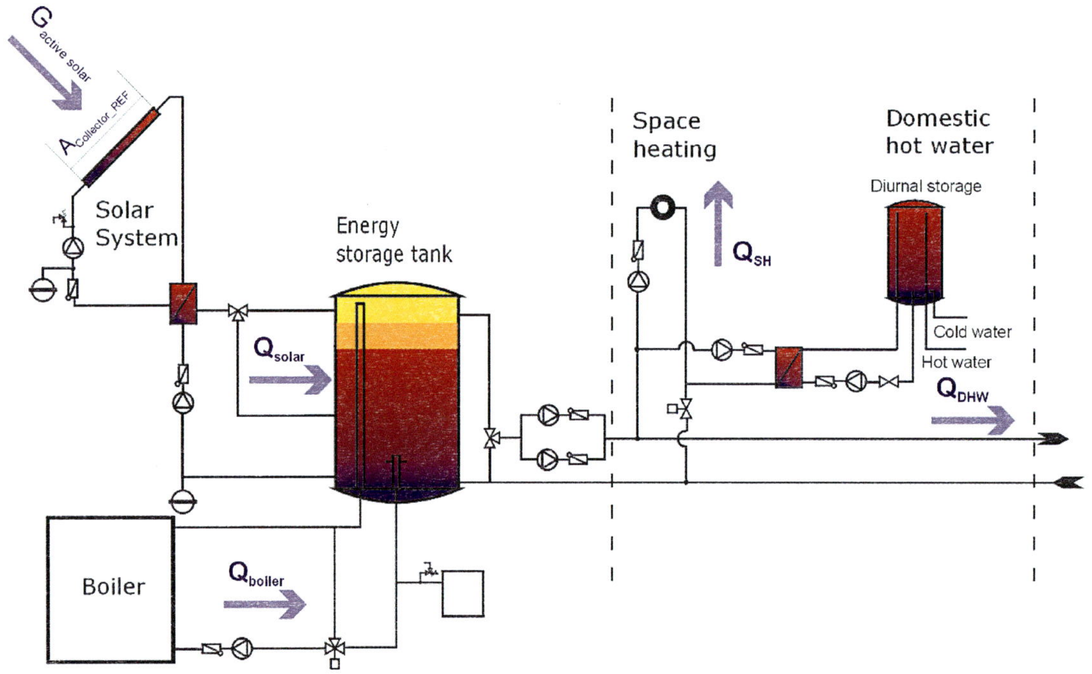

Figure 3.2. Allocation of the heat flow and reference values in the solar thermal system.

3.1.1 Solar fraction (f_{solar} or SF) and F_{save}

A commonly used solar system performance measure is the solar fraction, which identifies the fractional amount of the building heating energy needed is supplied by the solar thermal system. This Design Guide defines the solar fraction for solar supported heating networks as:

$$SF = \frac{Q_{Solar}}{Q_{Bioler} + Q_{Solar}} \quad (3.1)$$

where:

SF	solar fraction	[%]
Q_{Solar}	annual energy produced by collector loop (measured on secondary side)	[kWh/a]
Q_{Boiler}	annual heat input of the auxiliary heating system (boiler)	[kWh/a].

Thus, a building that requires 10.24 MBtu (3000 kWh) per year to generate hot water and obtains 6.83 MBtu/yr (2000 kWh/yr) from its solar system has a solar fraction of 67%. A very similar indicator is the "f-save" ratio. This indicator shows the thermal heating energy saved by the supply of solar energy as compared to the heating energy that would have been used by a non-solar thermal (reference) system for the same purpose. For example, when a non-solar heating system (the reference system) requires 10.24 MBtu/yr (3000 kWh/yr) to provide for a certain hot water use, and the solar thermal system requires 6.8 MBtu/yr (2000 kWh/yr) auxiliary heating, then the f-save of the solar thermal system is 33% (i.e., the fraction of auxiliary energy saved compared to the reference system). The formula for f-save (thermal) is:

$$f_{sav,thermal} = 1 - \frac{Q_{backup}}{Q_{boiler,ref}} \quad [-]$$

$f_{save,thermal}$ = *Auxiliary energy saved by the solar (sub-)system [kWh]*

Q_{backup} = *Energy used by the backup heater [kWh]*

$Q_{boiler,ref}$ = *Energy would have been used by a reference system without the solar thermal installation [kWh]*

For both indicators it must be specified whether direct energy (heating energy demand), primary energy, or secondary energy is used in their calculations. The major difference is what is included in the building heating energy use (the reference system) and the impact of thermal system losses and supporting equipment energy use such as by pumps, etc.

Higher solar fractions mean higher energy and CO_2-savings relative to the conventional energy source. However, one must also consider that the higher losses of the solar thermal system negatively affect the solar system efficiency. An economic optimum has to be found.

3.1.2 Solar system efficiency (SN)

In addition to the solar fraction, economic investigations of solar thermal systems are often affected by another parameter, the solar system efficiency (***SN***). The solar system efficiency describes the ratio between the annual amount of energy supplied to the heat storage unit and the global irradiation that strikes the collector surface:

$$SN = \frac{Q_{Solar}}{G_{active\_solar}} \tag{3.2}$$

where:

SN	= solar system efficiency	[%]
Q_{Solar}	= annual energy produced by collector loop (measured on secondary side)	[kWh/a]
G_{active_solar}	= annual global irradiation onto active solar collector area	[kWh/a].

3.1.3 Specific solar energy yield (SE)

The specific solar energy yield describes the annual amount of energy supplied to the heat storage unit from 11 sq ft (1 m^2) of collector surface area. Compared to other calculation results, the kind of surface (absorber, aperture or gross collector area) must always be indicated for the specific yield result.

$$SE = \frac{Q_{Solar}}{A_{collector\_REF}} \tag{3.3}$$

where:

SE	= specific annual solar energy yield	[kWh/m^2]
Q_{Solar}	= annual energy produced by collector loop (measured on secondary side)	[kWh/a]
$A_{collector\_REF}$	= collector area on which the solar yield refers to (gross, aperture or absorber area)	[m^2].

The specific solar energy yield is often said to be the crucial parameter for measuring the capacity of a solar energy system. For a correct interpretation of this parameter, the size of the system, the solar fraction and the system losses (storage and heat distribution losses) must be considered. Solar collector ratings provided by Solar Rating and Certification Corporation and the Florida Solar Energy Center use the gross area of the collector (not the net area) in their performance efficiency rating information. ASHRAE Standard 93-2010 (for testing to determine thermal performance of solar collectors) also uses the gross area of the collector.

3.1.4 Collector testing and certification organizations

Several organizations provide certification of solar collector and solar collector system performance in the United States, the most recognized of which are:

- Solar Rating and Certification Corporation (SRCC). This is the principle rating organization for solar domestic hot water collectors. The SRCC offers OG-100 certification for solar collectors and OG-300 for entire systems. OG-300 certification is for smaller residential systems. Website: http://www.solar-rating.org/ratings/ratings.htm
- Florida Solar Energy Center (FSEC). Mandated by the state of Florida to perform testing of solar energy products, the FSEC is a good source for performance characteristics on both solar thermal collectors and entire solar thermal systems. Although specifically geared towards the solar industry in Florida, the FSEC is a good resource for the Southeast region as a whole. FSEC Standard 101-09, which was revised May 2009, is the institute's solar collector certification. It supersedes FSEC Standard 102-05. Website: http://www.fsec.ucf.edu/en/industry/testing/index.htm
- North American Board of Certified Energy Practitioners (NABCEP). The NABCEP provides certification programs for solar electric and thermal system installers. A NABCEP-certified installer provides for an extra level of assurance as to the qualifications of the installer. However, because the solar thermal certification is a relatively recent development, NABCEP-certified installers are not especially common. Certified installers can be found on the website: http://www.nabcep.org/installer-locator-agreement

3.2 Types of hot water solar systems

3.2.1 Passive systems

"Passive" systems solar hot water systems do not have a pump or other moving parts. These heating systems rely on temperature changes in the water located in the solar collectors on the roof to move the water through the system. They are typically less expensive than systems having a pump (active systems) because they have no mechanical parts, but they are usually not as efficient. However, passive systems can be more reliable and may last longer. There are two basic types of passive systems: batch and thermosiphon.

Batch or Integrated collector-storage (ICS) systems. These systems work best in areas where temperatures rarely fall below freezing. They also work well in buildings with significant daytime and evening hot-water needs. Batch collectors (Figure 3.3) or ICS, use one or more black tanks or tubes in an insulated, glazed box. Cold water first passes through the solar collector and is preheated. The water then continues on to the conventional backup water heater, providing a reliable source of hot water. This type of collector should be installed only in mild-freeze climates because the outdoor pipes can freeze in severely cold weather.

Thermosiphon systems. Thermosiphon systems (Figure 3.4) move water through the system due to density differences (warm water rises as cooler water sinks). Neither pumps or electricity are used. However, the collector must be installed below the storage tank so that warm water can rise into the tank. These systems are reliable, but contractors must pay careful attention to the roof design because of the heavy storage tank. Although they are usually more expensive than ICS systems, they can be used in areas with less sunshine.

Passive solar water heating systems are used on individual buildings or for a single heating demand. They are not for central heating systems that service several buildings. They are also inefficient in cooler climates. Since the purpose of this design guide is to focus on systems that can serve multiple buildings, these systems will not be further discussed.

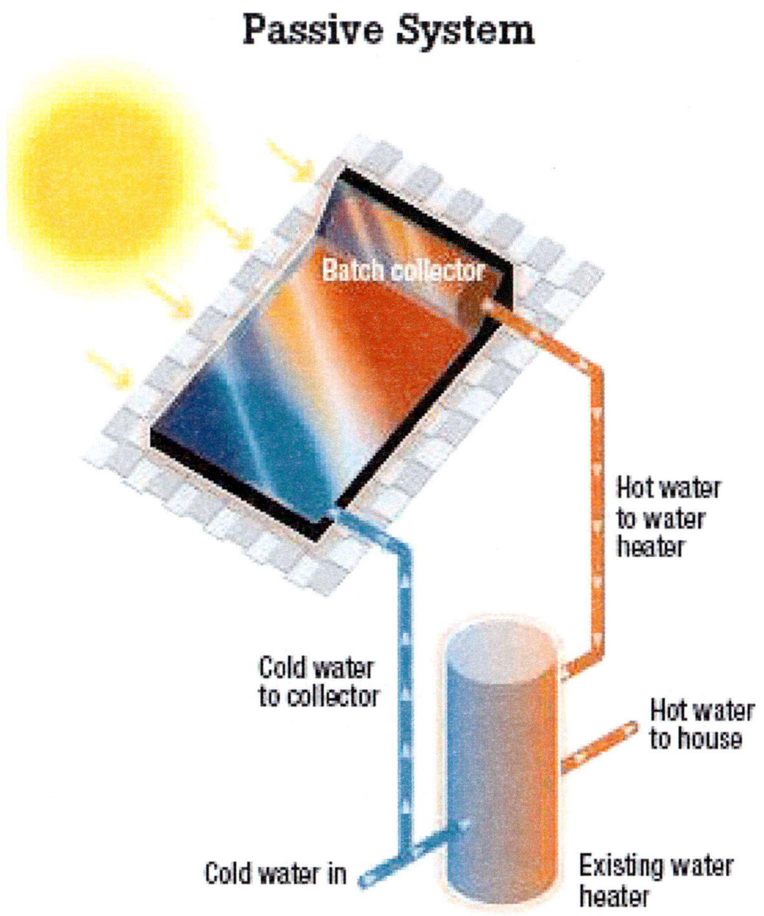

Figure 3.3. A simple passive solar water heating system with a batch collector.

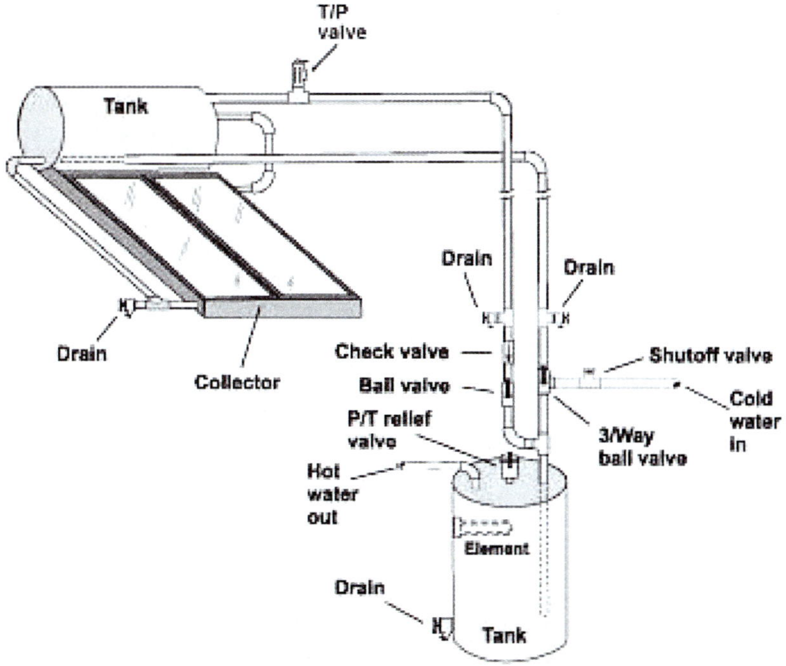

Figure 3.4. Schematic of a typical thermosiphon system.

3.2.2 Active systems

Solar water heating systems that rely on electric pumps to circulate fluid through the collector are called "active systems." Active systems are generally categorized into two types: direct and indirect, which simply means that water in the storage tank is either directly filled using the hot water flowing from the solar collectors (one loop) or indirectly using two water circulating loops separated by a heat exchanger. The latter type is normally used in locations where outdoor winter temperatures below freezing may occur. These systems use an anti-freeze solution such as a water glycol mixture as a heat transfer medium that circulates through the collectors to avoid freezing.

Direct circulating active system (no anti-freeze). These systems use pumps to transfer the sun's energy directly to potable water by circulating this water through the collector tubing and storage tank; no anti-freeze solution or heat exchanger is used. The pumps circulate water through the collectors, into the building, and back again. They work well in climates where it rarely freezes.

A direct active system (Figure 3.5) has one or more solar energy collectors installed and a nearby storage tank. The system uses a differential controller that senses temperature differences between water leaving the solar collector and the coldest water in the storage tank. When the water in the collector is about 15 to 20 °F (-9 to -7 °C) warmer than the water in the tank, the pump is turned on by the controller. When the temperature difference drops to about 3 to 5 °F (-16 to -15 °C), the pump is turned off, so the stored water always gains heat from the collector when the pump operates. A flush-type freeze protection valve installed near the collector provides freeze protection. Whenever temperatures approach freezing, the valve opens to let warm water flow through the collector. The collector should also allow for manual draining by closing the isolation valves (located at a height above the storage tank) and opening the drain valves. Automatic recirculation is another means of freeze protection. When the water in the collector reaches a temperature near freezing, the controller turns the pump on for a few minutes to warm the collector with water from the tank.

Another type of direct active solar water heating system is called "a drainback-system" (Figure 3.6), which is also designed for cold climates. This type of system typically uses regular water as a heat transfer fluid, and is designed to allow all of the water in the solar collector to "drain back" to a holding tank in a heated portion of a building. When no sunlight is available for heating, the solar pump turns off and the water flows into the drainback tank by means of gravity. Since these systems use water, they can be designed with or without a heat exchanger.

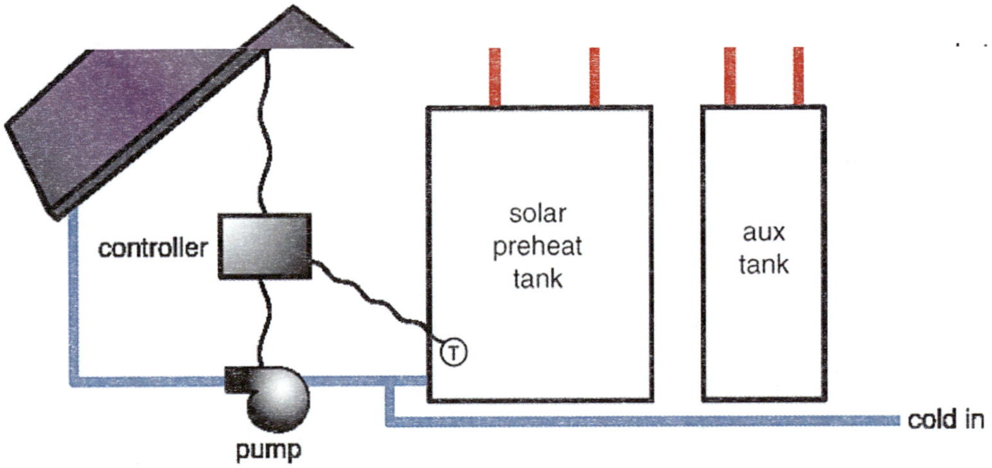

Figure 3.5. An active, direct solar water heating system. These systems offer no freeze protection, have minimal hard water tolerance, and have high maintenance requirements.

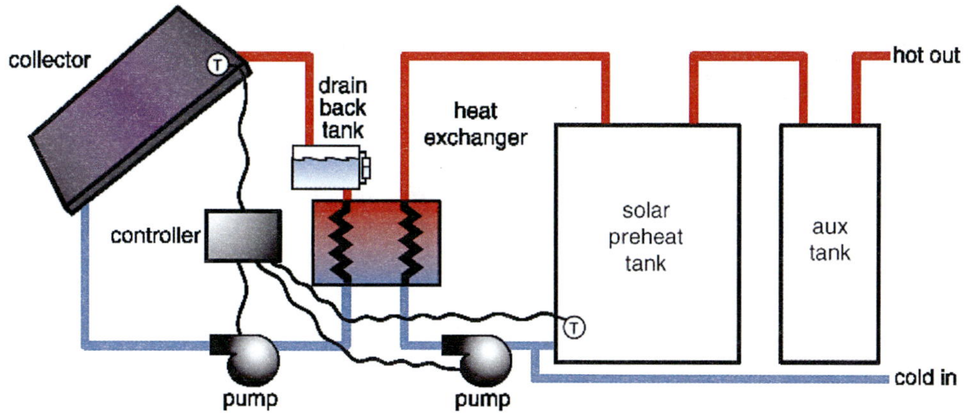

Figure 3.6. An active, drainback solar water heating system. These systems offer good freeze and overheat protection, tolerate hard water well, and have high maintenance requirements.

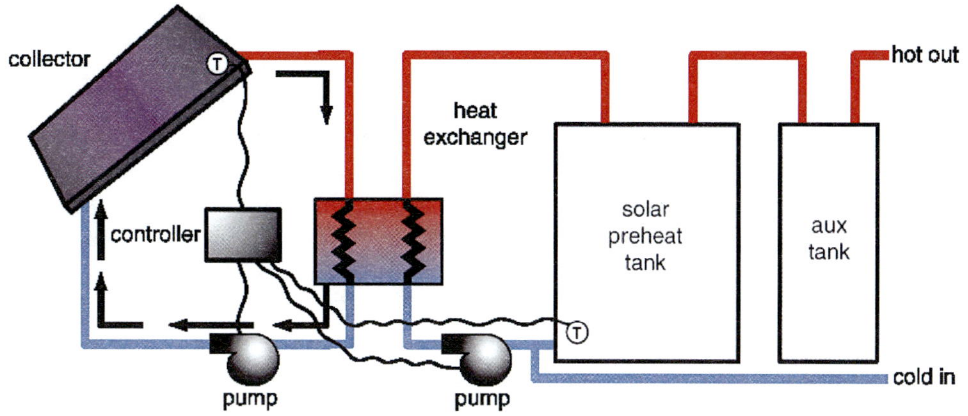

Figure 3.7. Schematic of an indirect active system that uses a heat exchanger to transfer heat from the collector to the water in the storage tank. These systems offer excellent freeze protection, tolerate hard water well, and have high maintenance requirements.

Indirect circulating active system (anti-freeze used). This system operates similar to the direct active system except that there are two circulating heat transfer fluid loops. In the first (solar primary) loop non-freezing, heat-transfer fluid such as a water-glycol mixture circulates through the collector field. A heat exchanger transfers the heat from the water-glycol mixture into the potable water. The heat exchanger can either be directly integrated into a storage (internal heat exchange) or the storage can be connected to a second loop (solar secondary loop) via an external plate heat exchanger (Figure 3.7). These systems are popular in climates that are prone to freezing temperatures.

If the solar collector is extremely well insulated and is not prone to freezing like an evacuated tube collector, water can be the heat transfer fluid. By using water the heated water from the collector can be sent directly to the storage tank; no heat exchanger is needed. Figure 3.8 shows the portion of this system before the storage tank (and, if used, before the heat exchanger).

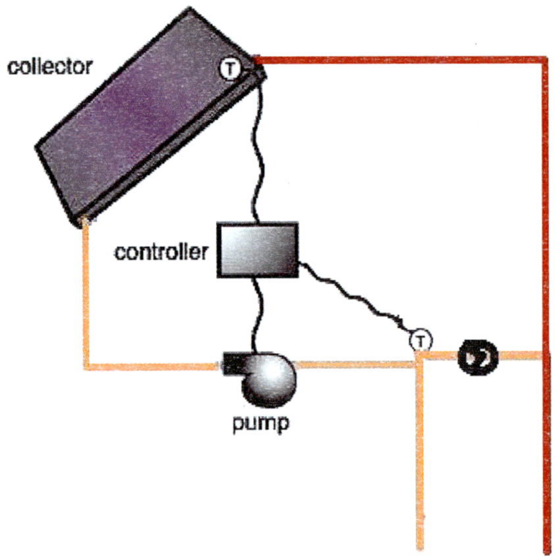

Figure 3.8. Schematic of a recirculating loop system. These systems require well insulated collectors such as evacuated tube to provide protection for freezing and overheating.

Indirect circulating solar water heating systems used on individual buildings or groups of buildings are consistent with the US Army's needs. They can easily be applied to central heating systems for domestic hot water and building heating. Since the purpose of this design guide is to focus on systems that can serve multiple buildings, these systems will be emphasized in further sections of this document.

3.3 Solar thermal energy collectors

The collector is a key part in a solar thermal heating installation. The following sections discuss the working principle of collectors, and the most common collector types. Most relevant for the application in domestic hot water and space heating systems are glazed flat plate and evacuated tube collectors. A collector should be selected based on the quantity and quality (temperature) of the demanded heat.

In principle, heat gain and loss mechanisms in different collector types are the same. In general, one differentiates between optical losses (reflection, absorption) and thermal heat losses due to heat transfer mechanisms (conduction, convection and radiation). Figure 3.9 shows the mechanisms for a flat plate collector (FPC).

3.3.1 General construction

Figure 3.9 shows how heat from the sun is collected by the absorber and is carried away by the fluid flowing through the tubes attached to the absorber. Since these collectors are located outside and normally in a cooler environment, the heated absorber can lose heat its surroundings. To reduce this heat loss, a cover is placed over the absorber and the sides and back of the absorber are insulated. The cover must allow solar radiation to penetrate and, since glass is typically used, some of this radiation will be reflected and radiated out to the atmosphere. The following sections will provide a brief overview and details of the solar collector parts.

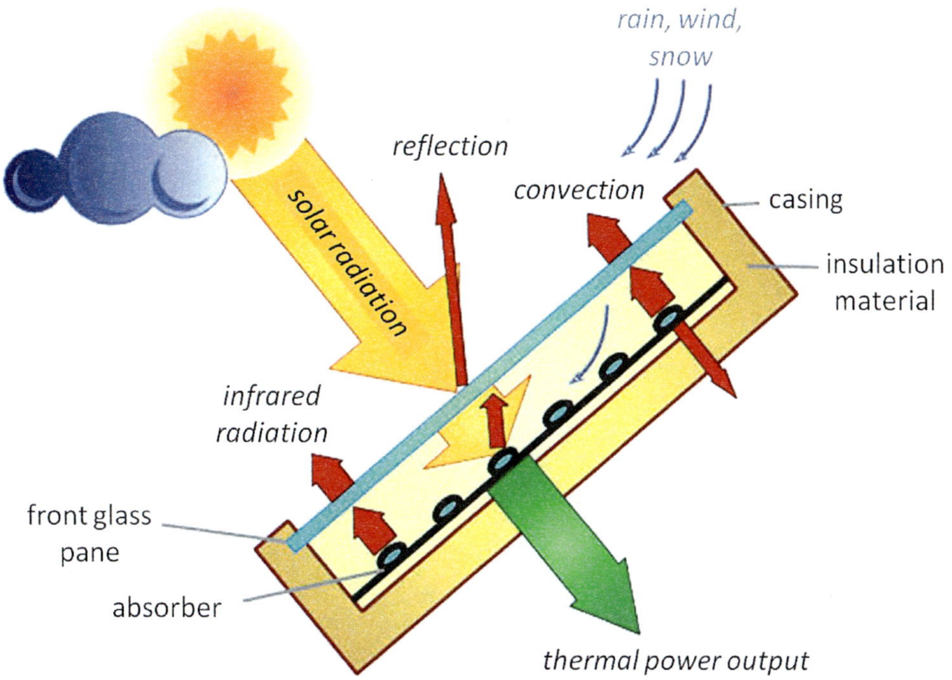

Figure 3.9. A Schematic of a flat plate collector (FPC) showing the heat gain and loss mechanisms that Play a role in determining the thermal efficiency of a collector (Regenerative Energiesysteme).

3.3.1.1 Absorber

The function of the absorber is to effectively convert solar radiation into heat. The absorber surface is often coated to maximize this energy collection. The absorber coating is thus designed with a high absorption coefficient, α, for the sun's radiation spectrum (typically $\alpha = 0.92$ to 0.96). Absorptivity is the fraction of incident sunlight captured (not reflected) by the absorber. The reflectance is the complement of the absorption and is given by: $\rho = 1 - \alpha$. For best performance, the absorber should have a low emission coefficient ε (typically, $\varepsilon = 0.05$ to 0.1) for infrared radiation to keep the losses from long wave radiation emission low as the collector heats up.

Emissivity is the ratio of radiant heat loss off the absorber relative to that of a perfectly black surface ("blackbody"). Most common materials, such as black paint, have an absorptivity equal to the emissivity, and the second law of thermodynamics requires that all materials have a=e at a given wavelength of incident light. However, special surface treatments (semiconductor coatings, blackened nickel layer) have an absorptivity in the short-wavelength solar spectrum that is much higher than emissivity in the long-wavelength infrared radiant heat loss spectrum. Such surfaces are called "selective surfaces" and improve the performance of solar collectors, especially when operating at elevated temperatures where radiant heat loss is more important.

Absorber coatings that possess high absorptivity and low reflectance are called "selective absorbers." Figure 3.10 shows this selective effect where the absorption/reflection characteristics of a selective surface are identified at various wavelengths of the solar spectrum. These wavelength values are taken with an atmospheric thickness of 1.5 of the thickness taken directly above (AM = 1.5). This is important since the atmosphere affects the spectral nature (wavelength distribution) of the solar radiation, and properties reported at AM=1 refer to one thickness of atmosphere. If the sun is not directly overhead, the sun's rays will have to go through more than one thickness of atmosphere. For example, AM = 1.5 corresponds to a zenith angle of around 48.2 degrees, and AM=0 refers to the wavelength outside of the earth's atmosphere.

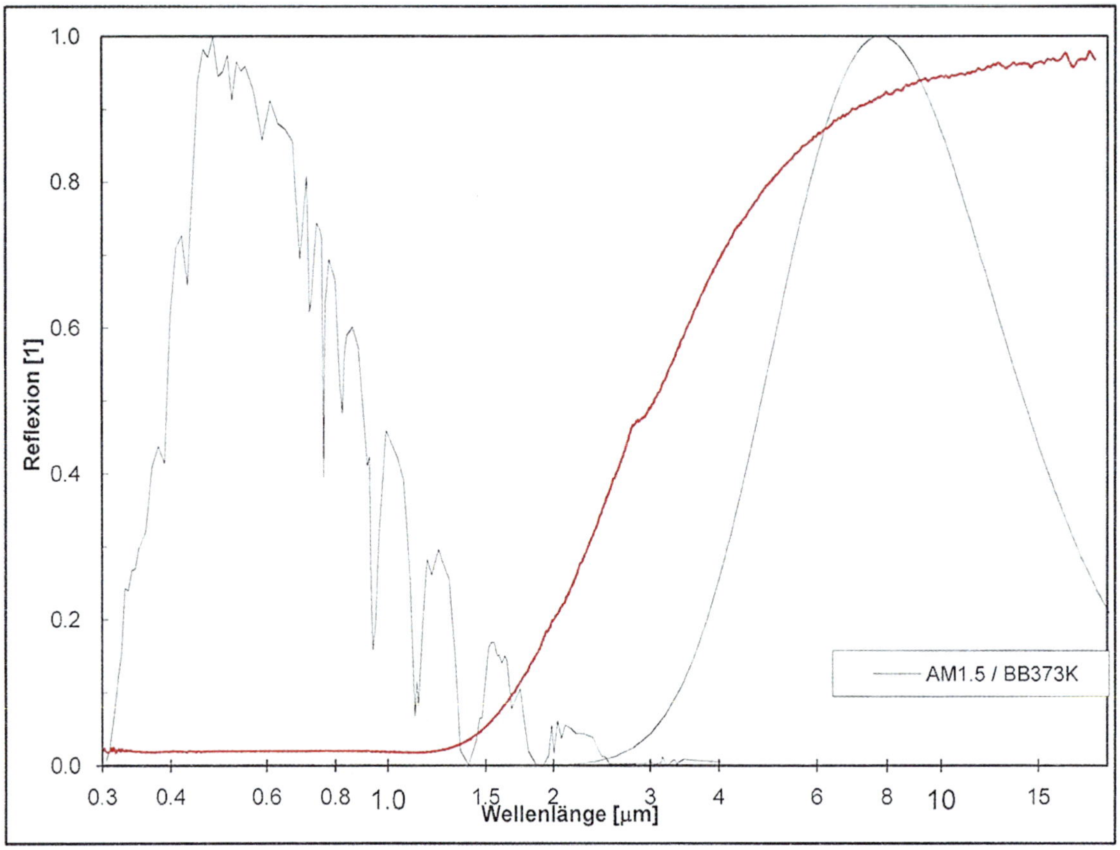

Figure 3.10. Spectral distribution over the wave length (Wellenlänge) of the solar radiation (AM 1.5) and of the thermal infrared radiation from an absorber at 212 °F (100 °C) (graph on the right). The spectral reflectivity (Reflection) of a selective absorber is indicated by the red line.

For spectral irradiance originating from the sun the solar constant (1.367 W/m^2) is defined as AM = 0. AM = 1 is defined as the spectral irradiance on a horizontal plane (zenith angle = 0°). AM = 1.5 is equal to a zenith angle of around 48.2 degrees and the global radiation accounts for 36,700 Btu/hr/sq ft (1000 W/m^2). (Tables of these standard spectra are given in ASTM G 173-03. The extraterrestrial spectral irradiance (i.e., that for AM0) is given in ASTM E 490-00a).

3.3.1.2 Transparent cover

The purpose of the transparent cover is to reduce the convection losses from the absorber, while allowing the maximum amount of radiation to reach the absorber. The cover must also provide the mechanical strength to protect the absorber from the environment.

Special solar glass with low iron content is used. It is occasionally called "water white glass" and its typical transmittance is τ = 0.89 to 0.91 for the wavelength range of the solar radiation. This can be enhanced to τ = 0.94 to 0.96 when anti reflective coatings are applied. This glass should be tempered to reduce breakage by impact.

3.3.1.3 Housing

The housing of a collector must provide the necessary mechanical strength to protect the absorber and the insulation to minimize heat loss to the environment. It must withstand wind and snow loads that occur in the area where the collector is installed. It also must be tight enough against rain penetration. These features need to be ensured over the entire lifetime of the system (20 to 25 yrs). Housings are typically made from aluminum sheet stock or extruded sections, galvanized and painted steel, molded or extruded plastic parts, or composite wood products.

3.3.1.4 Insulation

Insulation is added behind the absorber plate and on the sides of the collector to reduce thermal heat losses. The insulation must use a minimum of binders because it is intended for high temperatures (up to about 400 °F [204 °C] for flat plate collector stagnation); otherwise, the binders will outgas and form a film on the underside of the collector glazing blocking solar radiation. Common insulating materials include, for example, mineral fiber, ceramic fiber, glass fiberglass, and plastic foams. Sometimes polyurethane foam is used, though its resistance to temperature and moisture is limited so it should not be allowed to contact the absorber plate inside the collector. The insulation provides low heat conductivity, some mechanical strength, and temperature and fire resistance.

3.3.2 Types of solar thermal energy collectors

Figure 3.11 shows the four different types of solar hot water collectors. The type of collector chosen for a certain application depends mainly on the required operating temperature and the given ambient temperature range. Due to the design and simplicity of design each type has a maximum temperature that they are best suited to provide:

- Unglazed EPDM* collector - below 90 °F (32 °C)
- Flat plate - below 160 °F (71 °C)
- Evacuated tube - up to 350 °F (177 °C)
- Parabolic trough - up to 570 °F (299 °C).

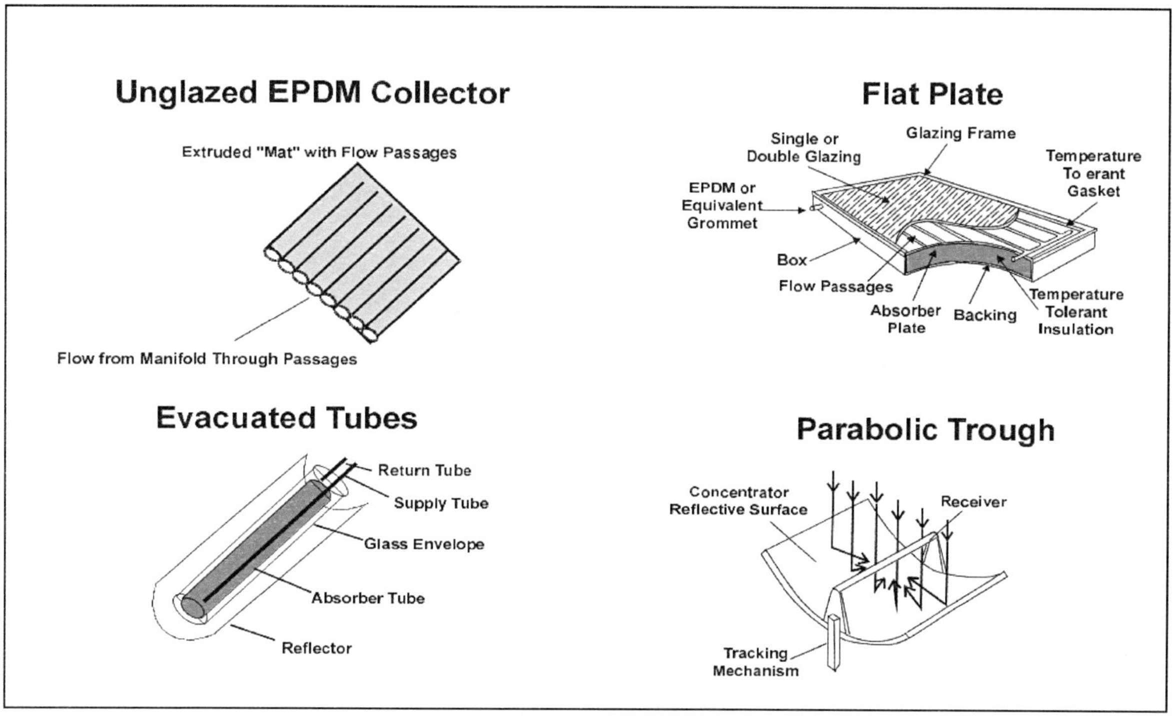

Figure 3.11. Types of solar thermal energy collectors.

* ethylene propylene diene M-class [rubber]

In the Army, the major hot water requirements are heating for domestic hot water, reheat for humidity control, and building heating. These requirements need a hot water source with a temperature of at least 140 °F (60 °C). This eliminates the unglazed EPDM collector from consideration. The ability of the parabolic trough greatly exceeds these requirements and thus would be a poor selection due to its high cost. This leaves the flat plate and the evacuated tube collectors as appropriate choices for Army applications. Both types of collector would be a good choice for most Army installations, but several factors could influence the selection:

- Cost (from RS Means Green Building Project Planning and Cost estimating, 3rd ed.)-
  - Flat plate - ~$17/sq ft = ~129€/m^2
  - Evacuated tube - ~$24/sq ft = ~182€/m^2
- Freeze Protection
  - Flat plate - Use non freeze liquid (glycol solution)
  - Evacuated tube - some are well insulated so they could use water as collector fluid with the strategy of cycling warm water into the collector from the storage tank if the collector fluid gets too cold.
- Stagnation Issues. Stagnation is caused when the flow through the collector stops and the solar energy heats the collector fluid to extremely high temperatures causing the collector fluid to boil. At what temperature this boiling occurs is dependent on the fluid and the operating pressure on the system. This boiling will push a portion of the collector fluid from the pipes in the collector and can hamper later collector performance. Section 3.4 (p 34) discusses this topic in more detail.

3.3.2.1 Unglazed flat plate

Unglazed flat plate collectors (Figure 3.12) are usually plastic collectors that are rolled out onto a roof and that are generally used for low temperature heating of such things as swimming pools or preheating of domestic hot water. Due to the absence of a glass cover they have no optical losses and therefore are most suitable for low temperature applications since heat losses increase more with higher temperatures compared to the other collector types. The manufacturers use plastic materials that reduce production and installation costs. Extensive testing and analysis have so far confirmed that the technology meets or exceeds reliability goals. They are generally less expensive, but less efficient than standard solar water heating collectors used throughout all seasons.

Figure 3.12. An unglazed solar mat-type solar collector made by FAFCO installed on a roof in California.

3.3.2.2 Glazed flat plate

Flat plate collectors (FPC) are essentially insulated boxes that have a flat dark plate absorber that is covered by a transparent cover (Figure 3.13). The solar energy heats the absorber and heat is carried away by a heat transfer fluid that flows through riser tubes that are connected to the absorber. The riser tubes are attached to the absorber in a parallel pattern or they meander from one side to the other.

The cover (usually a sheet of glass) is held in place by a frame above the absorber. The frame also seals the collector at the sides and at the back. It must provide mechanical strength and rain tightness, and must be designed to enable simple roof- and facade attachment or even integration into these building elements. The back and sides of the collector are insulated. Flat plate collectors are usually installed in stationary systems, i.e., they do not rotate to follow the path of the sun. The advantages of flat plate collectors are their simple, robust, low-maintenance design, and their large and effective aperture area.

Flat plate collectors are most commonly used for commercial or residential domestic hot water systems. These collectors generally increase water temperature to as much as 160 °F (71 °C). Special coatings on the absorber maximize absorption of sunlight and minimize re-radiation of heat. These collectors are prone to freezing and in climates where this can occur a mixture of about 60% water and 40% polypropylene glycol is used as the collector fluid (heat transfer medium).

3.3.2.2.1 Design considerations

Flat plate collectors similar to today's design have been manufactured for over 30 yrs and experience has been gained as to the proper materials to use for best performance and long life. The casing is typically made of aluminum. The absorber plate is made of copper or aluminum; steel is seldom used. To maximize the absorption of the solar energy the absorber plate is typically coated with black chrome, which is a selective covering providing good absorption and weak reflection of solar radiation.

Copper is normally used as the flow channel (tubing) through which the heat transfer fluid flows. It must be well bonded to the absorber plate for good heat transfer. The tubes are commonly placed in parallel rows (as shown in Figure 3.13) where the flow is released in a header at the top of the collector and is collected at the bottom.

Figure 3.13. Flat plate collector with selective coating on the absorber. The parallel lines indicate where the riser tubes are connected to the absorber by ultra-sonic welding.

Another tube arrangement is for the flow to meander across the surface of the absorber in a back and forth serpentine fashion. In this case the volume of heat transfer fluid spends more time on the collector surface and a greater temperature increase occurs. To obtain proper heat transfer from the absorber to the collector fluid the spacing between the runs of tubing cannot be too great and a tube interval of 4 to 5 in. (102 to 127 mm) is typical. In all cases, the tubes in a collector need to be placed so that the fluid can completely drain from the collector by gravity.

The housing around the absorption plate is mainly to minimize the heat loss to the environment and to provide a weather tight enclosure to prevent corrosion and other types of deterioration. Behind the absorption plate, rock or glass wool, or an insulating foam may be used an the insulating material. Typically a depth of 1-1/2 to 3 in. (38 to 76 mm) of insulating material is used. The insulating material must have the thermal stability to withstand the high temperatures that occur during times of collector stagnation. A glass cover is placed above the absorption plate that allows the solar radiation to pass through while limiting heat loss. Plastic covers deteriorate over time and are not recommended. Double pane glass covers retard the transparency to the solar radiation and thus are not commonly used. For sealing materials, EPDM or silicone rubber type materials should be used as the seal between the casing and the glass cover; adhesives should be silicon based and openings for pipes should be sealed with silicon based products.

3.3.2.2.2 Applications

Flat plate collectors are used mainly for producing domestic hot water and, in some cases, where building space heating is also accomplished. Standard flat plate collectors typically perform best providing hot water below 160 °F (70 °C). There are high performance flat plate collectors (those with a double, anti reflective cover) that perform well providing up to 200 °F (93 °C) hot water. These are seldom used due to their high cost. Above that temperature, the efficiency drops significantly due to the higher temperature difference between the collector fluid and the ambient air.

It is possible to reduce the thermal heat losses by avoiding convective losses such as by using vacuum tube collectors. The following section discusses this option.

3.3.2.3 Evacuated Tube

Evacuated tube collectors (Figure 3.14) can be designed to increase water/steam temperatures to as high as 350 °F (177 °C). They may use a variety of configurations, but they generally encase both the absorber surface and the tubes of heat transfer fluid in a vacuum sealed tubular glass for highly efficient insulation. Evacuated tube collectors are the most efficient collector type for cold climates with low level diffuse sunlight.

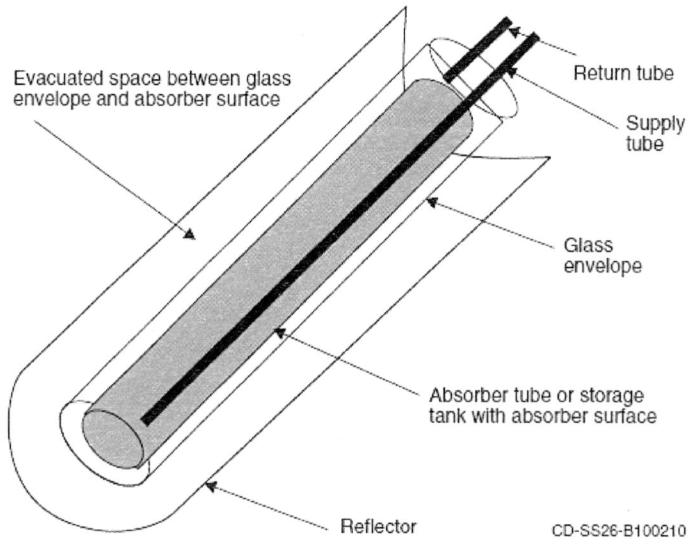

Figure 3.14. Evacuated-tube collector.

There are three types of evacuated tube collectors: (1) direct flow, (2) heat pipe, and (3) Sydney tube type. The direct flow type has the heat transfer fluid flowing through copper tubes attached to a absorber plate mounted inside the evacuated tube. The heat pipe type uses a heat pipe attached to the absorber plate. The heat pipe transfers the heating energy to the condensing section of the heat pipe where the collector fluid is warmed. This occurs in the header where the evacuated tubes are connected. The last type has an evacuated tube called a Sydney tube (Figure 3.15) that encapsulates a heat conductor sheet (absorber) with heat transfer fluid carrying tubes. The Sydney tube slides over the absorber section and locks into the collectors header forming a tight seal. Within the Sydney tube the space between inner and outer glass tube is evacuated. The selective coating is sputtered onto the outside of the inner glass tube. A heat conductor/transfer sheet is located inside the inner glass tube that conducts the heat from the glass into the U-form tubes carrying the heat transfer fluid. The Sydney tube type collector's performance can be enhanced through the use of a compound parabolic concentrator located behind each tube. This device will reflect the solar radiation that passes between each evacuated tube back to the underside of the cylindrical absorber in the collector tubes. There are various other construction methods like flat or round absorber, and single- or double-walled glass.

All evacuated tube collectors have the following in common:

- A collector consists of several evacuated glass tubes positioned in parallel and are joined by an insulated manifold at one end for the supply and removal of the heat transfer fluid (Figure 3.16).
- Due to the vacuum insulation (pressure < 10^{-2} Pa) heat loss caused by conduction and convection are minimal.
- The upper end of the tubes is connected to the "header."
- The tubes are circular to withstand the outside pressure.

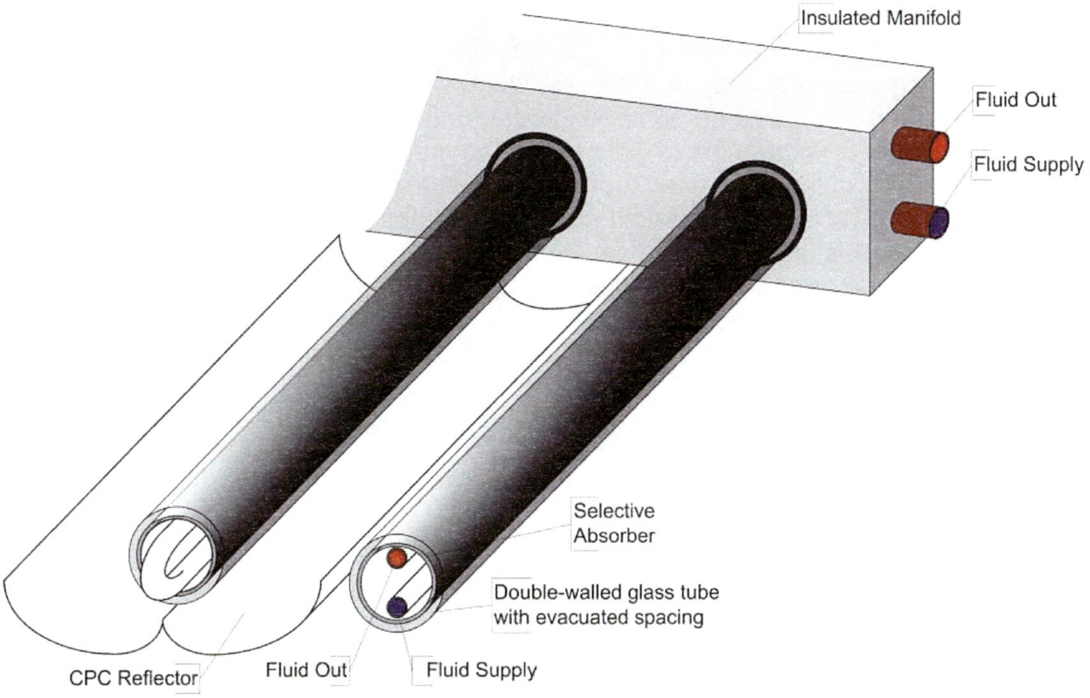

Figure 3.15. Basic elements of an evacuated Sydney tube collector. The ends of the tubes in the drawing are cut to show the internal tubing. On the left the tube is additionally equipped with an optional CPC (compound parabolic concentrating) reflector.

Figure 3.16. SunMaxx evacuated tube solar collectors on the roof of a commercial building.

3.3.2.3.1 Design considerations

Evacuated tube collectors have only insulated tubes and a pipe header to which the evacuated are connected. The collector fluid tubes use copper and typically black chrome is used as the selective absorber coating. The pipe header is insulated and has a protective cover.

3.3.2.3.2 Applications

This type of collector is used when there is a need for hotter water than would be necessary for domestic hot water heating. Hotter water is needed for applications that have cooling in the summer as a requirement and in some cases where building heating is a major need. Solar assisted cooling uses an absorption or adsorption chiller, which requires hot water temperatures in the range of 130 to 350 °F (55 to 180 °C).

An evacuated tube type collector may also be chosen as an alternative for a flat plate collector in areas where winter time freezing occurs. In this case, water would be used as the heat transfer fluid in the collector and warm water would be pumped into the outside piping and collector when freezing of those components is threatened. This would required a small amount of heated water due to the insulating quality of the evacuated tubes. As a result, the cost and inferior heat transfer characteristics of a water glycol mixture is avoided. Also the a hotter water could be produced in the collector providing a lower heat transfer fluid flow thereby reducing distribution pipe and storage tank sizes. Also, the heat exchanger between the collector and the storage tank could be avoided thus reducing the required leaving collector temperature. As a total system, the evacuated tube collector could have a total cost competitive with a flat plate collector system. The use of evacuated tube type collectors obviates most of the stagnation concerns associated with an anti-freeze heat transfer fluid.

3.3.2.4 Concentrating Collectors

These collectors use curved mirrors to focus sunlight onto a receiver tube (sometimes encased in an evacuated tube called CPC or compound parabolic collectors) running through the middle or focal point of the trough (Figure 3.17). They can heat their heat transfer fluid to temperatures as high as 570 °F (299 °C). Such high temperatures are needed for industrial uses and for making steam in electrical power generation. Because they use only direct-beam sunlight, parabolic-trough systems require tracking systems to keep them focused toward the sun and are best suited to areas with high direct solar radiation like the desert areas of the Southwest United States. These collector systems require large areas for installation, so they are usually ground mounted. They are also particularly susceptible to transmitting structural stress from wind loading and being ground mounted helps with the structural requirements.

Parabolic-trough collectors generally require greater maintenance and supervision and particularly benefit from economies of scale, so are generally used for larger systems. Because of their higher cost and greater maintenance needs this type of collector is not recommended for US Army heating needs in their standard buildings.

Figure 3.17. Parabolic trough collectors used to heat water at a large prison facility in Colorado.

3.3.3 Hot air collectors

Air collectors currently do not have a large market share (e.g., 0.5% in Germany in 2009). Nevertheless they can be considered as an alternative in certain situations (e.g., space heating, when an air heating system is used). As the type name indicates air collectors use air as the heat transfer medium (instead of water and glycol). This has some advantages:

- Air does not freeze or evaporate and air does not degrade when exposed to high temperatures. Freezing and stagnation thus does damage the system. Air collectors are usually intrinsically safe.
- Fresh air may be used directly as heat transfer fluid; a volume of air has no cost and is non-toxic. Leakage in the system does not cause damage to the system nor the environment.

On the other hand some disadvantages are:

- Air has lower heat transfer attributes and a lower heat capacity (a factor of 4 times lower compared to water and glycol).
- Higher driving power by a fan is needed for a comparable mass flow [kg/h] to a fluid pump.
- Larger cross sections for conduction pipes are necessary.
- If water is to be heated, an additional heat exchanger is needed.

3.3.4 Solar hot water collector efficiency

The efficiency of the solar collector is directly related to heat losses from the surface of the collector. Heat losses are predominantly governed by the thermal gradient between the temperature of the collector surface and the ambient temperature. Efficiency decreases when either the ambient temperature falls or when the collector temperature increases. This decrease in efficiency can be mitigated by increasing the insulation of the unit by sealing the unit in glass (for flat collectors), or providing a vacuum seal (for evacuated tube collectors). Figure 3.18 shows efficiency curves of these collectors. When comparing collector efficiencies, it is important to assume the same type of area (net vs. gross), and the same irradiation level.

The thermal performance of solar hot water collectors is characterized by:

- The power curve as shown in Figure 3.18, parameters: η_0, a_1 and a_2
- Incidence Angle Modifier (IAM) because of the optical efficiency of the collector
- Thermal capacity (C_{eff}), which is the measure of thermal response to heating and cooling.
- The quantity of heat input into the collector to heat it by -457.87 °F (1 °K). This information would be available from the collector manufacturer. This value is used in the solar collector simulation computer programs as it relates to the small time steps in the program to the estimated heat removed. The larger the $C_{eff,}$ the more energy that will be lost when switching off and on the solar heat transfer pump, which can happen as the weather changes during the day.

3.3.4.1 Power curve

For the power curve, collector performance is measured at different operating temperatures and with perpendicular insolation of G > 29,360 Btu/hr/sq ft (800 W/m²). The collector's performance is represented by:

$$\eta = \eta_0 - a_1 \cdot \frac{\Delta T}{G_{glob}} - a_2 \cdot \frac{\Delta T^2}{G_{glob}}$$

where:

η = Instantaneous efficiency of a collector at given operating conditions [-]

η_0 = Conversion factor at normal incidence of radiation [-]. This value determines the starting point of the collector efficiency curve, e.g., η_0 = 80% (at dT = 0 → no thermal, only optical losses!).

a1 = Linear heat loss coefficient [W/m²K]. This factor determines the starting downward slope of the collector efficiency curve (conduction + convection losses).

a2 = Quadratic heat loss coefficient [W/m²K²]. This factor determines the downward curvature of the collector efficiency curve and is assumed to cover all non-linear losses (losses due to radiation).

ΔT = The temperature difference between the mean fluid temperature in the collector and the ambient air temperature [K]

G_{glob} = Global irradiation intensity [W/m²].

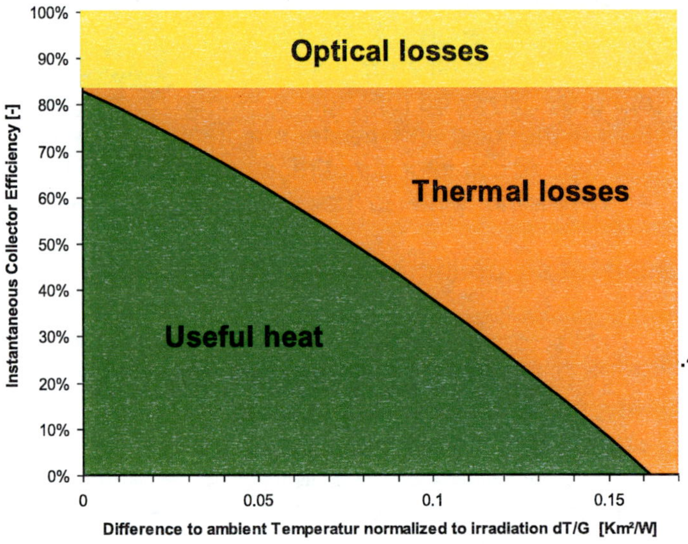

Figure 3.18. Typical solar collector efficiency curve with losses and useful heat indicated.

Figure 3.19 shows the power curves of four low* temperature collectors. A "rule of thumb" is to select a collector type that achieves an efficiency η ~ 50% for the working temperature range.

For use in water and low temperature space heating, both flat plate collectors (with solar glass and selective coating) and evacuated tube collectors are applicable. Both have certain advantages in specific applications especially when freezing, leaving water temperature, available space and installation cost are concerns. For applications with large collector fields, these two collector types should always be considered, and the final decision to select one technology over the other should be based on annual simulation results.

* Per definition low temperature collectors are applied in the range up to 176.0 °F (80 °C), medium temperature collectors up to 482.0 °F (250 °C) and above these: high temperature collectors.

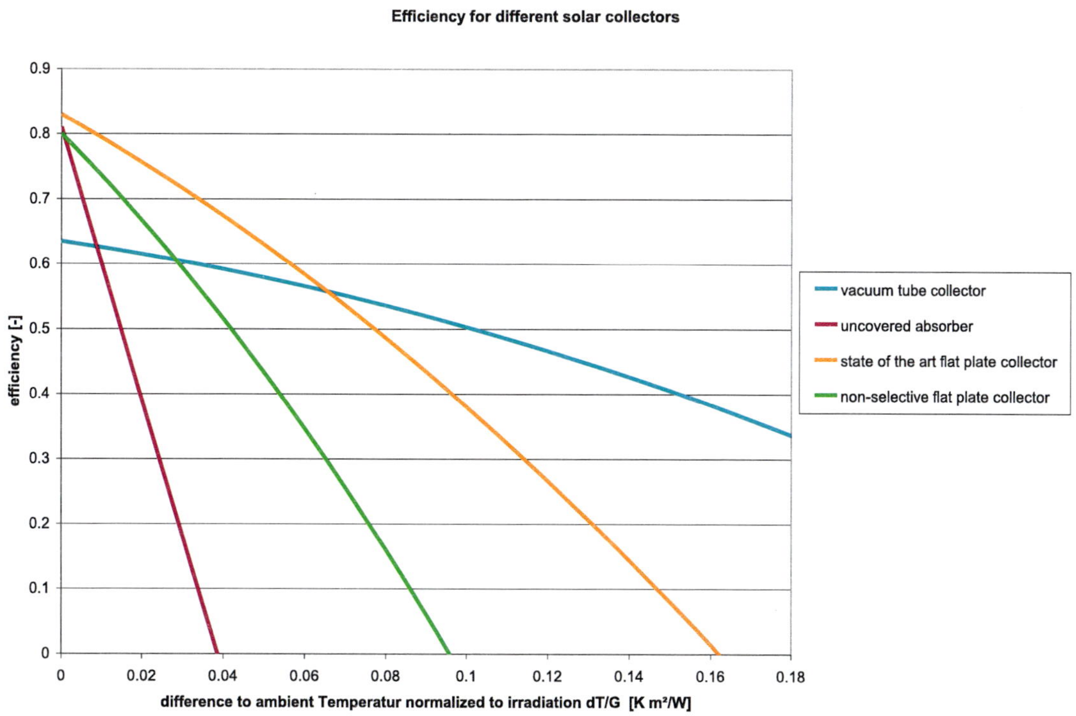

Figure 3.19. Power curves for four typical low temperature collectors.

3.3.4.2 Incidence angle modifier

The **IAM [-]** describes the modification of the conversion factor η_0 of the collector for non-perpendicular solar incidence angles. By definition, an IAM equal to 1 is for normal incidence. The IAM has a significant effect on the performance of stationary installed collectors as the incidence angle changes throughout the day and the year. Incidence angles less than 50 degrees do not have a significant effect on the solar thermal collector efficiency while an Incidence angle of 90 degrees is equal to a total reflection of the sun rays. Figure 3.20 shows the IAM curves of a typical FPA and a typical ETC.

The longitudinal IAM (i.e., in the direction parallel to the tubes for the ETCs) of the ETCs is similar to the flat plate collector's while the transversal IAM of most ETCs shows a characteristic increase at intermediate angles.

Collectors with a flat absorber surface, which includes some types of evacuated tubes, only have 100% efficiency at midday. Other evacuated tube collectors collect solar radiation in a perpendicular fashion over a longer period of the day since the collecting absorber surface is cylindrical. This feature can be enhanced by placing an optimally designed reflective compound parabolic concentrator (CPC) mirror behind the collectors, causing the sunlight to strike the collector at a perpendicular angle for a great percentage of the day. This provides most of the advantages of tracking systems while avoiding their high costs. The advantages of the CPC ETC include:

- longer usable daylight time
- more continuous power in the course of the day
- high target temperatures over the entire day
- higher daily and yearly energy yields.

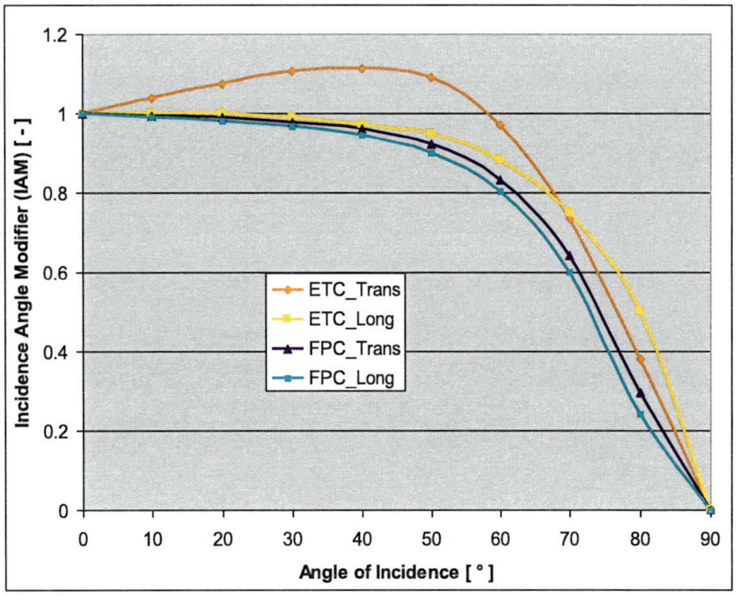

Figure 3.20. The graph shows the Incidence Angle Modifier (IAM) for evacuated tube and flat plate collectors in transversal and longitudinal directions across the collector.

3.3.4.3 Thermal capacity

The **thermal capacity** (C_{eff}) of the collector has an effect on the system performance. Every time the collector heats up, energy is absorbed by the collector. This energy is not fully recovered as useful energy. In simulations, this factor is taken into account when calculating the annual energy yield. The thermal capacity (C_{eff}) of the collector is expressed in kWh per m^2 collector area per degree K. The influence of this parameter is comparably small. Nevertheless, the lower the thermal capacity of the collector, the better.

3.4 Heat transfer fluid

As the solar collector heats up, the fluid in the collector increases in temperature. This fluid is then moved out of the collector so the heat can be extracted for some useful purpose. Larger solar systems either use the heat immediately, or use heat stored in a tank of heated fluid. When the solar system is located in a freezing climate, the fluid flowing through the collector is often an anti-freeze solution. When storing heated fluid in a tank, water is often the preferred medium (to reduce cost). If this is the case, a second fluids is used and flow through a heat exchanger is required to transfer the heat from the collector to the storage tank. There also may be a third fluid, potable water, which is used directly for domestic purposes, in which case another heat exchanger would be required. The heat transfer fluids used in a solar hot water system are very important and must meet a number of requirements to ensure good performance:

- a high heat capacity and conductivity allowing efficient heat transportation from the collector
- anti-corrosive-protection, if mixed or corrosion prone materials are present in the collector
- non-toxicity and environmental-friendliness
- low viscosity for easy pumping of the fluid
- low cost and availability.

Except where freezing is a concern, water is the fluid of choice in solar energy hot water systems. It has a low cost, is plentiful, and is compatible with the materials used in these systems. Water also has high heat capacity, good conductivity, a low viscosity, and can withstand the hot temperatures that are experienced during stagnation periods of time. Passive solar systems normally use water in the solar collector. Water is also often used in evacuated tube collectors. In freezing climates an anti-freeze liquid must be used in flat plate collectors.

Water used for domestic hot water is a potable water source that must be kept safe to drink and not be contaminated by chemicals. Because of this, large systems must have at least two fluid circulating systems, one that flows through the solar collector, and another that is heated and dispensed as domestic hot water. A heat exchanger is placed between the two piping systems for the movement of captured heat from the collector heat transfer fluid to the domestic water. Figure 3.6 shows such a system. Separating the collector fluid from the domestic water allows for water treatment to prevent corrosion of the piping and collector materials. To protect the potable water a double wall heat exchanger must be used when the heated fluid is not-potable. This separation also allows for an anti-freeze solution to be used in the solar collectors if needed.

3.4.1 Anti-freeze fluids

The typical anti-freeze solution is a mixture of water and propylene glycol, but a water/ethylene glycol solution, silicon oil, hydrocarbon oil, or refrigerant could also be used. With the water/propylene glycol fluid the percent glycol should be 40% or less. A 40% solution begins to freeze at –11 °F; below that temperature, an ice slurry develops that does not readily freeze solid causing pipes to burst. A glycol solution greater than 50% is not recommended due to a higher viscosity and lower heat capacity. Systems using glycol should be aware that it has a greater tendency to seep through piping joints than water and thus the piping system should be sealed with care and checked for leaks at scheduled intervals. Glycol is not compatible with zinc so galvanized pipes should not be used.

The use of automatic water makeup to heat transfer fluids selected to be an anti-freeze should be avoided. This is because the water makeup will dilute the anti-freeze mixture and making the fluid more likely to freeze when exposed to cold outdoor temperatures. The anti-freeze fluid should be periodically checked to assure proper performance.

3.4.2 Stagnation

The heat transfer fluid used in the collector must also withstand the high stagnation temperatures. Stagnation occurs when the heat transfer fluid stays in the collector too long and a high temperature is reached greater than the normal due to the heat from solar radiation. This could happen when there is a pump or control failure, when the heating demand of the users and the storage tank are satisfied or when the system is down for maintenance. Under normal operation the heat transfer fluid is under a pressure to avoid vaporization. Most collector systems are designed to operate at pressures below 125 psig (833 kPa). This is to allow the use of standard piping components (class 125) and avoid using more costly components. Since the solar collector is generally placed at a higher elevation than other parts of the system the static head of the fluid column must be added to the operating pressure in the collector when determining the pressure on components. This means pressures at or just below 75 psig (500kPa) are common in the solar collector. At this pressure, a 60/40 mixture of water/glycol will begin to vaporize at 320 °F (160 °C). If the pressure is dropped to 45 psig (310 kPa), then vaporization would begin at a temperature of 284 °F (140 °C). Vaporization of water at these pressures is 307 °F and 275 °F (153 and 135 °C), respectively.

When vaporization begins to occur, the resulting gas displaces the liquid in the collector and to some degree in the nearby piping. This displaced fluid should be directed to a recapturing tank so that it can be used again to fill the system when the temperature cools down. There are safety valves in the piping system that must be certified for the highest temperature that may occur. They should be placed on connections to the lines leading to the recapture tanks.

The disadvantages of frequent vaporizations are:

- Water/glycol mixtures will have a shorten life when exposed to at temperatures in the 300 °F – 320 °F (149 – 160 °C) range.
- Anti-corrosion additives and contaminants may stick to the interior of pipes and absorbers flow channels.
- Operating personnel need to spend time refilling the system.

All glycol systems should use the Pressure Stagnation Protection (PSP) method. This method allows over sizing of the pressure relief valve to 150 psi (1034 kPa), which allows the system pressure to rise with stagnation temperature. This protects the fluid from overheating and preserves the properties of the glycol by keeping it in a liquid form at all times.

Since glycols begin to break down and start to become corrosive when heated to temperatures greater than 240 °F (116 °C). One way to avoid this condition is to send excess heat to a nearby low priority heat user such as a swimming pool when the temperature in the storage tank is satisfied. Such a connection to a swimming pool is established using another heat exchanger with redundant, multiple pumps. Another solution may be to route external fluid to air heat exchangers, or in some cases, to a ventilated recooling device. For smaller applications (up to several hundred m^2) it is sufficient to design the solar loop expansion vessel in a way that both the additional volume of the heat transfer medium (due to the decreasing density), and the volume of the evaporated heat transfer medium (in case of stagnation) can be absorbed.

3.5 Piping arrangements

3.5.1 Collector piping alternatives

It is advised to choose collectors that feature a good emptying behavior during stagnation. This will reduce the strain on the heat transfer fluid and reduce the steam production of the collector field. During periods of stagnation, it is likely that steam will develop in the collectors. The steam will push the fluid out of the collectors. The fluid will later be sent back into the collector field when the collector field is cooled down again.

Tables 3.1 to 3.9 list a selection of collector designs. The pipe manifolds are shown in relation to the arrangement of the tubes attached to the collector absorbers. Both flat plate and evacuated tube collectors are addressed. An attempt is made to assess the different absorber piping designs with respect to:

- steam producing power (SPP) (a low SSP is a positive property with respect to a smaller expansion vessel needed)
- glycol strain (a low glycol strain is a positive property)
- air vent possibility (a good air vent behavior is a positive property).

For the designs listed in Tables 3.1 to 3.9, real measurements of the SPP were carried out on sample collectors. For the other designs, estimations are provided that are deduced from the experience made in the investigations. In these tables, note that:

- A question mark "?" in the tables means that even no rough estimation on the expected behavior can be given. The assessment is very rough and the given collection of piping designs is not complete.
- ETC = Evacuated tube collector
- FPC = Flat plate collector.

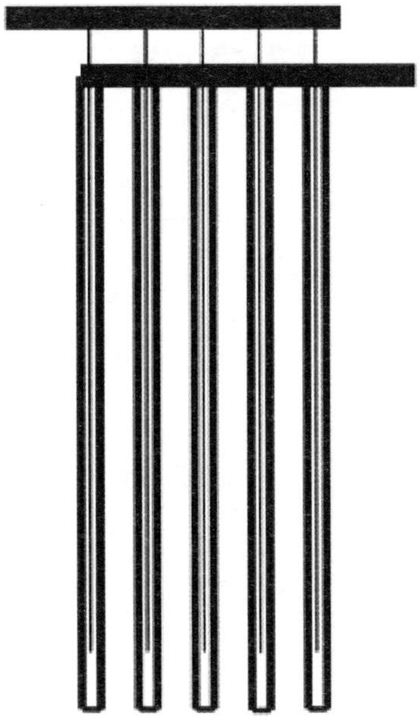

Table 3.1. Evacuated tube collector 1 (ETC), SPP 14,130 Btu/hr/sq ft (385 W/m^2).

SPP	High
Glycol strain	High
Air vent	Good

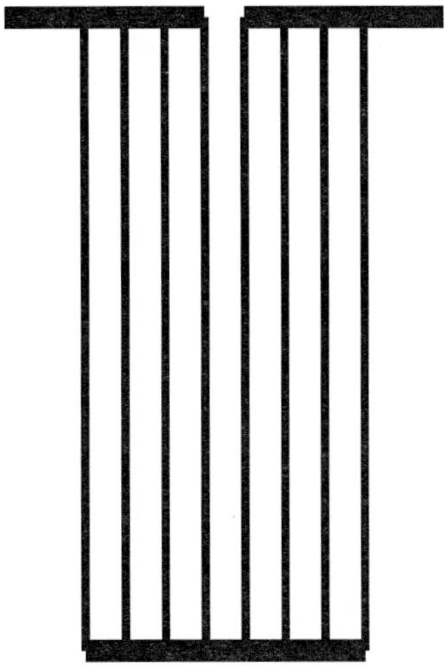

Table 3.2. Flat plate collector 1 (FPC), SPP 5,505 Btu/hr/sq ft (150 W/m^2).

SPP	High
Glycol strain	High
Air vent	Good

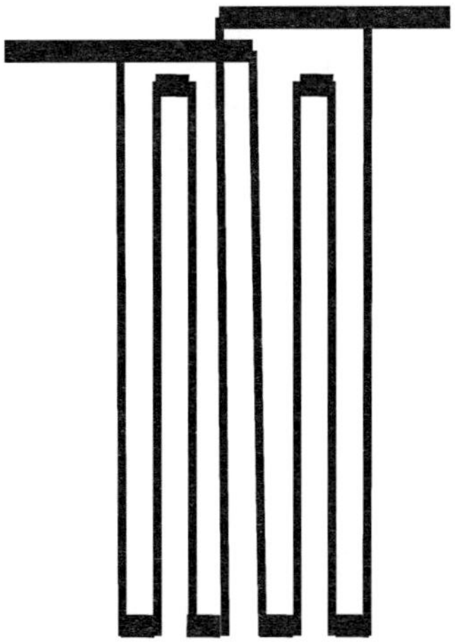

Table 3.3. *ETC 3*, SPP 2, 202 Btu/hr/sq ft (60 W/m^2), The collector does produce steam. Whether it is too much or not depends on the system.

SPP	?
Glycol strain	High
Air vent	?

Table 3.4. *FPC 2,* SPP 551 Btu/hr/sq ft (15 W/m²).

SPP	Low
Glycol strain	Low
Air vent	Good

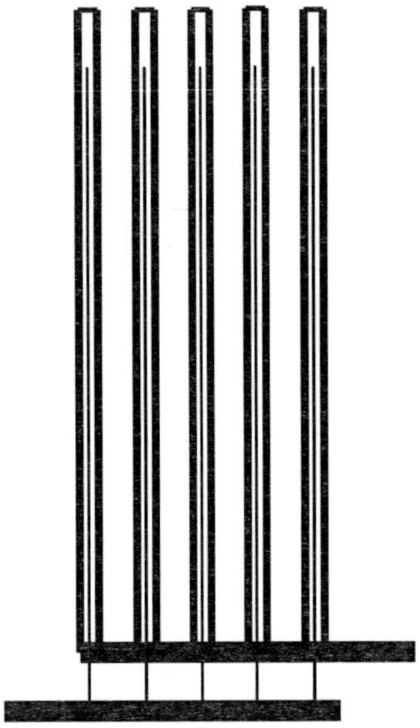

Table 3.5. *ETC 1 upside down.*

SPP	Low
Glycol strain	Low
Air vent	Bad

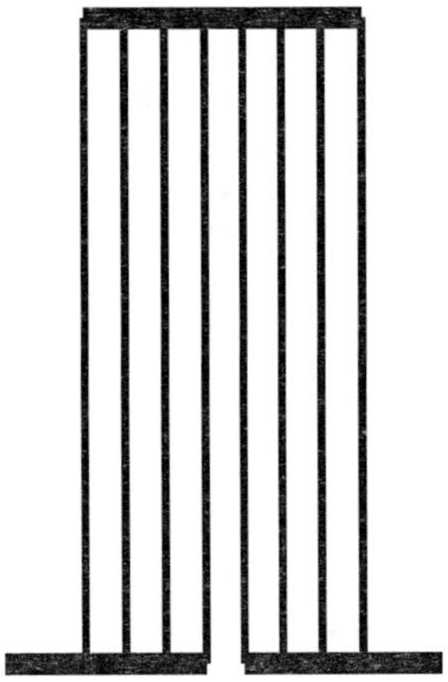

Table 3.6. *FPC.*

SPP	Low
Glycol strain	Low
Air vent	Bad

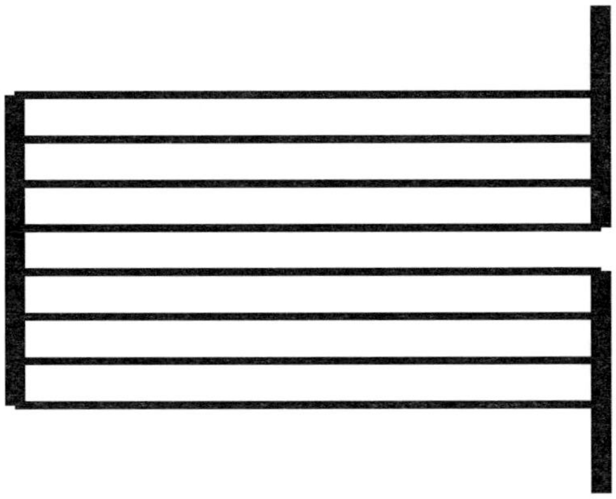

Table 3.7. *FPC, horizontal.*

SPP	Low
Glycol strain	Low
Air vent	?

Table 3.8. *FPC.*

SPP	?
Glycol strain	?
Air vent	Bad

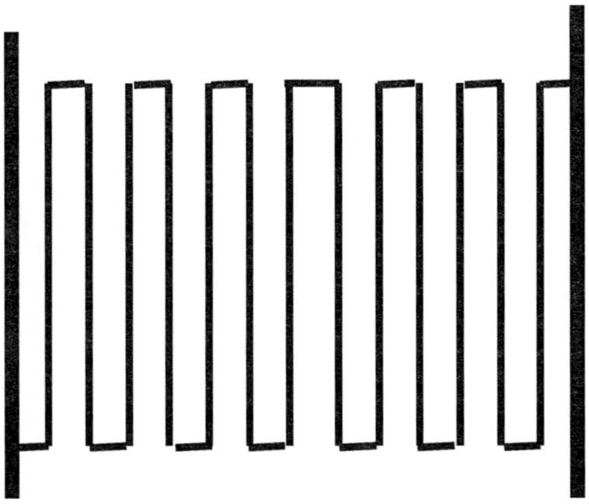

Table 3.9. *FPC, horizontal.*

SPP	?
Glycol strain	?
Air vent	?

There are two methods connecting a row of collectors together - with an external manifold or an internal manifold. The external manifold uses a supply and return header and all the collectors in the row have supply connections to the supply header and also have a flow outlet connected to the return header. The internal manifold type connection has the supply pipes of each collector connected to the next collector. The same applies to the return side of the collector piping. The piping of the collector is similar to that listed in Table 3.9. Internal manifolds offer the following advantages:

- less piping reducing costs and heat loss
- more attractive installation.

Disadvantages of internal manifolds are that they:

- make it more difficult to achieve a drainable system
- make it difficult to balance collector flow for a long row of collectors
- make it hard to remove a single collector
- create more significant expansion and contraction issues for a long row.

3.5.2 Solar System piping arrangements

Large solar systems serving multi-buildings typically have several heat transfer loops. First, there is one with the fluid flowing through the solar collector (solar primary loop). This one may contain an anti-freeze solution. Second, there is one that circulates from the heat exchanger with the collector fluid to the thermal storage tank (solar secondary loop). Third, may be a circulating fluid that heat the domestic hot water. Finally, there may be a circulating system going to a district or building heating system. Normally circulating systems use water as their heat transfer fluid. The types of piping systems can be characterized as:

- Solar supported **local heating system** with a two-pipe network:
  - two-Pipe Networks with Decentralized Domestic Hot Water Storages
  - two-Pipe Networks with Decentralized Heat Transfer Units
- Solar supported **local heating system** with a four-pipe network:
  - four-pipe networks with centralized energy storage and centralized domestic hot water storage
  - four-pipe networks with centralized energy storage and decentralized domestic hot water storage
- Solar supported **district heating network** with direct interconnection of a centralized solar thermal system:
  - solar thermal plant directly feeding the supply line of an existing district heating network
  - solar thermal plant directly preheating the return line of an existing district heating network.

3.5.2.1 Two pipe networks

In two-pipe networks, the heat supply to the heat sinks (buildings), including both domestic hot water and space heating, is by means of a pair of pipes. Domestic hot water is heated in a decentralized manner for the individual consumers using continuous flow water heaters, or by means of decentralized DHW storages using the charge-store principle. The two-pipe network shown in Figure 3.21 consists of centralized energy storage and decentralized heat transfer units for each building or unit connected to the network.

The energy storage is the central point for all heat flows and acts as a hydraulic gateway. To guarantee a reliable supply of heat with this design, it is essential that adequate reserves are permanently stored in the upper region of the energy storage (stand-by volume) so as to cover peak demand.

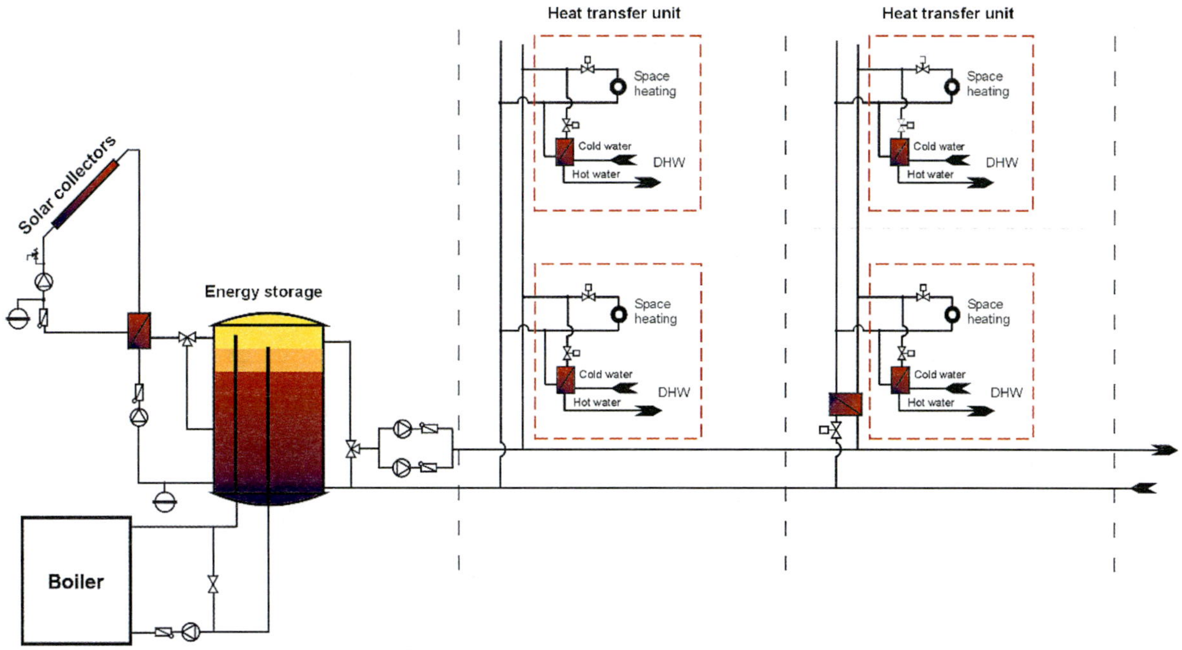

Figure 3.21. Solar supported two-pipe network with centralized energy storage and decentralized heat transfer units.

Domestic hot water is heated in a decentralized manner using continuous flow water heaters (usually via plate heat exchangers – no additional storage needed).

Two-pipe networks in combination with decentralized heat transfer units are ideally suited for use in compact unit blocks (**medium to high energy densities**). For less energy dense clusters of buildings (lower heating requirements per length of heat distribution pipe), a two-pipe network in combination with decentralized daily storages is preferable (Figure 3.22).

Solar-supported heating networks in combination with decentralized heat transfer units are also highly suitable for use in existing buildings. This includes buildings that are equipped with central space heating, but that also have a decentralized supply of domestic hot water (off-peak energy storage units). Whenever these energy storage units have to be renewed, they could then be replaced by decentralized heat transfer units. At the same time, improvements to the heat insulation (insulation of the building envelope, new windows) will mean that the space heating system can operate at lower temperatures.

Figure 3.23 shows a modified two-pipe network with decentralized energy storage for each building connected and additional decentralized solar thermal systems. The single consumers within the building are supplied via decentralized heat transfer units.

This concept is recommended for **local heating grids** with **high energy densities** (high heating requirements of buildings per length of heat distribution pipe), especially for new-built facilities that are constructed in several construction stages (modular enlargement).

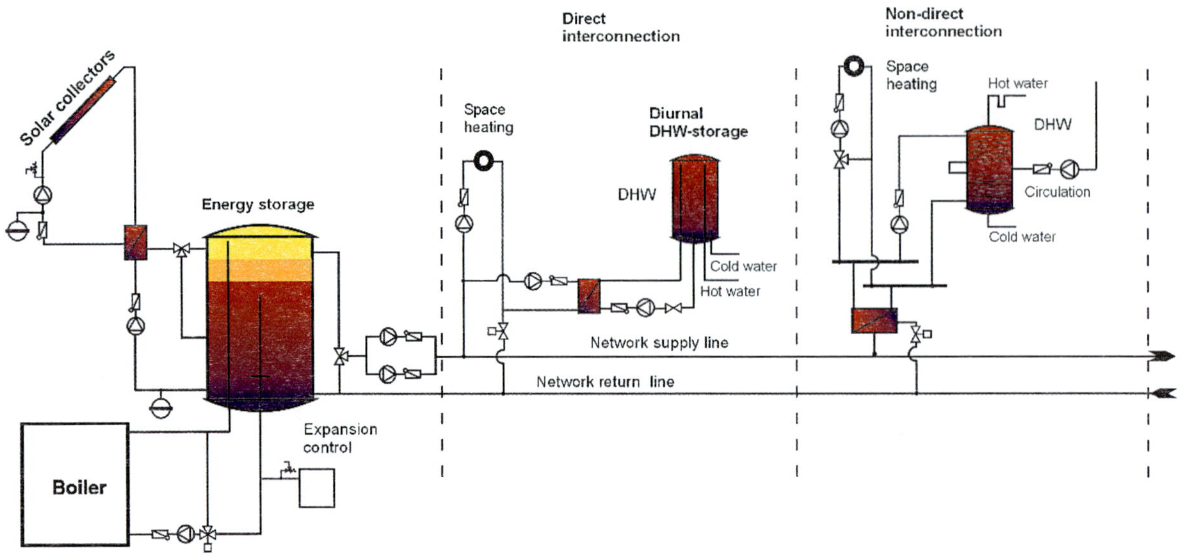

Figure 3.22. Solar supported two-pipe network with centralized energy storage and decentralized domestic hot water storages.

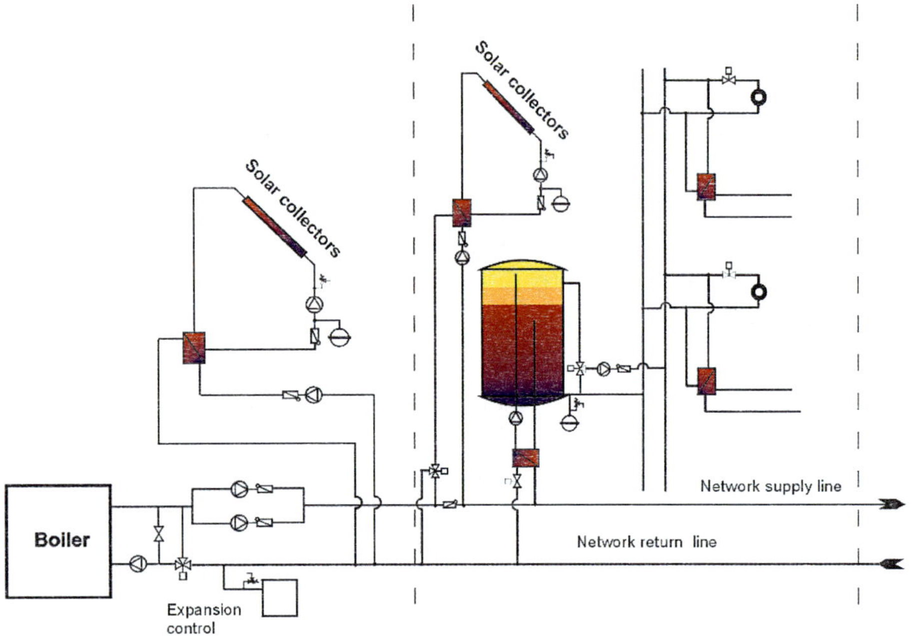

Figure 3.23. Modular expandable solar supported two-pipe network with decentralized energy storages for all buildings connected.

3.5.2.2 Four-pipe networks with centralized energy storage and centralized domestic hot water storage

In four-pipe networks, the heat is distributed through four pipes. In addition to flow and return lines for the space heating system, four-pipe networks also have two pipes for the supply of domestic hot water (distribution pipe for domestic hot water and circulation line).

The four-pipe network shown in Figure 3.24 consists of centralized energy storage and centralized domestic hot water storage. The energy storage is the central point for all heat flows and acts as a hydraulic gateway. Domestic hot water is heated in a centralized manner using the charge-store principle.

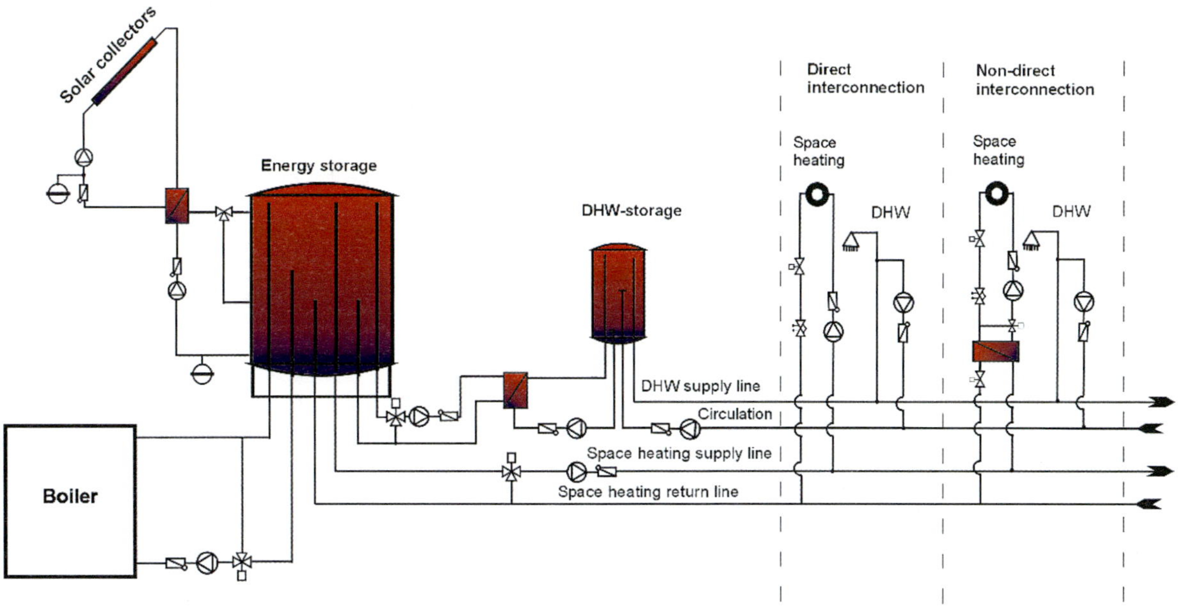

Figure 3.24. Solar supported four-pipe network with centralized energy storage and centralized hot water storage.

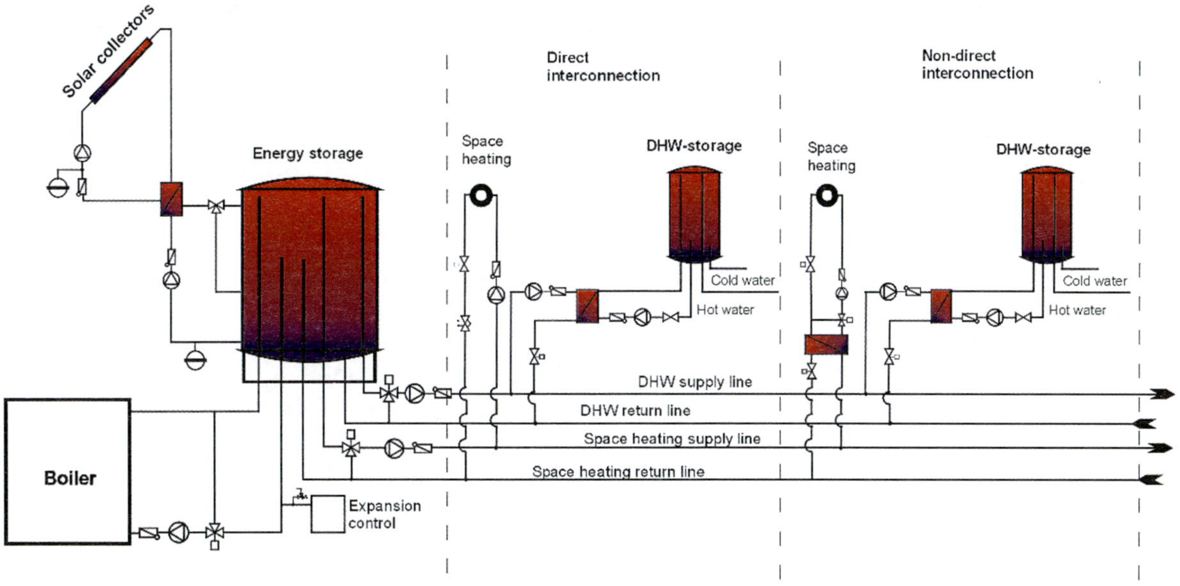

Figure 3.25. Solar supported four-pipe network with centralized energy storage and decentralized hot water storage.

Due to the high distribution losses in the DHW supply and circulation lines, this concept is especially recommended for **local heating grids** with **high energy densities** (high heating requirements per length of heat distribution pipe) such as Army New-Recruit Troop Training Sites.

The four-pipe network shown in (Figure 3.25) consists of centralized energy storage and decentralized domestic hot water storages in the buildings.

The energy supply for space heating and domestic hot water is performed by two pairs of distribution pipes using the heating water as the heat transfer medium within the whole heating network.

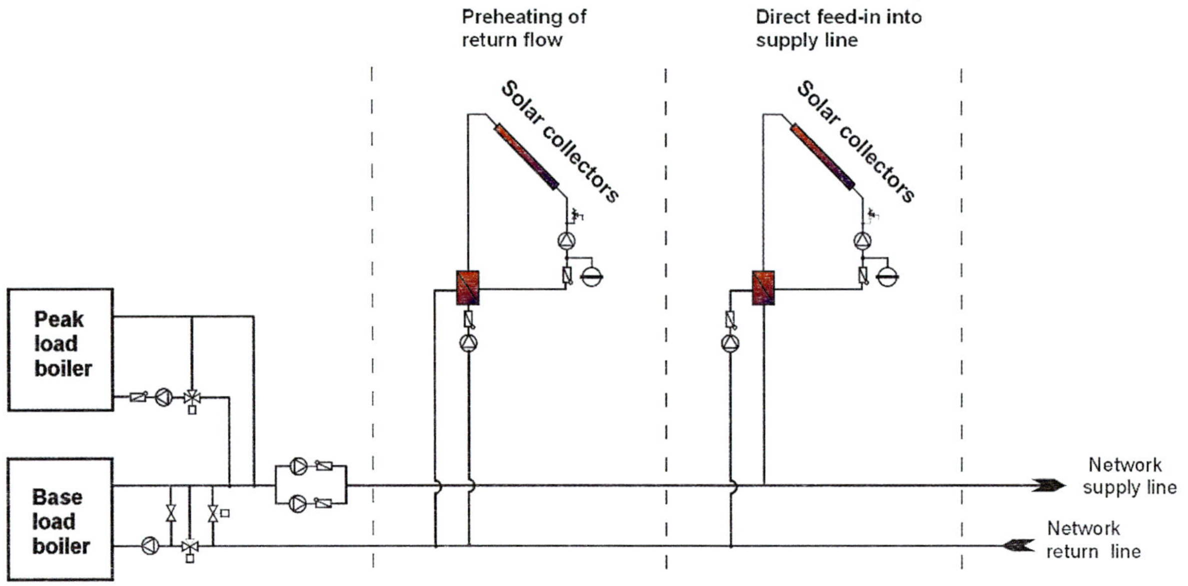

Figure 3.26. Solar supported district heating network with direct interconnection of a central solar thermal system.

Due to its higher specific investment costs, this concept is especially recommended for **local heating grids** with **high energy densities** (high heating requirements per length of heat distribution pipe) such as Army New-Recruit Troop Training Sites.

The direct interconnection of a solar thermal system to an existing district heating network (Figure 3.26) is applied to provide some base load energy directly to the grid. In general, two different applications are most commonly used:

- solar thermal system directly preheating the return line of an existing district heating network
- solar thermal system directly feeding the supply line of an existing district heating network.

Applications of this kind are commonly designed based on the available space and the existing dimensions of the district heating branch on site, and not on the actual load in a specific building. The solar collectors can either be roof- or ground-mounted. The majority of these systems can be operated without additional storage as they use the district heating network as storage (as long as they provide a small amount of heat in comparison to the total load in the district heating system).

3.5.3 Balancing fluid flow in collector field

A large solar hot water system will flow through many solar collectors. The flow through each solar collector should have basically the same pressure drop. This will ensure that the system is balanced such that each collector is receiving the same flow rate of heat transfer fluid. Thus the fluid temperature increase of each collector will be equal to the others. With several collectors, a reversed return piping system is used to achieve the equal pressure drop. This assumes each collector is arranged to have parallel fluid flow. In some systems, the collectors are arranged for flow in series. Here the fluid goes from one collector to the next collector picking up heat along the way. Figure 3.27 shows the parallel and series flow arrangements. There is a limit to the number of collectors that can be arranged in series due to the pressure drop of the flow through so many collectors.

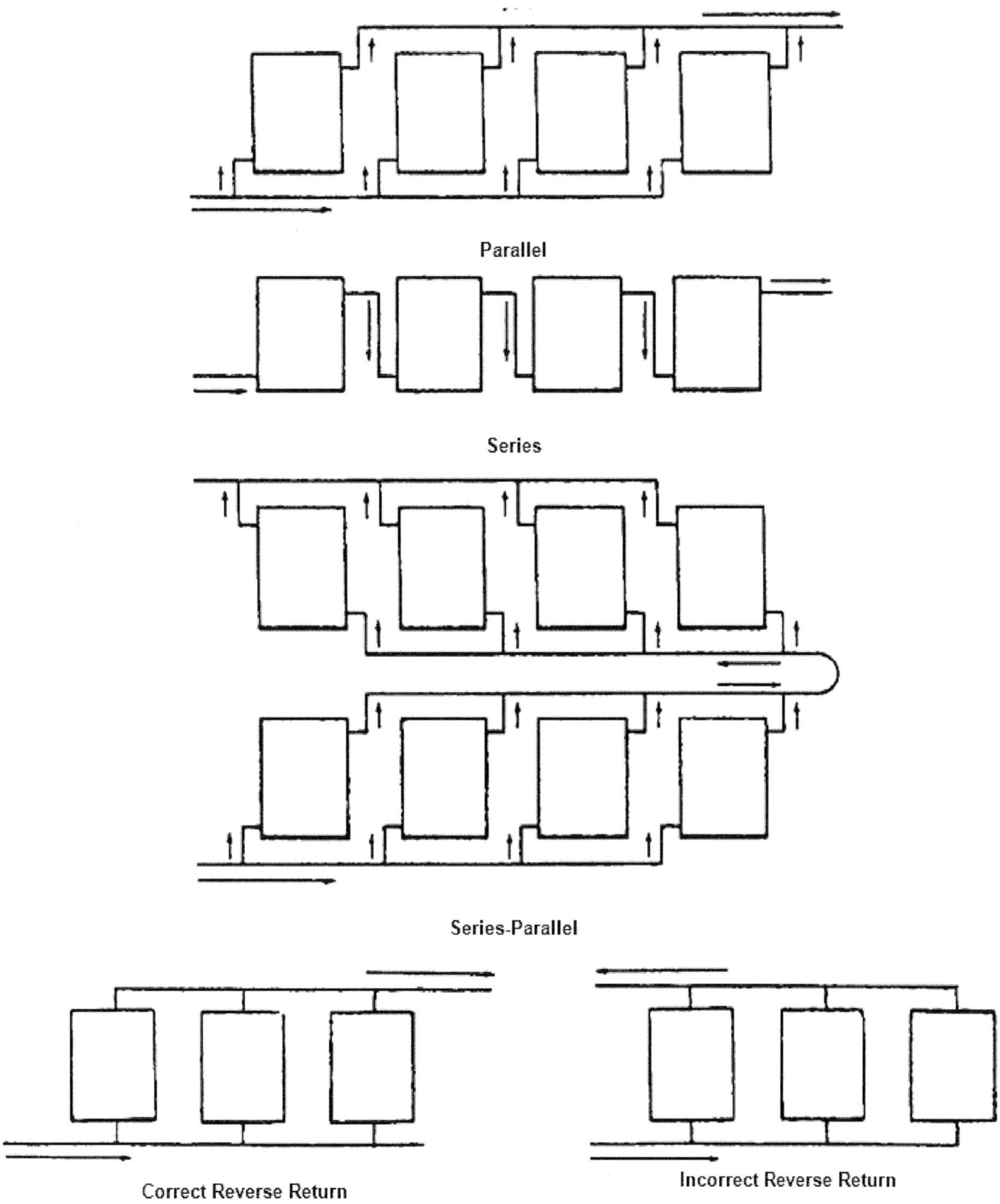

Figure 3.27. Piping arrangements of collectors for balanced flow.

In large collector systems a combination of parallel and series flow is used. An arrangement of collectors that have series flow are placed in a group or zone. Each group is then arranged to have parallel flow with the other collector groups. Figure 3.27 shows such arrangement. The collector layout could also have a group collectors with parallel flow placed in series with another collector group having parallel flow.

The design objective is for the pressure drop of each group to be equal. This is accomplished using a reverse return piping layout (Figure 3.27). The amount of piping should be kept at a minimum to keep installation costs low and to minimize the system resistance to flow, which helps keep the required pump pressures at acceptable levels. A rule of thumb is for the pressure drop per pipe length in the collector to be slightly more than three times the pressure drop per pipe length in the general piping system

With these large solar thermal systems it is advised to design the collector loop system as a low flow system. That means that the flow rate is ~0.37 gal/sq ft*h (15 l/m^2). By contrast, in the so called high flow type the flow rate is in the range of 1.23–1.72 gal/sq ft*hr (50–70 l/m^2*hr). This type system is well suited for small (less than 161sq ft [15 m^2]) solar DHW systems and for applications having a small temperature rise such as solar cooling applications. It is possible with a low flow system to connect more collectors that are piped for series flow. The flow through an individual collector is then less compared to a high flow operation mode and a higher temperature rise is possible. The benefits of a low flow system are its:

- lower investment costs due to smaller pipe sizes required
- lower piping lengths, as more collectors are connected in series (lower investment) and reduced heat losses
- smaller pump requirements (lower investment), which use less pump energy (lower operation costs) due to lower volume flow
- requirement for less fluid in the solar loop, and consequently less glycol (lower investment)
- quicker response to achieving the target temperature in heat storage tanks including those using stratified charging
- ability to achieve a useful temperature in a single flow cycle a greater percentage of the time.

Table 3.10 lists the range of specific mass flow rates of the various operating modes of the solar installations and the differences in the total mass flow rate in the feed and return piping of an assumed collector area of 10,760 sq ft (1000 m^2)· This example shows that larger manifold pipe sizes and higher electric pump power are required when the high-flow operating mode is used in large-scale solar thermal systems. The costs of installing and operating this type of system would therefore soon exceed acceptable limits.

3.5.4 Freeze protection

When water freezes, it expands. Frozen water in pipes is likely to rupture them. This is a major cause of breakdown of solar heating systems. Water thus must be prevented from freezing inside the collector loop. The following options will protect against freezing of the piping system:

- Provide drain-back pipe design.
- Use an antifreeze solution in the outdoor collector and piping, a water-glycol mixture.
- Add heat from the storage tank to the outside piping and collector field.

Each solution has its benefits and drawbacks. Some general aspects are discussed here. The freeze protection implemented is usually not a free choice, but depends on the selected collector technology and supplier; it is usually based on environmental conditions, operating temperature, maintenance, costs, and local availability.

Table 3.10. Comparison of mass flow rates for high flow, low flow and matched flow systems.

Designation	Range of the specific mass flow rate		Mass flow rate for a collector area of e.g., 1,000 m^2	
Low-Flow	5 – 20 kg/(m^2·h)	0.12 – 0.49 gal/(sq ft)	5,000 kg/h to 20,000 kg/h	11,025 to 44,100 lb/h
High-Flow	50 – 70 kg/(m^2·h)	1.23 – 1.72 gal/(sq ft)	50,000 kg/h to 70,000 kg/h	110,250 to 154,350 lb/h
Low-Flow — speed-controlled	5 – 20 kg/(m^2·h)	0.12 – 0.49 gal/(sq ft h)	5,000 kg/h to 20,000 kg/h	11,025 to 44,100 lb/h

In a **drain-back system design** the water drains from the collector when the collector pump is stopped. This requires that the collector is constructed and the piping is installed so that no water remains in any part that could freeze. A continuous downward gradient for all the draining pipes is needed to avoid locations where pipes sag and where water can be trapped; this can be difficult to achieve. After a system draindown, the system pump is started to fill the pipes and solar collector with water. This means that the pump must be able to push the circulating fluid up the piping system when starting. The extra amount of pump pressure needed is equal to the height difference between the top of the collector and the position of the drain-back storage vessel. The drain-back approach can also be used during stagnation periods, preventing the formation of steam, and avoiding the resulting expansion and high pressures within the collector. Refer to Section 3.5 for various collector piping arrangements.

Please note: Common practice of solar system designers in Europe is to use a water glycol mixture in drain back systems nevertheless. This way it is ensured that heat transfer fluid does not freeze if draining does not work perfectly. The remaining advantage of drainback systems is: Overheating of oversized collector fields does not necessarily lead to problems during stagnation.

When using a **water-glycol mixture** the freezing temperature of the mixture is lowered (lower than water alone). The mixing ratio determines the lowest operating temperature. It is important to follow the manufacturer's or installation engineer's recommendations. The different types of glycol that could be used vary in their properties. See Section 3.4.1 (p 35) for more information. The pipes still need to be pitched to drains so that the collector field can be emptied for maintenance even when a anti-freeze collector fluid is used.

The main drawbacks of using a glycol antifreeze water mixture are that:

- The heat capacity of the fluid is reduced while the viscosity is increased. This causes a reduction in efficiency of the collector field. In addition, for a comparable mass flow more pump energy is needed.
- Glycol deteriorates when heated up to common collector stagnation temperatures that occur in evacuated tube collectors. Deteriorated glycol forms solid particles that can block and even destroy the collector loop. The glycol mixture then must be replaced.
- Care must be taken when an antifreeze solution is used to heat domestic water (which is a potable water source). The two fluids must be kept separate with a leak detection device at the point of heat exchange. This affects the efficiency of the heat transfer.

The third option is to use water in a closed, always filled loop, and to heat up the collector loop by turning on the pump when near freezing temperatures are sensed. This means that heat from the storage tank is used to prevent the collector and piping from freezing. This has the obvious drawback of wasting captured heat and using pump power. Therefore, the viability strongly depends on the site with the specific occurring ambient temperatures in winter. It would probably be the logical choice at locations where freezing temperatures are not normally experienced. This option also requires a reliable power supply because if no power is available, this option will fail to protect the collector loop.

3.5.5 Air elimination

When the piping system is filled with the heat transfer fluid air is pushed out of the pipes as the fluid enters. This air must be removed for proper flow to occur. To remove this air, valved openings in the pipe system are installed at high points. These are called "air vents"; their valves are opened to allow air to escape. Usually a 3/8-in. (9.52 mm) ball valve is used. This is a manual operation and when fluid begins to be released at these high points the valve is closed with the captured air is removed. This air removal exercise must be done not only at the initial fluid fill, but also a short time after operation begins and then on a routine basis thereafter. Gases are released from the fluid as it is heated and trapped air can be moving through the system as bubbles. With a fluid velocity greater than 1.3 ft/second (40cm/second), these gas bubbles will move with the fluid until they reach the air vents in the system and there they can be released. The air vents consist of a short section of an expanded pipe section with a valved pipe outlet placed on the top of the expanded pipe section. The larger pipe volume allows the gas to collect without disturbing fluid flow and thus the gases can be removed when the air vent is opened.

3.5.6 Heat loss

The collector loop piping system between collectors and storage tank must be insulated to avoid energy losses. The thickness of insulation on pipes of up to 1 in. (22mm) (outside diameter) should be at least 0.8 in. (20 mm), pipes up to 1-1/2 in. (42mm) should have at least 1.18 in. (30 mm) insulation.* Mineral wool is a widely used insulation of solar system installations. Any pipes and insulation material that are exposed (to weather and animals) should be resistant to damage (e.g., UV degradation and removal by birds) as not to fail. Zinc coated steel, aluminum, or stainless steel are all common materials used for the pipe insulation covering.

3.6 Storage tank

3.6.1 Tank configuration and use

A challenge in applying renewable energies is often the mismatch between the time energy is needed and the time energy is available. Thus storage tanks are a necessary part of any hot water system since they couple the timing of the intermittent solar resource with the timing of the hot water load.

For systems that provide heat for domestic hot water, 1 to 2 gal (3.8–7.5 L) of storage water per square foot of collector area are generally adequate. The storage fluid can either be potable water or non-potable water if a load side heat exchanger is used. For small systems, storage is most often in the form of glass-lined steel tanks. Solar heated water may be stored in a "one-tank" system, or it may be stored in a separate tank that feeds into the tank of a conventional gas or electric water heater (a "two-tank" system). Whether one or two tanks are used, solar energy heats the water before use. On sunny days, a typical solar system can raise water to 140 °F (60 °C).

Bigger commercial solar hot water systems are basically the same as those used for homes, except that the thermal storage tank, heat exchanger, and piping are larger. The storage tanks in these applications are commonly steel tanks with an enameled interior coating. The sizes of these components are proportional to the size of the collector array. Most systems include a backup energy source such as an electric heating element or are connected to a gas or fuel fired central heating system that will heat the water in the tank if it falls below a minimum temperature setting, enabling the system to work year-round in all climates.

If the solar hot water system provides for some of the building heat, a larger storage tank may be advisable. Figure 3.28 shows a breakdown of common storage types.

---

* See EN 12976 Appendix B. In addition the thermal conductivity (λ) of the insulation should be equal or less than 0.2 Btu/hr-sq ft °F (0.035 W/mK).

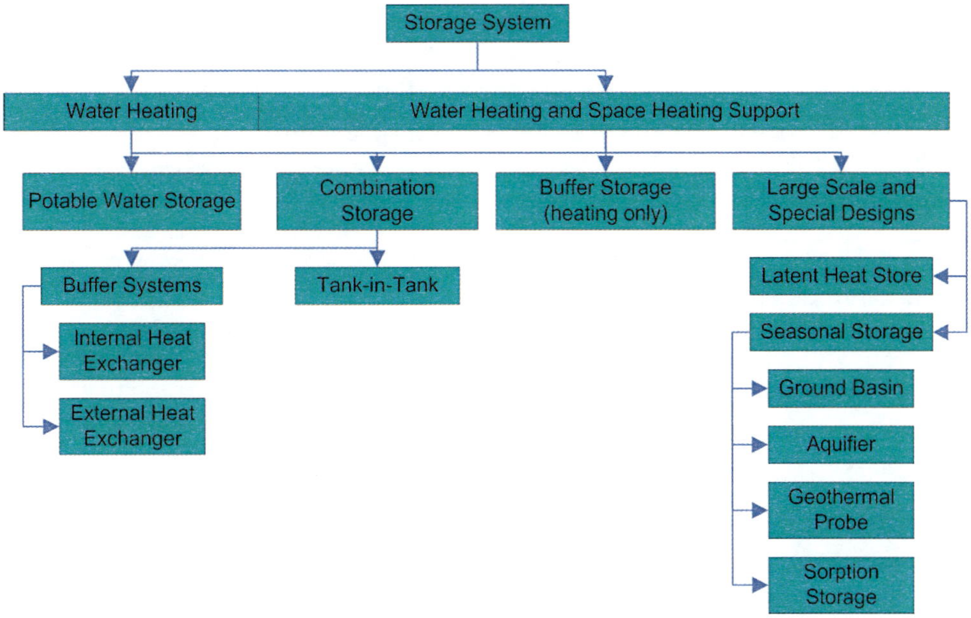

Figure 3.28. Overview of storage system types and their application.

Most large systems have stratified storage tanks where cooler temperatures are at the tank bottom and the hotter temperatures are at the top. The cooler fluid is drawn off the tank bottom and is sent to the collector system for heating. It may go directly to the collector or be used to cool the heat transfer fluid that is then sent to the collectors. Using this cooler water, increases the collector efficiency. The heat transfer fluid is held in the collector until it reaches the desired hot temperature. This is accomplished by stopping the fluid flow or by slowing the flow in the collector until the desired temperature is reached.

Thermal losses from the storage tank are a significant part of the heat balance of the solar thermal system. The losses are proportional to the surface area of the storage tank. Because the volume of a solid body (e.g., cylindrical storage tanks) increases faster than its surface, larger storage tanks have a lower heat loss per volume than smaller tanks. Combining multiple storage tanks into one large storage tank is thus beneficial regarding reducing storage heat losses. All storage tanks need to be insulated to reduce the amount of heat lost from the system. It is good practice to limit the heat losses to 10% of that in the storage tank over 24 hrs. An insulation value of R-16 is the minimum insulation required. If the tank is placed outdoors the insulation should have a weather proof cover.

The storage tank temperature must satisfy the required service temperature and quantity. For DHW, this would be 140 °F (60 °C). For other heating needs the temperature could be hotter. A hotter temperature than the service requirements in the storage tank allows for greater storage of heat, but reduces the collector efficiency. This consideration for the storage tank normally results in a design that provides a stratified water condition in the tank with the hot water on the top and the cold water at the bottom. Stratification of the water is naturally achieved and maintained if mixing of the water is minimized since the cooler water has a higher density and tends to stay at the tank bottom. For good stratification, the tank should be as tall as possible, which means a high, narrow tank is best. Such a tank could be divided into several shorter tanks plumbed in series where the outlet at the top of the first is connected to the inlet near the bottom of the second and later tanks similarly piped. Up to four tanks are often installed in this manner for large systems. If one large tank is used, there may be a pipe connection in the middle section that allows for the entry of water at a temperature warmer than the cold water at the bottom, but not as hot as that in the upper part of the tank. Using this technique minimizes the mixing of the stored water and helps keep the water column stratified.

For larger storage tanks it is common not to store drinking water in the whole storage volume, but to use an internal or external heat exchanger through which drinking water flows for heating (see Figure 3.29). This avoids Legionella growth at mid temperatures, eliminating the need to heat to higher temperatures, more than 140 °F (60 °C).

The most commonly used storage tank systems are used for short-term storage, and are designed to store surplus solar energy for 1–2 days (diurnal storages). This limits the possible solar fraction* (*SF*) to 10 – 20% of the total heat demand (space heating + DHW) supplied by the solar system, but provides the lowest system investment cost.

Long-term storage tank systems can compensate for seasonal fluctuations in solar irradiation between winter and summer, which can include solar fractions in the range between 40 and 70%. The negative effects are higher system investment costs and higher heat losses of the system. Between short and long-term storages, weekly or medium-term storages have been realized in Austria.

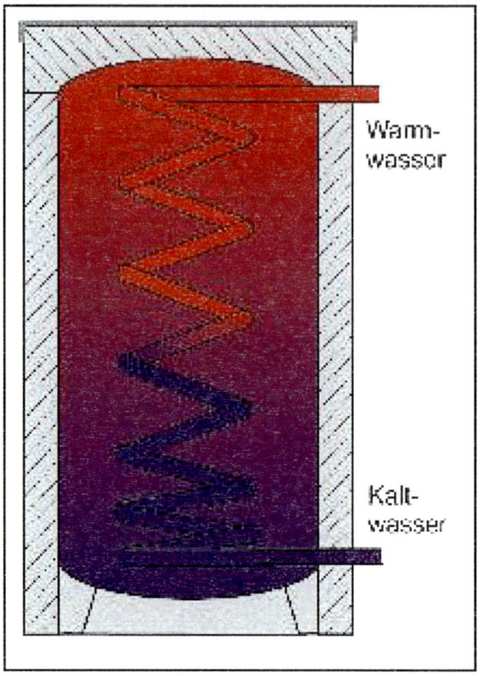

Figure 3.29. Example of internal heat exchanger in a storage tank for heating domestic hot water.

3.6.2 Legionella considerations

One should exercise caution to avoid the risk of the growth of the *Legionella Pneumophila* bacterium. These bacteria are naturally found in water, but at low concentrations. They can multiply quickly when the water temperature is between 86 and 113 °F (30 and 45 °C).† Their numbers start to diminish at temperatures above 122 °F (50 °C), and almost instantly die above 150 °F (66 °C). The greatest health danger is not in drinking contaminated water, but in inhaling contaminated dispersed water droplets, which can happen when taking a shower. Storing large quantities of water at temperatures in the range of 86 to 113 °F (30 to 45 °C)should be avoided when it is intended to be used (from a tap) later. Local legal requirements may apply to this issue and should be taken into account.

Table 3.11. Techno-economic comparison of different storage concepts for local heating networks†

Type of Storage Connected to System	Short-Term (Diurnal Storage)		Medium-Term (Weekly Storage)		Long-Term (Seasonal Storage)	
Solar thermal generated heat for:	DHW		DHW + space heating		DHW + space heating	
Solar fraction (in% of the total heating demand)	10 – 20%		30 – 40%		40 – 70%	
Collector area (m^2) per 1076 sq ft (100 m^2) heated floor area	6 - 12 m^2	65 –129 sq ft	12 - 30 m^2	129– 323 sq ft	30 - 120 m^2	323– 1,291 sq ft
Specific storage capacity (L/m^2 collector area)	50 – 70 l/m^2	1.23 – 1.72 gal/sq ft	200 – 400 l/m^2	4.91 – 9.82 gal/sq ft	2,000 - 4,000 l/m^2	49.10 – 98.20 gal/sq ft
Specific solar thermal system costs (€/m^2 heated floor area)	20 – 25 €/m^2	$2.65 – $3.31 /sq ft	30 – 50 €/m^2	$3.97 – $6.62/sq ft	90 - 150 €/m^2	11.91 – 19.85 $/sq ft
†A tempering valve must be used to prevent scalding.						

* Within this report solar fraction SF is defined as the ratio between the energy produced via the solar thermal system and the total energy demanded for DHW + space heating

† All temperatures mentioned here are indicative only; refer to local requirements for actual temperatures.

The typical approach in the United States to avoiding the Legionella bacteria issues is to store the hot water at 140 °F (60 °C). Other design options are to limit the quantity of usable hot water by means of a heat exchanger or tank-in-tank construction, where the temperature of the delivered water is 140 °F (60 °C). In Europe, other options include periodically reheating the water above 150 °F (66 °C) (setting the controller for the auxiliary heater to do this when solar heat alone cannot achieve this), or instantly heating the water just before use to above 185 °F (85 °C).*

3.7 Heat exchangers

In the solar hot water system heat exchangers transfer the heat from one moving fluid to another. To achieve the temperature transfer the fluid from the collector will be hotter than the leaving temperature of the secondary fluid. This higher temperature from the collector has a negative effect on system performance. To keep this temperature difference as small as possible a plate-and-frame heat exchanger is used. Where this temperature difference is not a large concern a shell-and-tube heat exchanger can be used.

When the solar thermal system is used to heat domestic water there must be a guarantee that the heat transfer fluid is kept isolated from the potable water source. This is normally accomplished by separating the two fluids by a vented open space, or by using a third fluid that could be monitored for fluid leakage. The double wall or additional heat exchanger also protects the solar collector fluid from being diluted by water and becoming prone to freezing. (Refer to Section 5.3.5 [p 82]).

All heat exchangers have a small pressure drop of the fluid passing through them. This must be considered in selecting the pumps. Their materials of construction need to be compatible with the heat transfer fluids and with other materials in the piping system. They also need to be able to withstand the temperatures and pressures that will be experienced in the system. To achieve good heat transfer and to adhere to the other requirements, stainless steel and copper are the normal materials used in these heat exchangers.

The pipe design at the heat exchanger that transfers energy from the collector to the storage tank may need a by-pass pipe circuit around the heat exchanger for the collector heat transfer fluid if an antifreeze solution is used. The antifreeze solution on cold days during system startup could be several degrees below freezing and thus could freeze the water in the heat exchanger from the storage tank when flow begins. The pipe bypass would be opened when the collector pump is started. The valves would be closed to allow flow through the heat exchanger when there is usable heat in the heat transfer fluid (a temperature ~80 °F [27 °C]). This action will also avoid cooling the heat exchanger during the initial flow of the collector heat transfer fluid that was downstream of the solar collectors.

A comparison between the two types of heat exchangers will show the following results. Plate-and-frame heat exchangers have small passageways and thus have a higher pressure drop and are more prone to flow blockage due to contaminants in the circulating fluids. They have a quicker reaction time and typically take less space in the piping system. A shell-and-tube heat exchanger can handle fluids having more contamination. These heat exchangers are easier to clean and have a lower pressure drop for the fluids flowing through them.

Selection of the size of the heat exchanger depends on the amount of heat that must be transferred and the desired leaving temperature. An increased heat transfer surface area is needed for improved performance. Note that heat exchanger ratings are established using water as the heat

* A tempering valve must be used to prevent scalding.

transfer fluids. With the use of an anti-freeze fluid, heat exchange performance will decrease and the exchanger surface area will need to increase. A water-glycol mixture has a higher viscosity and lower heat capacity. Heat exchanger manufactures can provide useful information to aid selection.

3.8 Pumps

Pumps move the heat transfer fluids through their piping circuits. The pump associated with the collector fluid will be exposed to the highest temperature and the greatest pressure. To minimize the high temperature in this loop, the pump is placed in the pipe going to the collectors. But even here there will be short term periods when the pump is handling temperatures greater than the design collector discharge temperature. Such a time is when the hot heat transfer fluid has been forced out of the collector into the recapture tank due to stagnation. If using an anti-freeze solution as the collector heat transfer fluid, the pump components will need to be compatible with the water/glycol mixture. These considerations will affect the materials used in the construction of the pump.

The pump will also need to operate through the flow range that is designed for the system. In some solar thermal systems, the heat transfer fluid flow is slowed when the solar radiation is not at its peak. At this lower flow the pump must still provide the required pressure to obtain fluid movement through the pipe system. Having clean fluid strainers/filters, a minimum of valves, and properly operating sensors will aid in keeping the system pressure drop as low as possible. In large systems, the piping system flow is modeled to define the pump pressure needed for proper operation. It is important to take into account that glycol/water mixtures have a different viscosity than water alone. The manufacturer of the heat transfer medium should provide the necessary data. The energy needed to pump the heat transfer fluid is considered parasitic energy; it must be kept as low as possible. The pump should thus not be oversized.

3.9 Expansion tank

As the heat transfer fluid increases in temperature its volume also increases thus an expansion tank is used to capture the addition fluid as it is pushed out of the piping loop. The expansion tank also helps prevent the loss of heat transfer fluid that would escape the piping system through the safety valves during high temperature periods. When the heat transfer cools, fluid in the expansion tank can then flow back into the system.

Most expansion tanks are a steel tank with a rubber membrane inside. On one side of the membrane is a gas such as nitrogen, which is under a pressure equal to the system operating pressure. On the other side of the membrane is the space where the extra heat transfer fluid goes when expansion of the fluid occurs. The membrane material will begin to deteriorate when exposed to temperatures above 160 °F (71 °C). To protect the expansion tank from such high temperatures, they should be connected to the piping that goes to the collector. This will be where the coolest fluid will be found. If temperatures exceeding 160 °F (71 °C) are expected in this pipe then a buffer tank or auxiliary reserve tank should be placed between piping loop and the expansion tank. This is a normal steel tank filled with heat transfer fluid. Since it is not circulating in the collector piping system loop, it will be somewhat cooler. When expansion occurs, the extra hot fluid will enter the auxiliary reserve tank causing the residing cooler fluid then to enter the expansion tank protecting its membrane.

The membrane expansion vessel may need to be designed to compensate not only the additional volume of heated up fluid, but also the volume of steam that is produced by the collector field. That can be more than just the volume of the collectors. Water in pipes above the collector field, for example, can drain into the collectors after they are already empty and then start to produce even more steam.

3.10 Back-up/supplemental heater

Each solar thermal system must have a back-up heating supply. Often gas or electricity and sometimes biomass (wood) are used as a heat source. Due to the 20+ year lifecycle of heating systems, including solar thermal systems, possible changes in energy supply options should also be considered.

It is important to control the solar energy system to provide the required temperature to the building energy systems. So if the domestic hot water must be 140 °F (60 °C), the heat transfer fluid must be hot enough to deliver that temperature. If it is not, then the supplemental heating system will be energized causing additional energy costs to occur. Running the supplemental heater is often very inefficient for this use since it would be cycling on/off and would not be operating at its normal firing rate.

3.11 Controls

The controller controls the flow of the heat transfer fluid in the collectors by modifying the pump operation. Normally the pump is just turned on/off in small systems. The most common pump controller used in solar thermal systems is the "differential controller." This controller requires two different temperature settings, one for "on" (upper band) and one for "off" (lower band). The system temperatures are measured on an absorber in a collector (usually one whose flow is a short distance to the storage tank) and in the storage tank (near the tank outlet to the collectors) or in this discharge pipe adjacent to the tank. If the collector temperature exceeds the storage tank temperature plus the upper band the pump will turn on. If the collector temperature drops below the store temperature plus the lower band the pump will turn off. A common upper band ("on") temperature difference is 41–46 °F (5–8°C) and for the lower band ("off") 36–39 °F (2–4 °C). Proper location of these sensors is very important. Good placement guarantees that the pump only runs when this is beneficial for collecting solar energy that can be stored in the tank. Failing to measure the temperature difference correctly will affect the overall system performance significantly; this requires care and verification at the installation.

The controller also must turn off the pump when the maximum storage temperature (e.g., 194 °F (90 °C) is reached, or when the collector exceeds the maximum flow temperature (e.g., 68.00 °F [20 °C]).

Some pump-controller combinations offer different power settings (often step-wise) to be able to match the mass flow rate with the amount of solar radiation available. This limits the temperature increase in the solar circuit and prevents unnecessary pumping.

More advanced controllers offer additional functionality: insolation and heat measurement, data-logging, error diagnostics, and auxiliary heater (boiler) regulation, and a graphical user interface.

4. DOD Installation Solar Hot Water Applications

4.1 Areas of potential Army use

There are a number of Army installations in the United States. They break down into several general types: Introductory troop training, General troop installations, Proving Grounds, Depots, and Arsenals. The installations that emphasize the training and support of troops have a large number of housing units and corresponding dining facilities. Both of these types of buildings have a high domestic hot water use and are good candidates for solar hot systems. All Army installations have recreational and health care facilities, which also could be large domestic hot water users. The administration, maintenance, training, storage, manufacturing, and service facilities are typically low domestic hot water users and thus are poor candidates for a central solar hot water system to heat their domestic water.

Air-conditioning system reheat for humidity control is another year-round heating energy use and thus is a candidate for a solar hot water system. Humidity control is required in some administration, data processing, communication, and health care buildings. To accomplish this control, the air in the air-conditioning system is chilled to a low temperature for moisture removal. This air is normally too cold to introduce into occupied spaces so it is reheated to a more comfortable temperature. The energy to accomplish this reheat is a good candidate for the heat provided by a solar hot water system.

Building profiles found at the different types of Army installations are:

- Army Introductory Troop Training Sites (Fort Benning, Fort Jackson, and Fort Sill)
    - High density barracks and lodging
    - Dining facilities
    - Administration buildings
    - Company Operation Facility (COF)
    - Maintenance facilities (motor pools, aircraft wash areas, car washes)
    - Recreational (Fitness centers, pools)
    - Training
    - Healthcare
    - Service (Fire and police stations, PX, AAFES)
    - Storage
    - Central heating plants
- Army Troop Locations (Fort Bragg, Fort Campbell, etc.)
    - Barracks and lodging
    - Dining facilities
    - Administration buildings
    - Company Operation Facility (COF)
    - Maintenance facilities (motor pools, aircraft wash areas, car washes)
    - Recreational (fitness centers, pools)
    - Training
    - Healthcare
    - Service (fire and police stations, PX, AAFES)
    - Storage
- Army Proving Grounds
    - Research and manufacturing buildings
    - Lodging
    - Storage facilities
    - Administration buildings

  - Recreational (fitness centers, pools)
  - Healthcare
  - Maintenance facilities (motor pools, aircraft wash areas, car washes)
  - Service (fire and police stations, PX, AAFES)
- Army Depots
  - Research and manufacturing buildings
  - Lodging
  - Storage facilities
  - Administration buildings
  - Recreational (fitness centers, pools)
  - Maintenance facilities (motor pools, aircraft wash areas, car washes)
  - Service (fire and police stations, PX, AAFES)
- Army Arsenals and Ammunition Plants
  - Manufacturing buildings
  - Lodging
  - Storage facilities
  - Administration buildings
  - Recreational (fitness centers, pools)
  - Maintenance facilities (motor pools, aircraft wash areas, car washes)
  - Healthcare
  - Service (fire and police stations, PX, AAFES).

The size of these installations can be small with as few as 50 occupied buildings. Others are very large, having a population more than 70,000 people. Some are in very cold climates and others in extremely hot/humid climates. A better description of each can be obtained by doing a internet web search on a specific installation.

There is no "average size" Army installation; they have grown over time to serve a number of different needs. Perhaps the best way to approach Army building groups is to look at what the Army is doing now. The Army has recently standardized on a grouping of buildings to support the supervising, housing, and training of combat troops. The troops are organized into a Brigade size fighting unit and thus is called a Brigade Combat Team (BCT). The group of buildings are arranged in a manner shown in Figure 4.1. Looking at this layout, the BCT consist of Barracks (light blue) with a Dining Facility (purple) in the middle. Then the Headquarter Building (orange) is close by on a through street. South of this street are green buildings called Company Operation Facilities (COF) and red buildings called Tactical Equipment Maintenance Facilities (TEMF). The total building floor space is about 1,402,000 sq ft (130,386 m^2), which is divided into Barracks with 567,000 sq ft (52,731 m^2), Dining Facility with 31,000 sq ft (2,883 m^2), Tactical Equipment Maintenance with 229,000 sq ft (21,297 m^2), Company Operation Facilities with 447,000 sq ft (41,571 m^2), and 129,000 sq ft (11,997 m^2) for the Headquarters building.

The buildings that have the major DHW heating requirements are the barracks and the dining facility. Some healthcare and recreational facilities also have a high DHW heating need. The rest of the buildings have a very small DHW heating requirement that is most often satisfied with small local heaters. As can be seen below there are 11 barracks buildings and one dining facility in the BCT cluster. The average barracks size is 51,500 sq ft (4790 m^2). Each barracks will house ~260 soldiers. The dining facility is 31,000 sq ft (2880 m^2).

Figure 4.1. BCT building grouping at Fort Bliss.

The heating of these buildings can be from a central source such as a boiler plant that serves the entire installation or a group of buildings within the installation. Another method often seen is individual boilers in each building with one for heating the building and the other for generating DHW. Even with a central boiler system heating the DHW is often by local building boiler equipment.

Few Army installations have a cogeneration system where both electricity and heating hot water are generated in the central plant. Most Army central boiler plants provide heating only and thus it can be desirable to turn this system off during the nonheating season. About half the central systems are steam distributed at a pressure in the range of 60 to 100 psi (414 to 689 kPa). The others are hot water with temperatures as hot as 400 °F (204 °C).

The most common system is the use of local heating systems within a building. In this case DHW is generated using a hot water boiler. The next most likely situation is a central heating system that is used to heat the buildings, but the DHW is generated by a hot water boiler as above.

4.2 Building hot water demands

The barracks and dining facilities are large DHW users. The estimated DHW needs for a barracks building is 30 gal (114 L) of hot water per soldier per day at a temperature of 140 °F (60 °C). Most soldiers take two showers per day and one of them is in the morning, from 7:30 am to 8:00 am. To handle this peak usage DHW is stored in a tank having the size of ~22.5 gal (85 L) per soldier. Figure 4.2 graphs the heating rate throughout a typical week for a person's DHW use in a barracks. Such a high DHW heating energy use makes a barracks a good candidate for a central solar hot water system.

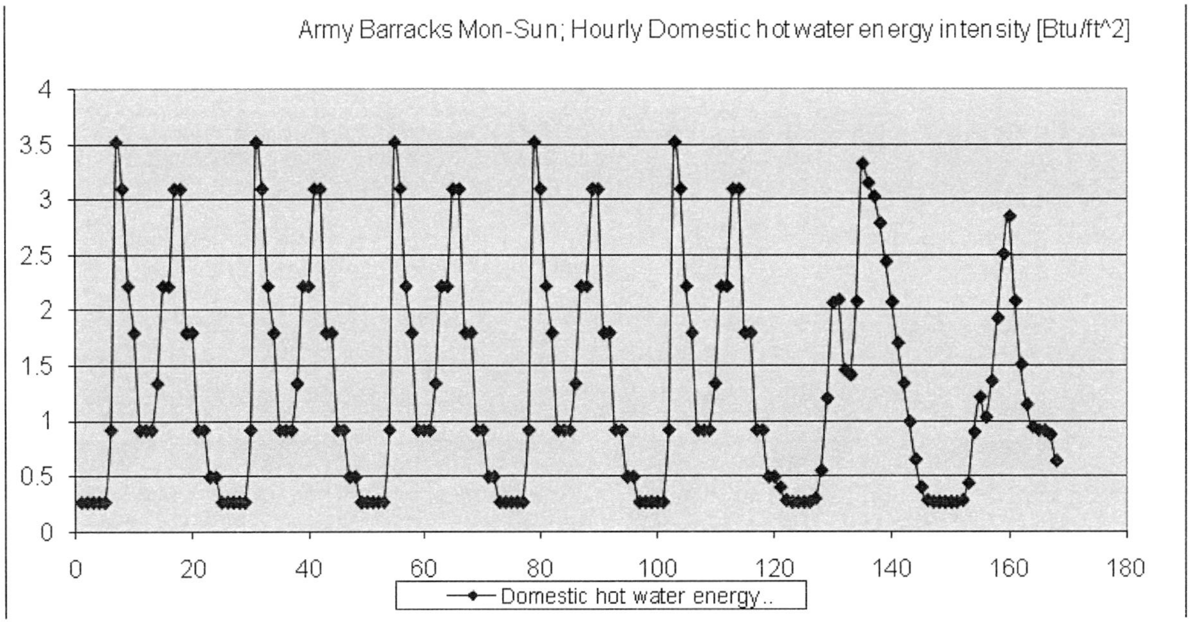

Figure 4.2. Weekly barracks domestic hot water heating profile, Btu/sq ft by hour.

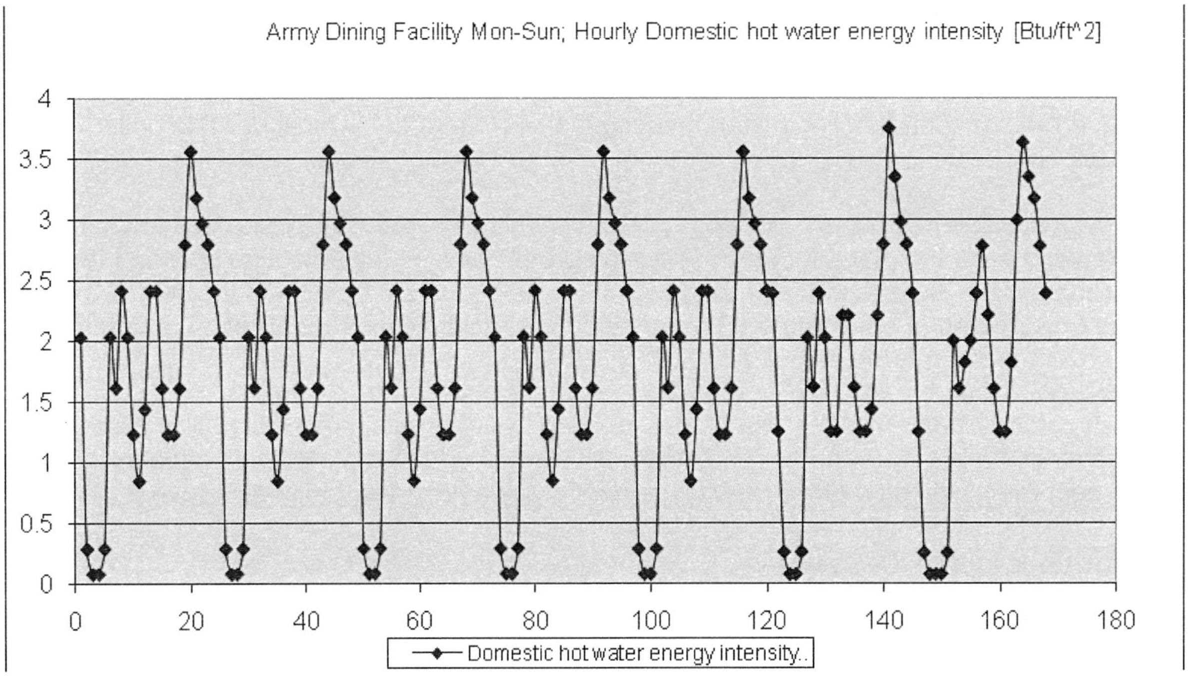

Figure 4.3. Weekly domestic hot water heating profile, Btu/sq ft by hour for a dining facility.

The other building in the cluster that should be included in the central solar hot water system is the dining facility. Figure 4.3 graphs DHW heating energy needs. Typically, these facilities are occupied from 5:30 am to 7:30 pm. Three meals are served per day - breakfast (7:30 to 9:00 am), lunch (11:00 am to 1:00 pm) and supper (5:00 pm to 6:30 pm). The schedule will change slightly on the weekend. The hot water use of this facility is ~2.4 gal (9 L) per meal served. Assume that a dining facility of this size would serve 7500 meals each day.

The office buildings have a hot water use of ~1 gal (3.79 L) per person per day. If the occupancy of the building is not known, use 100 sq ft (9.3 m^2) per person.

Another source for domestic hot water use is provided by NREL (shown below). Some guideline for determining hot water loads for sizing commercial SHW systems are:

- **Apartments:** 20 gal (76 L) /day of 120 °F (49 °C) set temperature water per bedroom
- **Hotels/Motels:** 15 gal (57 L)/day of 125 °F (51.67 °C) set temperature water per room
- **Laundries**: 20 gal/10 lb (76 L/4.5kg). wash of 130 °F (54 °C) set temperature water
- **Restaurants**: 24 gal (91 L)/day of 140 °F (60 °C) set temperature water per 10 full meals served
- **Retirement Homes**: 18 gal (68 L)/day of 120 °F (49 °C) set temperature per room
- **Office Buildings (without showers):** 1 gal (3.79 L)/day of 120 °F (49 °C) set temperature water per person.

4.3 Basic solar system design

Selecting the right solar water heating system for a federal facility will depend on three key factors: climate, budget, and water usage needs. Solar water heating systems are economical, especially in commercial buildings where the energy used to heat water is significant. There are a number of technologies available to heat water efficiently. However, before implementing these technologies, it is important to first reduce hot water use with water-saving fixtures and appliances.

Solar water heating systems can be used throughout the United States on any building with a south-facing roof or unshaded grounds for installation of a collector. In addition, reliable off-the-shelf systems can be selected from the Directory of the Solar Rating and Certification Corporation at: www.solar-rating.org/ratings/ratings.htm

The Arizona Solar Center Albuquerque, NM uses system sizing estimates based on climate:

- Sunbelt—use 1sq ft (0.09m^2) of collector per 2 gal (7.61 L) of tank capacity (daily usage).
- Southeast and Mountain states—use 1 sq ft (0.09 m^2) of collector per 1.5 gal (5.71 L) of tank capacity.
- Midwest and Atlantic states—use 1 sq ft (0.09 m^2) of collector per 1.0 gal (3.79 L) of tank capacity.
- New England and the Northwest—use 1 sq ft (0.09 m^2) of collector per 0.75 gal (2.81 L) of tank capacity.

Estimates will be affected by water temperature, consumption amount, and the solar resource available at the site.

4.3.1 Simple system calculation

A simple evaluation procedure can help to determine if solar water heating is appropriate. Traditional solar hot water heating systems are most cost effective in facilities with:

- Constant water heating load throughout the week and year; housing units and dining facilities are good candidates.
- High fuel costs to heat water; this is area specific.
- Sunny climates (which helps, but is not required); this is area specific.

The economic viability of a solar system depends on:

- amount of annual sunshine
- heating energy requirements throughout the year
- cost of the solar system
- price of conventional fuels (are the utility rates high in your area?)
- financing and incentives available (for 3rd party investors)
- what temperature of hot water is required (e.g., swimming pool vs. laundry)
- annual operation and maintenance costs.

The step-by-step method listed in Table 4.1 can be used to estimate the solar system size and its cost effectiveness. Table 4.2 lists general data for water heating loads for building types, solar resource, etc. Numbers are included for a sample building in Denver, CO and assume that the building uses electricity to heat water. The sample building group is four barracks, each housing 120 soldiers for a total of 480 soldiers. With a hot water use rate of 30 gal (114 L)/day/person, the hot water use would be 14,400 gal (54,000 L)/day. The heating energy to make this hot water is:

Q = 14,400 gal/day x (140 °F - 60 °F) x 1 Btu/lb/°F x 8.3 lb/gal /3413 Btu/kWh = 2800 kWh)/day

= 9,556 MBtu/day

4.3.2 Equation 1 – Solar water system size

Estimate collector size using the following equation:

$$A_c = \frac{L}{\eta_{solar} I_{max}} = 700\ \text{kWh/day} / (0.25 \times 6.1\ \frac{\text{kWh}}{\text{m}^2}/\text{day}) = 459\text{m}^2\ [4939\ \text{sq ft}]$$

where:

A_c =collector area [m^2]

L =Daily Load [kWh/day]

η_{solar} =efficiency of solar system (assumed to be 0.40)

I_{max} =maximum daily solar radiation [kWh/m^2/day] (Imax in the equation above means the system is designed to meet the load on the sunniest day of the year, which eliminates excess capacity and optimizes economic performance).

The paper "Annual System Efficiencies for Solar Water Heating" by Craig Christensen of National Renewable Energy Laboratory and Greg Barker of Mountain Energy Partners presents calculated efficiency for domestic solar water heaters that varies between 26.4% and 44.3% depending on location and hot water load with an average of 40.2% for all locations and load profiles.

Table 4.1. Method to estimate solar system size and cost effectiveness.

Evaluation Method	Sample Building	
Estimate daily water heating load	9,556,400 Btu/day	2,800 kWh/day
Determine the solar resource (kWh/day)	1,934.86 Btu/sq ft Maximum 1,744.55 Btu/sq ft/day Minimum	6.1 kWh/m^2/day Maximum 5.5 kWh/m^2/day Minimum
Calculate solar system size - for water heating load on the sunniest day - undersize rather than oversize the system	12,352.48 sq ft	(1,148 $m^{2)}$
Calculate annual energy savings	3,576.82 Btu/yr	(1.048,000 kWh/yr)
Calculate annual cost savings	$88,000/yr	
Estimate system cost	$746,000	
Calculate savings-to-investment ratio	2.8	
Calculate simple payback period	8.5 yrs	

Table 4.2. Typical daily water heating loads.

Barracks	**30 gal/day/person**	**(114 L/day/person)**
Motel	15 gal/day/person	(57 L/day/person)
Hospital	18 gal/day/person	(68 L/day/person)
Office	1 gal/day/person	(3.79 L/day/person)
Food Service	2.4 gal/meal	(9 L/meal)
Residence	40 gal/day/person	151 L/day/person
School	1.8 gal/day/student	6.8 L/day/student

4.3.3 Equation 2 – Annual energy savings

Annual energy savings (electricity in this example) can be estimated using the following equation:

$E_s = A_c\, I_{ave}\, \eta_{solar} 365 / \eta_{boiler}$ = 1148 m^2 X 5.5 kWh/m^2/day X 0.40 X 365/0.88 = 1048,000 kWh/yr

= 3,576,824 MBtu/yr

Where:

E_s = annual energy savings [kWh/yr]

I_{ave} = average solar radiation [kWh/m^2/day]

η_{boiler} = auxiliary heater efficiency.*

Typical auxiliary heater efficiencies are:

- Gas: 0.43 to 0.86, assume 0.57
- Elec: 0.77 to 0.97, assume 0.88
- Heat pump: assume 2.0
- Prop: 0.42 to 0.86, assume 0.57
- Oil: 0.51 to 0.66, assume 0.52.

4.3.4 Equation 3 – Annual cost savings

Annual cost savings can be estimated using the following equation:

$S = E_s\, C_e$ = 1048,000 kWh/yr X 0.084/kWh = $88,000/yr

= 3,576,824 MBtu/yr X $0.0246/MBtu = $88,000/yr

where:

S = annual cost savings [$/yr]

C_e = cost of auxiliary energy

Typically:

Electricity:	$0.084/kWh[= $0.0246/MBtu
Natural Gas:	$0.020/kWh = $0.0059/MBtu
Propane:	$0.040/kWh = $0.012/MBtu
Oil:	$0.025/kWh = $0.0073/MBtu.

4.3.5 Equation 4 – Solar system cost

Solar system cost can be estimated using the information in Figure 4.4 and the following equation:

$C = c_{solar}\, A_c$ = $650/$m^2$ X 1148 m^2 = $746,000

($C = c_{solar}\, A_c$ = $60.41/sq ft X 12,349 sq ft = $746,000)

where:

c_{solar} = per-unit-area cost of installed solar system [$/$m^2$], typically:
$650/$m^2$ ($60.41/sq ft) for large system
$1600/$m^2$ ($148.70/sq ft) for small systems
$1080/$m^2$ ($100/sq ft) for medium size system.

* Gas Appliance Manufacturer's Association.

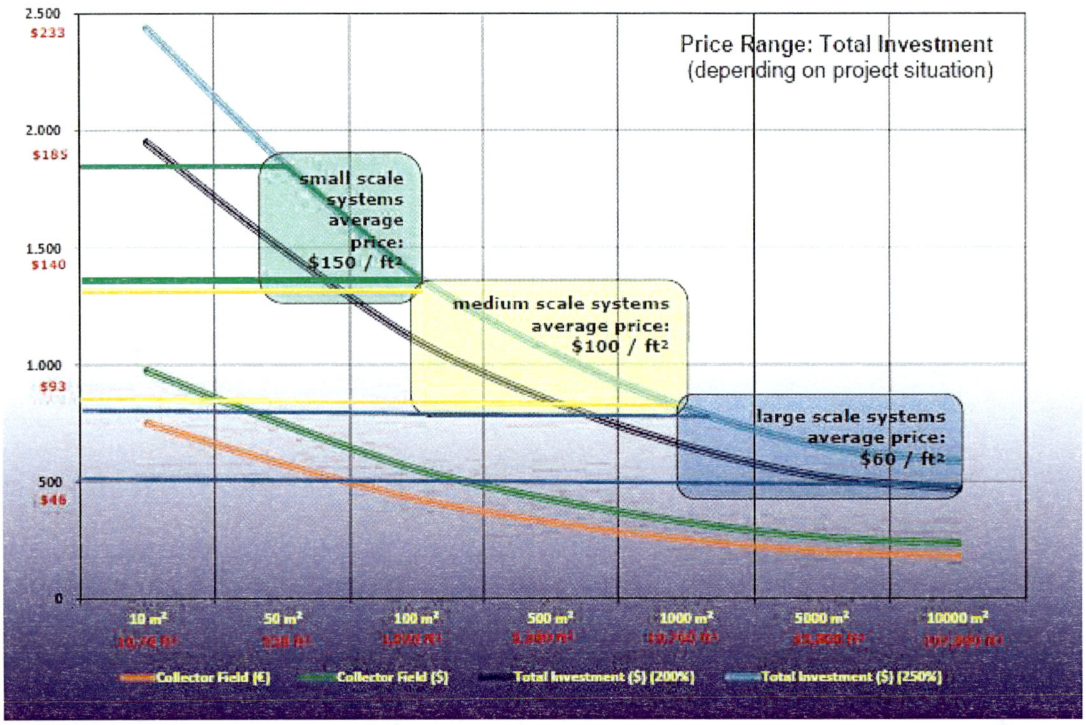

Figure 4.4. Solar system costs.

4.3.6 Equation 5 – Savings-to-investment ratio

The savings-to-investment ratio can be calculated using the following equation:

SIR = S * pwf / C = $88,000/yr X 24/ $746,000 = 2.8 (project is cost effective if SIR>1)

where:

pwf = present worth factor for future savings stream
 = 24 for 40-yr lifetime and 3% real discount rate (specified by NIST for 2009).

4.3.7 Equation 6 – Simple payback period

SPB = C / S = $746,000/$88,000 = 8.5 yrs

Another system sizing analysis would be:

For a location receiving 2,220 – 2,854 Btu/sq ft (7–9 kWh/m^2) on a high-radiation day a system that will provide heat for a domestic hot water system located will have a degree of utilization of ~50%. The result is 1,110 to 1,427 Btu/sq ft (3.5 to 4.5 kWh/m^2) of collector area provided as useful heat for generating the hot water. The value of 4.0 kWh/m^2 is equal to 13,650 Btu/m^2 or 1,270 Btu/sq ft of collector. This amount of heat can raise 19.2 gal water/m^2 from 55 to 140 °F. A sq ft of collector area can heat 1.8 gal. A solar storage system for this application should be in the range of 60 to 70% of the normal daily use. If the daily use is 1.8 gal/sq ft, then the storage tank size would be ~1.2 gal per sq ft

4.3.8 More accurate sizing approaches

Consider a typical solar system providing heat to maintain domestic hot water at 140 °F (60 °C) in the domestic hot water storage tank. There is an heat exchanger between the solar system storage tank and domestic hot water system. To achieve a DHW temperature of 140 °F (60 °C), the incoming stored water temperature must be 5 °F (2.7 °C) warmer or 145 °F (63 °C) as it enters the heat exchanger, assuming a plate-and-frame type. The design temperature in the storage could be as warm as 15 °F (-9 °C) to provide a few degrees for temperature loss and extra storage capacity. There is another heat exchanger between the water storage system circulation and the collector fluid circulation, which means the collector fluid should be a hot as 160 °F (71 °C) entering the plate-and-frame heat exchanger.

Thus the solar collector leaving temperature will be a minimum 160 °F (71 °C), which will establish the collector efficiency and the amount of heating energy the collector will produce per sq ft area. Knowing the domestic hot water demands and the incoming cold water temperature this will establish the energy demands for heating the domestic water. The size of the solar collector can be determined from this information. In sizing the collector system, an important consideration in the design is the selection of peak solar radiation. As the peak value chosen is reduced closer to an average solar radiation value for a specific installation, there will be periods when the collector will generate heat that the DHC system cannot consume. At these times, the collector system will have no place to send the heat. The flow through the collector will stop and the collector system will go into stagnation with the temperature of the heat transfer fluid going above the design 160 °F (71 °C). If it goes high enough the heat transfer fluid will begin to vaporize.

In Section 4.3.1 (p 59) "rules of thumb" were used to determine the solar collector size. A more accurate evaluation for determining the size of a solar hot water system would use values specific to the collectors that are planned to be used in the system. The Solar Rating and Certification Corporation (SRCC) provides such Information and efficiency ratings for collectors available in the United States. This independent organization tests and provides a certification rating of various solar collectors and solar water heating systems. Equipment that has certified and rated by SRCC must show their certification label that provides the products performance rating. Information from the SRCC is available through URL: www.solar-rating.org/ratings/ratings.htm

The SRCC performance ratings in Btu/sq ft/day are provided under three different solar weather conditions - clear, mildly cloudy, and cloudy skies. Table 4.3 lists the five levels of service. The ratings also include durability and efficiency. This information can be used to compare different types of collectors. Although these ratings give the heat output for collectors, they do not provide cost information. Look at the amount of heat (BTUs) the collector delivers per day relative to its cost.

To compare two panels, first look at the service category (A through E) that represents water use. Then look at the output of the three solar weather conditions for each panel. To determine the collectors that produce the most heat for the least cost, figure the price per square foot of the panels by dividing the panel price by the panel area. When comparing panels, some may perform better in sunny conditions and some will perform better under cloudy conditions. Unless the system will be located in an area with lots of cloudy days such as the Pacific Northwest, it is more important for a collector to do better under sunny conditions than under cloudy conditions because there is more heat to capture on sunny and partly sunny days than on cloudy days.

Table 4.3. Solar collector categories.

Use Category	Temperature difference	Typical Application
A	-9 °F (-5 °C)	Swimming pools and solar assisted heat pumps
B	9 °F (5 °C)	Swimming pools and solar assisted heat pumps
C	36 °F (20 °C)	Service hot water and space heating - air systems
D	90 °F (50 °C)	Service hot water, space heating - liquid systems and space cooling
E	144 F (80 °C)	Space heating - liquid systems, space cooling and industrial process heating

SRCC also tests and provides performance ratings for small hot water systems. Information is provided on the storage tanks size and the total system performance for locations that can be selected from a pull-down list. The computer program *TRNSYS* is used to determine the values, which provide the Solar Energy factor (SEF), equivalent Solar Fraction, and equivalent solar savings (QSOLAR).

An example solar collector performance report is provided in the Appendices.

The performance of a solar hot water system (its "ability to capture solar radiation" and its "ability to deliver hot water") depends on the configuration of the system (collector area, controller setting, storage volume etc.), the current state of the system (i.e., temperature of the hot water in the storage tank) and factors from its environment (notably the ambient temperature and insolation patterns over the year). Indicators (*SF, SN, and SE*) that are used to measure the performance of solar thermal systems are discussed in Section 3.1 (p 15). Another indicator is the standby hot water volume, which is the amount of water at the desired delivery temperature that the system can deliver at any time without any instant additional heating required for this delivery. With the following situation, the standby hot water volume is 83 (= 50x[65-15]/[45-15]) liters of water (22 gal) without any additional instant heating.

- Standby storage tank: 26 gal (100 L)
- Backup fraction: 13 gal (50 L)
- Set temperature of the backup: 149 °F (65 °C)
- Tapping temperature: 113 °F (45 °C)
- Cold water feed: 59 °F (15 °C).

The performance parameters *SF, SN* and standby hot water volume, together with the costs for the heat form a triangle; that by maximizing one, others may be affected negatively. For example:

- Maximizing for *SF* means that the collector area and the store volume accordingly will be maximized as to make more solar energy available, reducing *SN*, increasing the costs.
- Maximizing for *SN* means that the collector area will be undersized, because then excess available energy is avoided, but less solar energy will be available for the user reducing.
- Maximizing for standby hot water volume means that a larger fraction of the store will be at a higher temperature, which requires more auxiliary heating (lower *SF*) and also has a negative effect on the efficiency of the collectors.´

Optimization thus involves a prioritizing and compromising (Figure 4.5).

Independent evaluation of single components of a solar system provides limited predictability of the performance of the system as a whole (i.e., yield). With larger components (in particular the collector area and the storage tank) the "principle of diminishing returns" applies. This means that although the yield will increase when the collector area or storage tank volume is increased, the marginal increase becomes smaller with increasing area and volume. In addition, maximizing one component over others also has a strongly diminishing effect. A diminishing effect usually also means an increase in energy costs (increase of marginal and average $/kWh). Simulations with various component sizes aid in the design of an (cost) optimal system.

There are design tools available for sizing the solar collector system. The National Renewable Energy Laboratory (NREL) developed the Federal Renewable Energy Screening Assistant (FRESA) software that can help facility managers determine if their building is a possible candidate for a solar water heating system. This Windows-based software tool screens federal renewable energy projects for economic feasibility and evaluates renewable technologies including solar water heating systems, photovoltaic, and wind energy systems. The Federal Energy Management Program is developing a new version, but information about the current version is available at: www.wbdg.org/tools/fresa.php

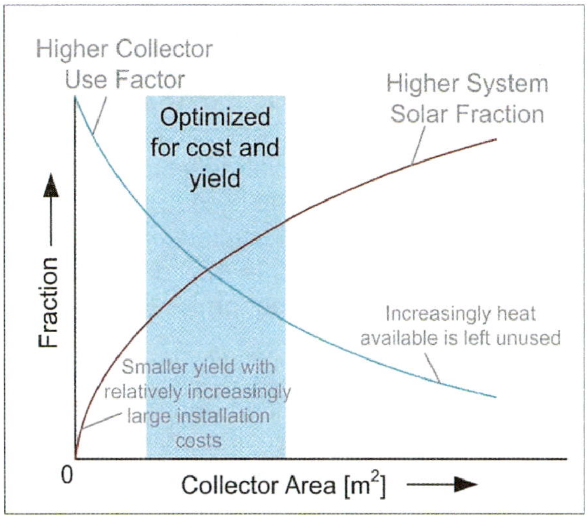

Figure 4.5. Sizing is a compromise between costs and yield.

A somewhat more detailed screening tool is provided by the Canadian Retscreen at www.retscreen.net/. This computer program is available as a free download and it will model glazed, unglazed and evaporative cooling solar hot water collectors. It uses f-chart to calculate the estimated collector area required, annual solar energy yield, solar system efficiency, solar fraction and pumping energy needed. The analysis is based on the input of average monthly solar radiation, outdoor temperature and relative humidity and wind speed. The building's hot water use is input for an average day over the year. This use cannot be varied. The results of this simulation compares favorably with other year-long evaluations. If a more detailed engineering and economic analysis is required, consider using the following software programs

- F-CHART, correlation method, available from the University of Wisconsin, available at http://sel.me.wisc.edu/fchart/new_fchart.html.
- TRNSYS, software, available from the University of Wisconsin, available at http://sel.me.wisc.edu/trnsys/.

F-Chart is available for the purchase price of $400 for a single user. It has the capability of modeling flat plate, evacuated tubes, CPC and 1 or 2 axis tracking collectors. It can handle water storage heating, building storage heating, domestic water heating, and integral collector storage heating. The output can provide a life cycle analysis with cash flow.

Other computer programs that can be used for sizing the solar system are T*SOL and Velasoris Poly Sun. The Federal Energy Management Program (FEMP) Help Line (800-DOE-EREC) provides manuals and software for detailed economic evaluation and for the Energy Savings Performance Contracting Program, which allows federal facilities to repay contractors for solar water heating systems through bills for energy savings instead of paying for initial construction.

4.4 Cost effectiveness

4.4.1 Cost of centralized versus de-centralized systems

Cost of Collectors Vs System Cost. When comparing solar thermal system alternatives: system size, configuration, types of collectors, etc., all cost components over the full lifecycle of the system should to be taken into account. This requires detailed information on the system hardware and results of system's performance simulation.

Advanced technology and production economies of scale have led to significant cost reductions in solar hot water collectors. The value of shipped low-temperature collectors was $1.89/sq ft (~14 €/m^2) in 2008. The average cost of thermosyphon systems with the storage integral to the collector was $24.27/sq ft (~183 €/m^2) ; the price of flat-plate collectors was $17.40/sq ft (~131 €/m^2); the price of evacuated tube solar collectors was $25.69/sq ft (~194 €/m^2); and the price of parabolic trough solar collectors was $11.96/sq ft (90 €/m^2) These values are based on collector factory revenue divided by output, so retail prices would roughly double (not including labor for installation), and the installed system price with all the other components is on the order of $75 to $225/sq ft (565 to 1696 €/m^2) depending on project size and location. New construction systems usually have better economics than retrofit projects because of reduced installation expenses.

Independent evaluation of a single system component cost and its performance, e.g., solar collector, provides limited information for the whole system evaluation. Increasing the size of a single component (e.g., the collector area or the storage tank volume) results in "diminishing returns." This means that, although the yield will increase when the collector area or storage tank volume is increased, the marginal increase becomes smaller with increasing area and volume. Also, maximizing one component over others has a strongly diminishing effect, which usually results in increased energy costs ($/kWh). Simulation of the system with variation of component sizes aid the design and energy cost optimization. Figure 4.6 shows approximate composition of the large system first cost.

Large central solar water heating systems are normally more cost effective due to economies of scale in installation, operation and maintenance compared to several small systems. This includes higher efficiencies possible* for backup heating systems in centralized systems. Table 4.4 lists the approximate effect of the solar thermal system size on investment cost generated from analysis of case studies described in Section 4.5 (p 71).

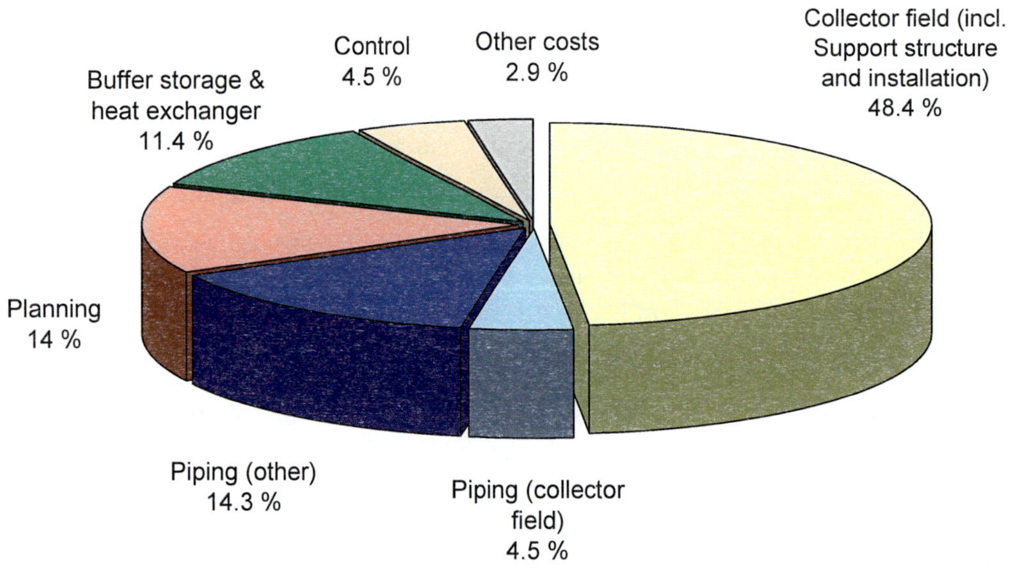

Figure 4.6. Distribution of costs for a large solar thermal system installation (Regenerative Energiesysteme).

* This depends on the specific configuration, but in particular equipment costs and boiler operating efficiencies are meant here.

Table 4.4. Economies of scale.

Price Ranges - Total Investment Solar System		
Large scale systems >10,000 sq ft (930m^2)	Medium scale systems 1,000 – 10,000 sq ft 93 – 930 m^2	Small scale Systems 500 – 1000 sq ft 46 – 93 m^2
av. \$50/sq ft ~377 €/m^2)	av. \$100/sq ft ~754 €/m^2)	av. \$150/sq ft 1,130.55 €/m^2

The values in Table 4.4 represent a first approximation and do not apply to every large building or cluster of buildings. The main criterion is the heat demand of the building or a cluster of buildings and their density (required distribution system).

4.4.1.1 System efficiency factor

Losses in heat distribution increase with distance and have an additional negative effect on overall system efficiency. For example, for large centralized systems the required storage temperature and delivery temperatures increase (above the operating temperature) due to heat losses through the walls of a storage tank and in the distribution system. Simulations of specific system is required to determine the overall performance (e.g., distribution losses as a fraction of the energy supplied).

4.4.1.2 Operation

In the particular situations with high peak loads and long periods when there is no hot water use in some buildings (e.g., in barracks during soldiers' deployment) central system will simplify adjustments needed to control such situations and potentially reduce the damage to the system. In the case of compounds with multiple barracks, large central SWH system can be more easily installed, maintained, and operated. Even finding a location for the solar collectors may be simpler. Large SWH systems also have the advantage in that the influence of individual users is minimal on their operation.

Small SWH systems require periodic checking and maintenance. Scheduling of maintenance for many distributed systems may be an issue of prioritization. On the other hand, maintenance of large central systems is more critical since a larger number of users could be affected and thus system monitoring and preventative maintenance are required activities to assure good performance.

4.4.1.3 Storage

Thermal losses from the storage tank have a significant effect on the efficiency of the solar thermal system. Heat is lost via the surface. The capacity of a sensible storage tank (e.g., water steel tanks) is defined by its volume. The volume grows with the 3rd potency of its circumference. The surface grows with the 2nd potency. Thus storage tank losses are reduced when one large store is used in comparison with many small storage tanks.

4.4.1.4 Back-up /auxiliary heating

Each solar thermal system must have a back-up heating supply. Oil, gas, electricity, or biomass is used as a heat source. Due to the 20+ year lifecycle of heating systems, including solar thermal systems, possible changes in energy supply options should also be considered. This could, for example, relate to future auxiliary (back-up) heating options from biomass (bio-waste) and waste heat supplies. In the case centralized systems, these options are often more easily integrated later.

4.4.2 Large scale SWH cost effectiveness

Monetary savings from installing a solar water heater depends on a variety of factors, including climate, the amount of hot water used at your location, the cost of conventional fuels, and system performance. Using these parameters for a large solar hot water system that could be applied to a cluster of buildings, the cost effectiveness can be illustrated with regional coloring presented on the maps of the United States shown in Figures 4.7 to 4.10. The information presented in these figures are for solar hot water systems that are replacing electrical or gas heating of hot water. An average efficiency of 40% for the solar system is assumed and a life of 40 yrs is used in the analysis. Two solar system costs are used: \$50 per sq ft of collector, and \$75 per sq ft (~377 €/m^2 and 565 €/m^2).

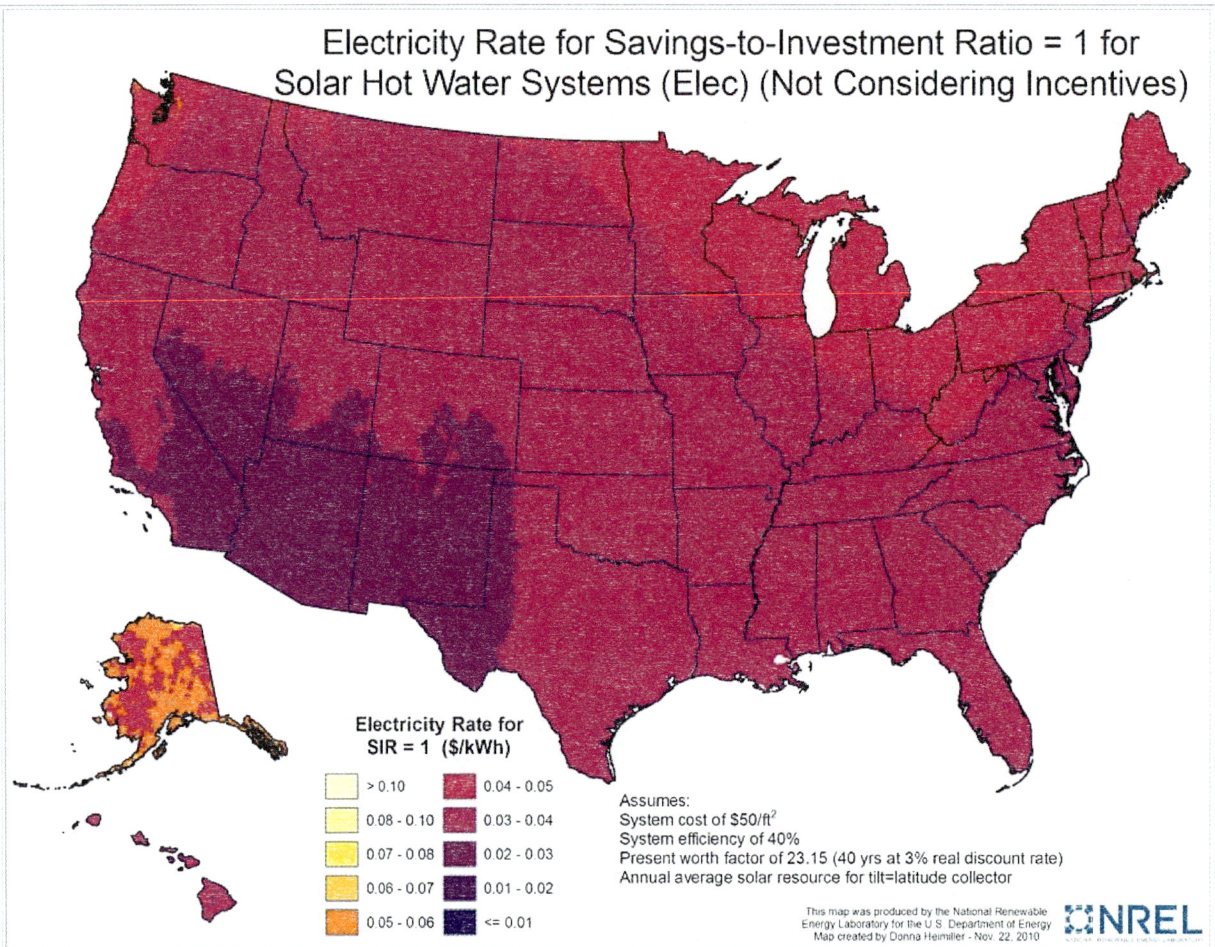

Figure 4.7. Cost effectiveness of solar hot water systems priced at \$50/sq ft (~377 €/m^2) replacing electrical heating use.

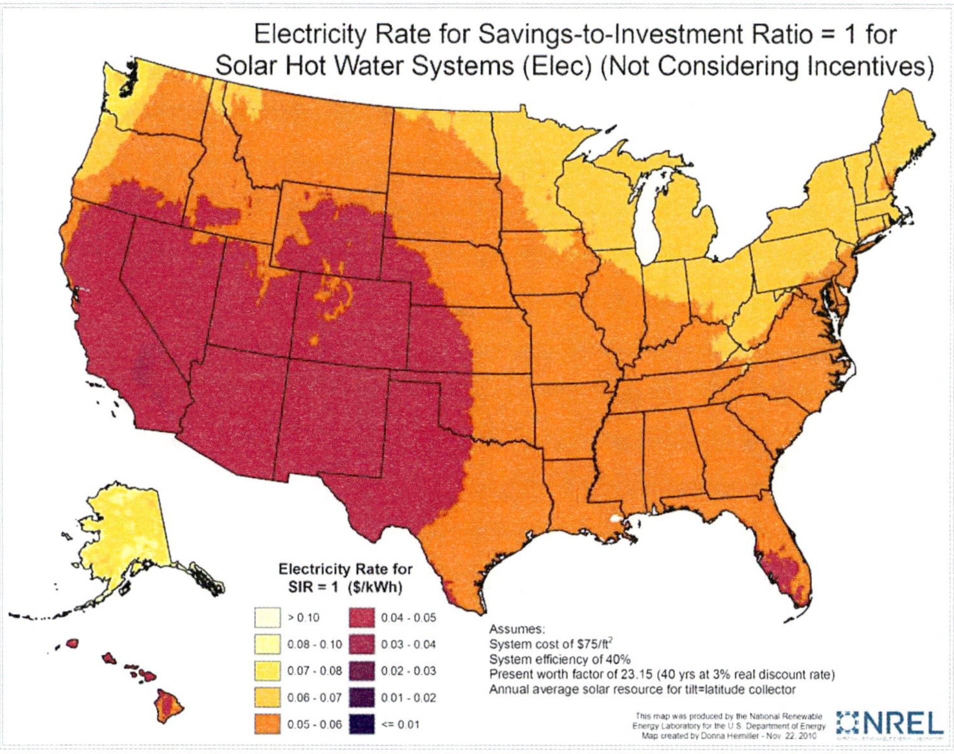

Figure 4.8. Cost effectiveness of solar hot water systems priced at $75/sq ft (~565 €/m^2) replacing electrical heating use.

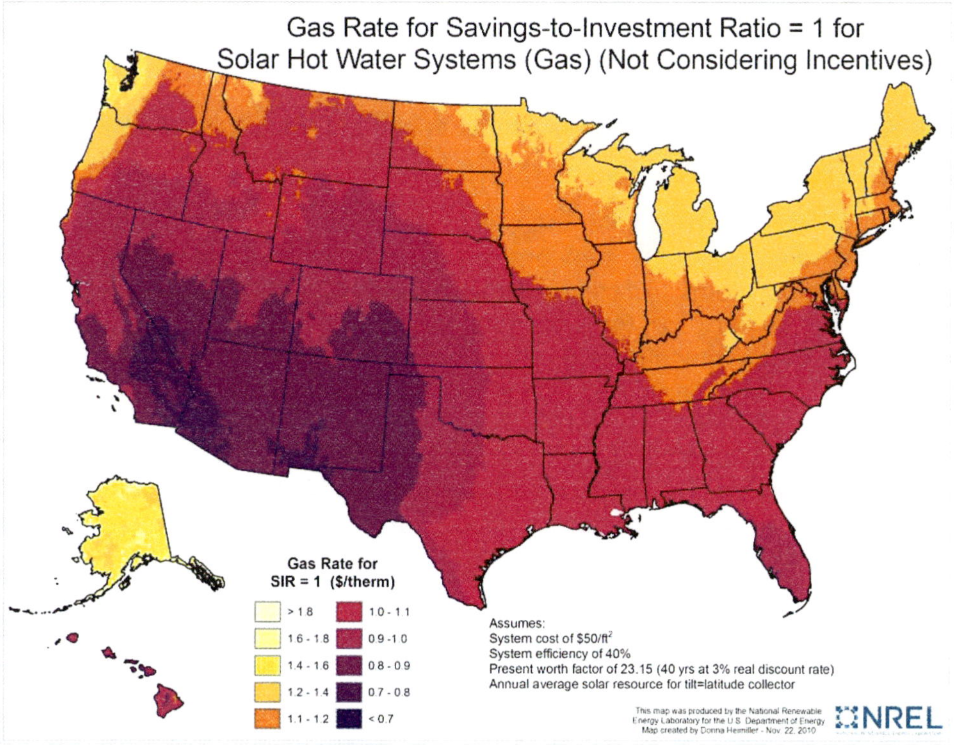

Figure 4.9. Cost effectiveness of solar hot water systems priced at $50/sq ft (~377 €/m^2) replacing gas-fired heating use.

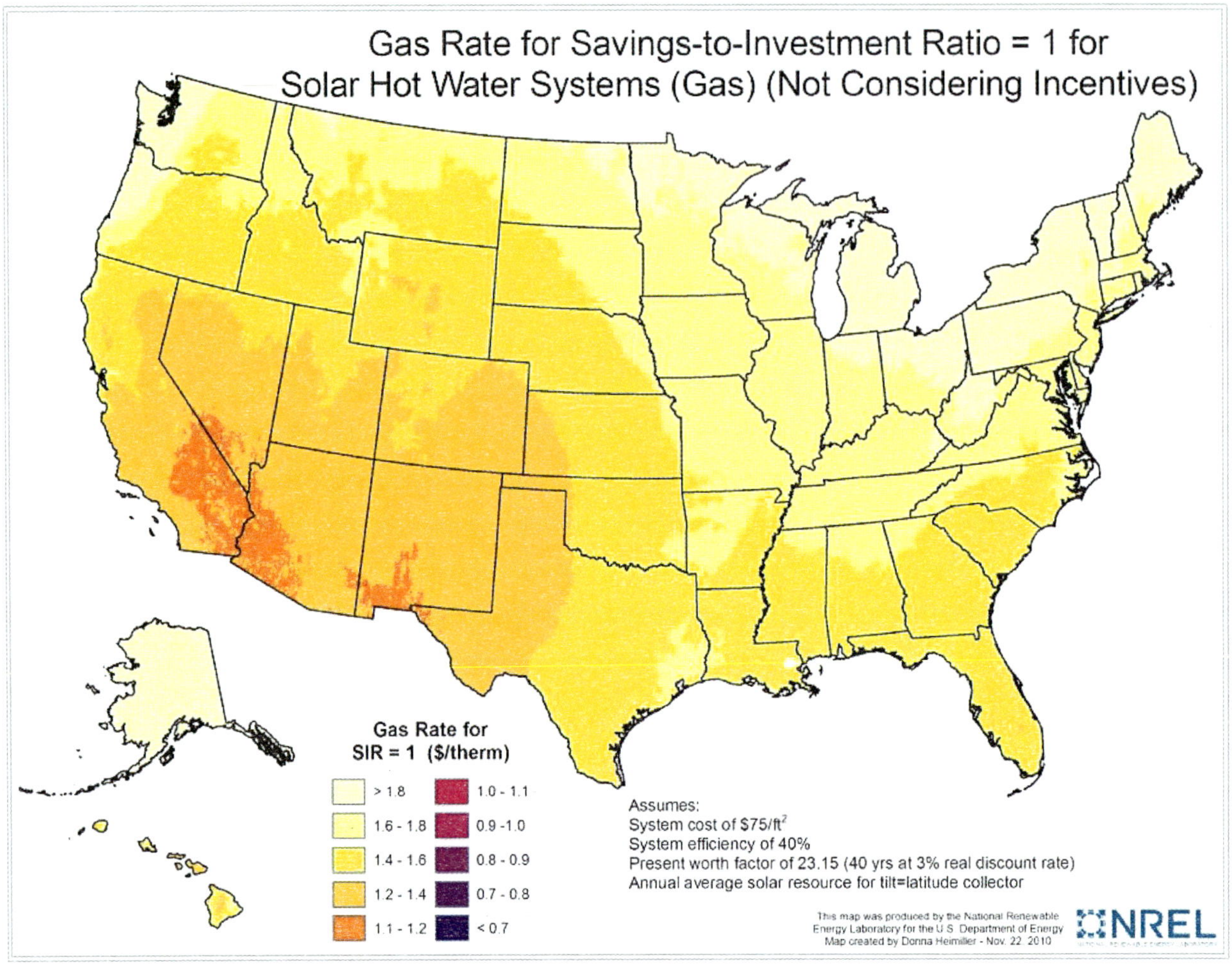

Figure 4.10. Cost effectiveness of solar hot water systems priced at $75/sq ft (~565 €/m^2) replacing gas-fired heating use.

This Guide is focused on large centralized solar thermal systems with short and medium-term storages connected to heating networks and they provide much lower specific system costs than decentralized small-scale solar systems. As can be seen in Figure 4.11, the cost/benefit-ratio (investment cost/ energy savings per year) for large solar systems with collector areas > 1,076 sq ft (100 m^2) is about half that of small, decentralized systems, and can even be reduced by more than 20% when using large scale systems combined with seasonal storages.

A general rule of thumb for federal facilities is that a renewable energy installation should pay for itself within about 10–15 yrs. System life spans can be as much as 30 yrs, which means a facility can look forward to as much as 20 yrs of "free energy."

4.5 Case studies

Appendix A provides includes descriptions of a number of case studies, which illustrate the application of solar hot water collector systems that supply buildings with hot water and heat.

There are very few large solar collector systems in the United States. In Europe the use of such systems is more commonplace. Table 4.5 lists the selected systems size, cost, and performance values.

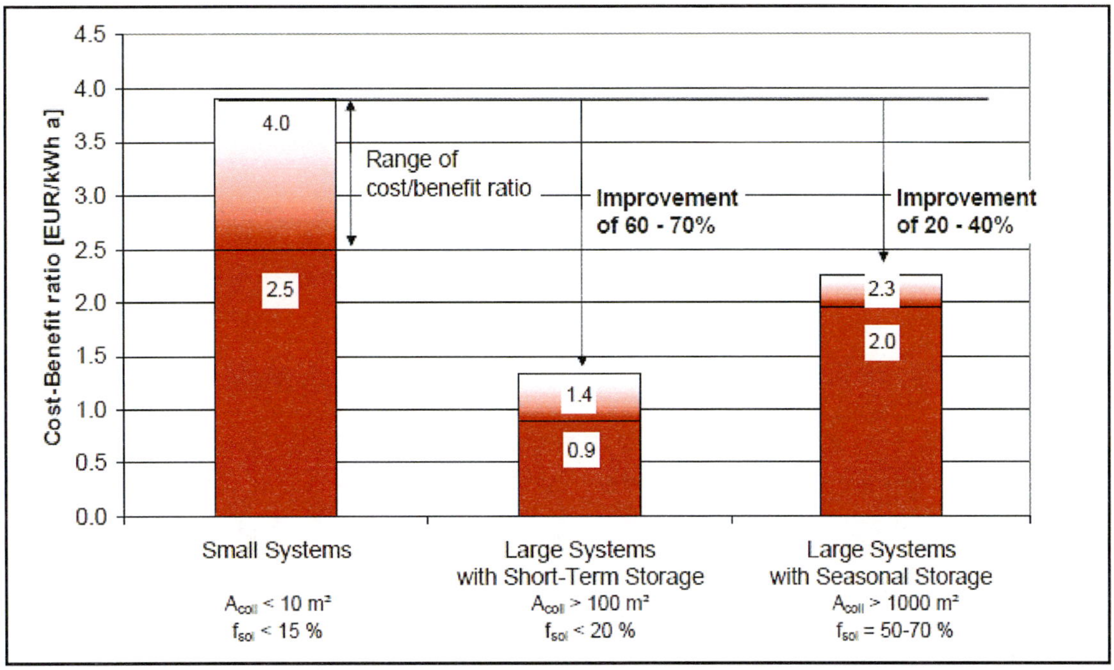

Figure 4.11. Cost/Benefit ratios of small decentralized solar thermal systems vs. large solar thermal systems with different storage capacities connected to heating networks.

Table 4.5. Summary of selected solar system case studies.

Location	Type Collector	Size sq ft (m²)	Storage Vol. gal (L)	Cost	Solar Energy Collected, MBtu/yr/sq ft (kWh/m²)	System Temp., out/in °F (°C)
Austria - AEE						
Gneis Moos, Salzburg	Flat Plate	4,412 (410)	26,420 (99,999)	$218,530	119.9 (377)	149(65) / 86–95 (30–35)
Wasserwerk Andritz	Flat Plate	41,481 (3,857)	17,067 (64,599)	$1,950,000	131.6 (415)	167–248 (75–120) / 140 (60)
UPC arena Graz-Liebenau	Flat Plate	15,139 (1,407)	-	$223,860	114.2 (360)	167–248 (75–120) / 140 (60)
Demark- ARCON						
Ulsted, Denmark	Flat Plate	53,929 (5,015)		$1,700,000	377.1 (1,189)	167– 248 (75–120) / 140 (60)
Strandby, Denmark	Flat Plate	86,769 (8,069)	396,423 (1,500,461)	$2,900,000	314.0 (990)	194–185 (90–85) / 68–50 (20–10)
Frederikshavn, Denmark	Flat Plate	1,614 (150)	1,320 (4,996)	$50,000	149.1 (470)	140 (60)/60 (16)
Skørping, Denmark	Flat Plate	5,918 (550)	396,423 (1,500,461)	$190,000	124.7 (393)	149 (65) / 77 (25)
Braedstrup, Demark	Flat Plate	86,769 (8,069)	528,564 (2,000,614)	$2,500,000	125.9 (397)	158–194 (70–90) / 95–101 (35–38)
Frankfurt/Main, Germany	Flat Plate	2,712 (252)	775 (10,503)	182,838E	83.1 (262)	DHW
Old Slaughterhouse Speyer, Germany	Flat Plate	5,864 (545)	26,428 (100,029)	357,020E	59.4 (187)	153 (67) / 90–99 (32–37)
Residential area Speyer, Germany	Flat Plate	3,077 (286)	6,607 (25,007)	189,200E	92.9 (293)	160–180 (71–82) / 126–140 (52–60)
Residential area Nordemey, Germany	Flat Plate	2,098 (195)	2,640 (9,992)	208,851E	97.9 (309)	149–176 (65–80) / 132–135 (56–57)

Location	Type Collector	Size sq ft (m^2)	Storage Vol. gal (L)	Cost	Solar Energy Collected, MBtu/yr/sq ft (kWh/m^2)	System Temp., out/in °F (°C)
Residential area Hennigsdorf, Germany	Flat Plate	9,218 (857)	10,570 (40,007)	549,570E	129.1 (407)	149–203 (65–95) / 97–106 (36–41)
Residential area Heilbronn, Germany	Flat Plate	4,049 (376)	11,100 (42,013)	234,561E	100.3 (316)	155–161 (68–72) / 113–123 (45–51)
Apt. Bldgs. Hannover, Germany	Flat Plate	1,333 (123)	1,586 (6,003)	99,410E	99.8 (315)	162–175 (72–79) / 117–135 (47–57)
Residential area Stuttgart, Germany	Flat Plate	16,613 (1,545)	23,780 (90,007)	797,788E	96.7 (305)	171 (77) / 122 (50)
Saint Paul, MN, USA	Flat Plate	21,034 (1,956)	System uses district heating distribution system as tank; solar storage volume: 1200 gal (4542 L)	$2,200,000	(Less than 1 year's data available)	180 °F – 190 °F (82 – 88 °C)/ 160 °F (71.1 °C) (design temperatures)
Paradigma paper						
Trade Park, Housing Estate Ritter, Karlsbad, Germany	Vac Tube	667 (62)	1,585 (5,999)	$52,500	168.8 (532)	140–194 (60–90) / 77–144 (25–60)
Festo, Esslingen, Germany	Vac Tube	14,310 (1,330)	4,491 (16,998)	$825,000	124 (391)	176–203 (80–95) / 167–185 (75–85)
Cooney Island, New York	Vac Tube	1,761 (163)	3,963 (14,999)		203.4 (641)	140–194 (60–90) / 77–140 (25–60)
Alta Leipziger, Oberunsel, Germany	Evac Tube	1,268 (117)	1,849 (6,998)	$135,000	145.3 (458)	149–194 (65–90) / 95–158 (35–70)
Panoramasauna, Holzweiler, Germany	Evac Tube	1,057 (98)	0	$84,000	193.4 (610)	158–194 (70–90) / 149–176 (65–80)
Wohnheim Langendamm, Nienburg, Germany	Evac Tube	505 (46)	1,321 (4,999)	$38,000	162.2 (511)	140–194 (60–90) / 95–158 (35–70)
Kraftwerk, Halle, Germany	Evac Tube	241,024 (22,415)	9,511,200 (35,999,892)	$12,900,000	124.6 (393)	176–203 (80–95) / 131–149 (55–65)
Wels, Austria	Evac Tube	39,629 (3,685)	0	$3,000,000	146.4 (461)	194– 239 (90–115) /167–221 (75–105)
AWO Rastede, Oldenburg, Germany	Evac Tube	1,054 (98)	0	$112,500	174.4 (550)	167–185 (75–85) / 140–158 (60–70)
METRO Istanbul, Turkey	Evac Tube	11,063 (1,028)	3,963 (14,999)	$760,000	200.1 (631)	176–203 (80–95) / 167–185 (75–85)
USA - NREL						
Edison, NJ	Evac Tube	150 (13)	280 (1,059)	$26,000	121.7 (384)	180 (82) / 60 (16)
Philadelphia, PA	Evac Tube	576 (53)	None	$58,000	248.3 (783)	140 (60) / 60 (16)
Phoenix, AZ	Parabolic	17,040 (1,584)	23,000 (87,055)	$650,000	3.964 (12)	140 (60) / 60 (16)
Modesto, CA	Parabolic	57,969 (5,391)	None		14,600 (46,016)	460 (238) / 420 (216)

5. Design Considerations

5.1 Collector site placement

In an US Army installation, solar collectors can be placed on building roofs or on the ground adjacent to the buildings they would serve. Rooftop is limited by useful area and maintenance is a greater issue than with ground-placed systems. Access to roof mounted systems is more difficult; stairs or ladders need to be climbed, and space for safe movement between the collectors and the roof edge must be provided and maintained. Ground placement of the collectors can displace green spaces that are desired in building clusters, thus limiting the number of buildings to a given ground area. A compromise may be to place large collector systems above parking spaces in the soldier parking lots so the collectors can shade vehicles, and still remain close to the buildings and to the ground level for ease of maintenance. Figure 5.1 shows solar collectors located in a parking area.

When using a roof placement make sure that the predicted life of the roof is about equal to or longer than that of the solar system. Avoid placing a solar system having a life of 20 yrs on a roof scheduled for replacement in a shorter time.

Increasing the collector field distance to the mechanical room and storage tank increases losses and should always be minimized. The maximum distance should be less than 600 ft (1823 m) one way.

Figure 5.1. Solar hot water collectors place above automobiles where they are parked.

5.2 Structural (foundation)

The support of the rooftop solar collector becomes part of the building's structural system. For small collectors serving a single wood frame building the collector system can typically be secured directly into the roof rafters. For larger installations more typical of that on US Army barracks, dining facilities, and other buildings, an engineered substructure is generally required to meet local codes and to support the collector arrays. The design object would be to use less material and fewer roof penetrations. Considerations of the supporting structure design are:

- Penetrations though the roof for connection to the existing structure must be water tight using properly sealing methods. Apply insulation where needed.
- Dynamic loads of wind and snowfall (see Figure 5.2) must be included in the structural analysis
- Expansion and contraction of system components must be considered.
- Where geographically required, seismic loads shall also be included in the structural design.
- The system design must provide access for maintenance and safe movement near roof edges.

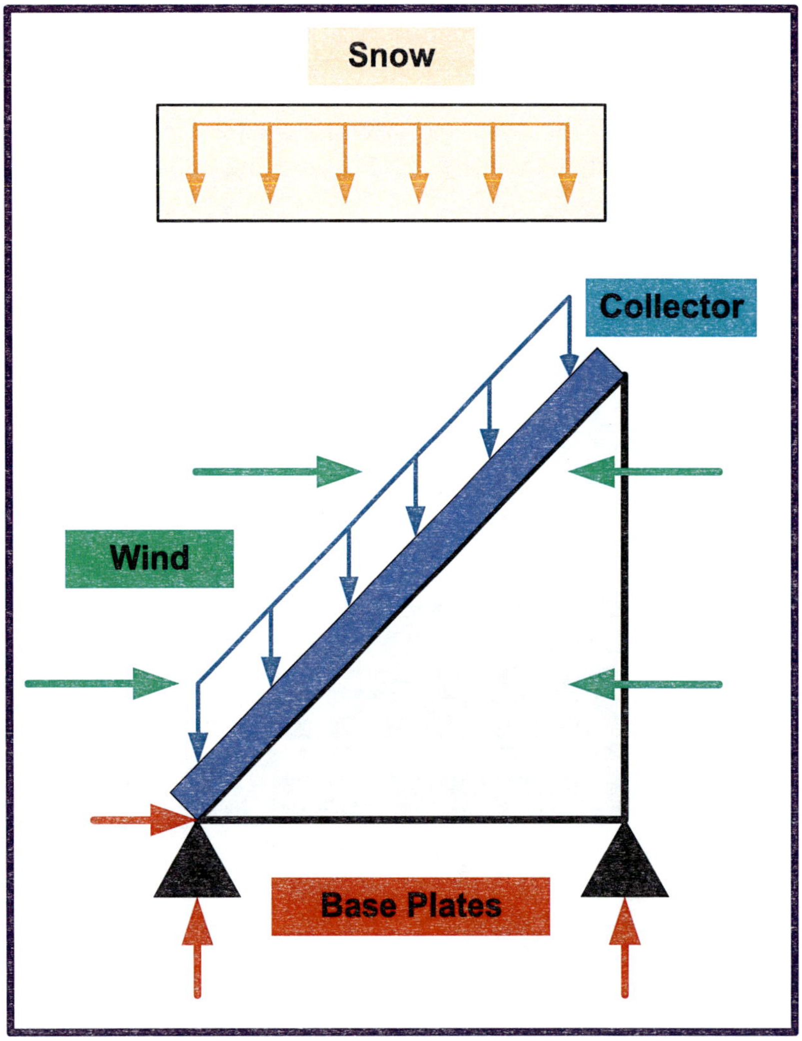

Figure 5.2. Forces on solar collector that determine structural requirements.

A major concern with the installation of solar collectors is the wind impact on the structure. Increased wind loading on the support substructure can be offset by increasing the securing mounting points and reducing the tilt of the solar collector. A taller solar collector has a greater wind resistance and thus a higher wind load.

A typical engineered substructure supporting the solar collectors consists of beams or open web joists mounted on supports attached to the building structure below the roof. These reasonable spaced attachments are optimized with the cost of the rest of the supporting structure. The cost on a flat roof is ~$10/sq ft (~75.4€/m^2). If the use of reasonably spaced support points is not desired, the support structure must span longer distances up to the width of the roof. The cost of using the longer span can approach 2.5 times the structure having intermediate supports. The frame that supports the solar collectors should be made using a non-corrosive metal such as aluminum.

Collectors mounted on sloped roofs typically take the roof slope as their tilt angle. This should be between 20 and 50 degrees. Slopes within this range on a sun facing roof will have only a slight reduction in performance when compared with the optimum tilt angle. These collectors may be attached a few inches above the roof to allow for rain water to flow underneath. Another style of placing solar collectors on a sloped roof is to integrate them into the roof surface. Figure 5.3 shows how these collectors form part of the roof replacing the roof materials being used. Pipe connects directly to the collectors from the ceiling space below. In-roof placement typically requires roofs with a slope of not less than 20 degrees to avoid standing water on the glazing, which may void the manufacturers water tightness guarantee.

Collectors can also be placed in a vertical wall (Figure 5.4). Flat plate collectors attached to a full surface of the building façade can eliminate the need for insulation and a weatherproof covering for the affected wall area, thus providing avoided cost savings that will offset the reduction in solar energy collection performance.

Collectors placed on flat roofs or the ground need to be arranged so that the optimum performance can be achieved. The materials used for constructing the structural supports need to be protected from corrosion. The use of steel hardware and fasteners in contact with aluminum collector frames and copper piping create a high likelihood of corrosion. Separating dissimilar materials with fluorocarbon polymer, phenolic, or neoprene rubber materials is recommended,

Figure 5.3. Collectors integrated into the roof as a structural element.

Figure 5.4. Collectors integrated into the façade as a structural element.

Placement distance between rows must be such that shading of one row by the next row behind is minimized. A slight amount of shading in the winter when the sun is lowest in the horizon can be permitted if the space for collectors is limited or piping costs are a concern since the solar radiation lost is a small percentage of the total annual amount. In areas where snowfall is normal a space of at least 12 in (30.5 cm). between the roof and the lowest collector part should be provided to allow for the collection of snow that has slid off the collectors. There also should be space for safe human movement between the collectors and around the end of collector rows. The space between the roof edge and the collector row must satisfy local building codes. Figure 5.5 shows a diagram of the space between collector rows (located on the ground or flat roofs) when the collectors are placed one behind the other on a horizontal surface.

When placing the collector on a roof, care must be taken not to damage the roof surface or interfere with other roof functions. Placement should allow for proper operation roof drains, HVAC and exhaust systems, plumbing vents, flues, or chimneys and antennas. Space for maintenance of the roof and those system placed on the roof shall also be provided. If the roof can handle the addition load of the solar collectors, then the collector supports can be attached to concrete slabs placed on the roof. To avoid harming the roofing below them these pads need to be placed on protective mats. Roof penetrations for structural connections and piping need to be made water tight. This is typically done using sleeves that surround the connecting structural or piping components, and that pass through the roof. A sealant is placed in the sleeve to form a watertight barrier and the top of the sleeve is ~3 in (77 mm). above the probable roof water level. The sleeve is appropriately flashed around for a good weather tight roof penetration. Figures 5.6 to 5.8 show several methods of supporting solar collectors on the ground or on a flat roof. It is advised that the roofing company that installed the roof be used to perform the roof work required to install the solar collector to preserve the remaining roof warranty.

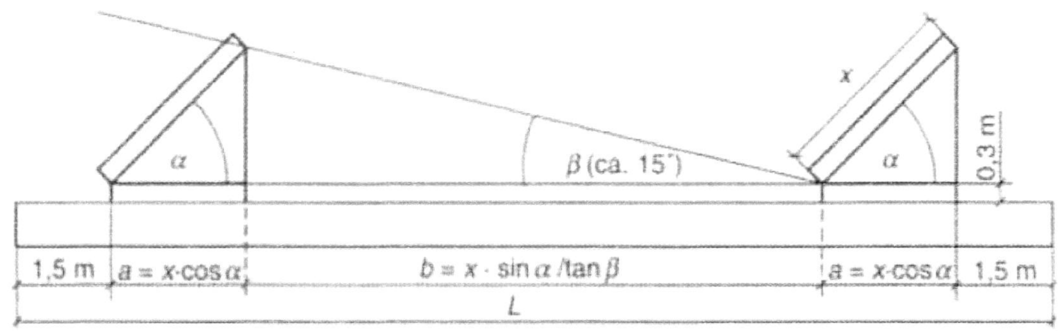

Figure 5.5. Spacing between collector rows.

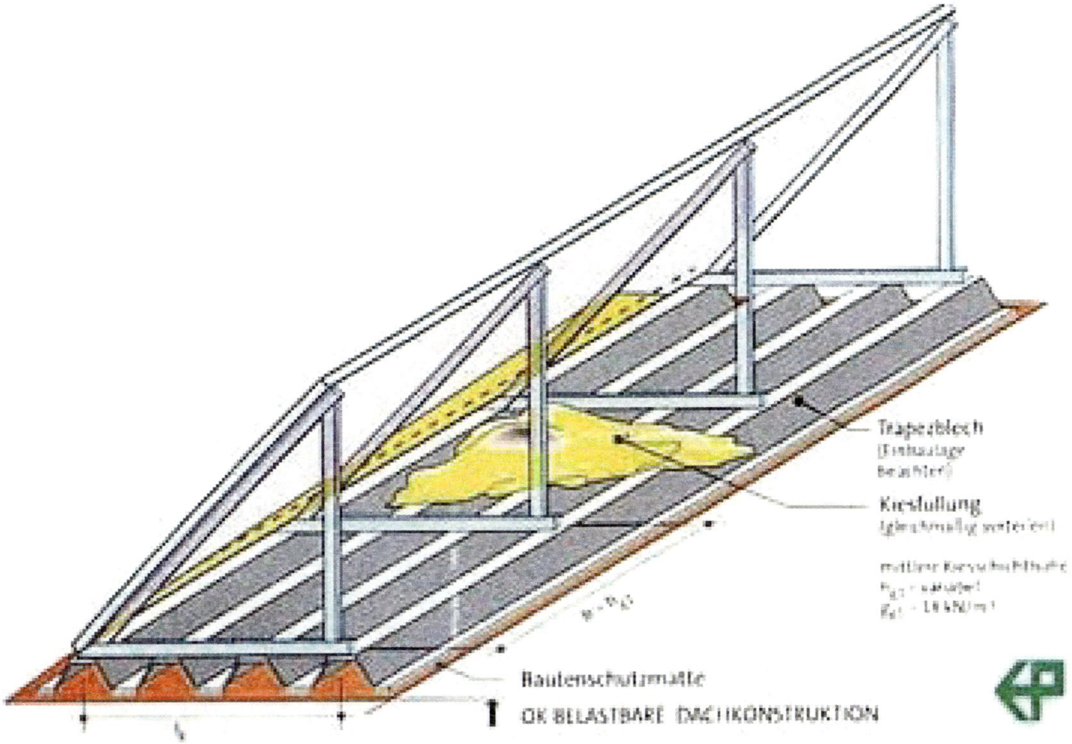

Figure 5.6. Collectors mounted on a trapezoidal sheet filled with a rock material.

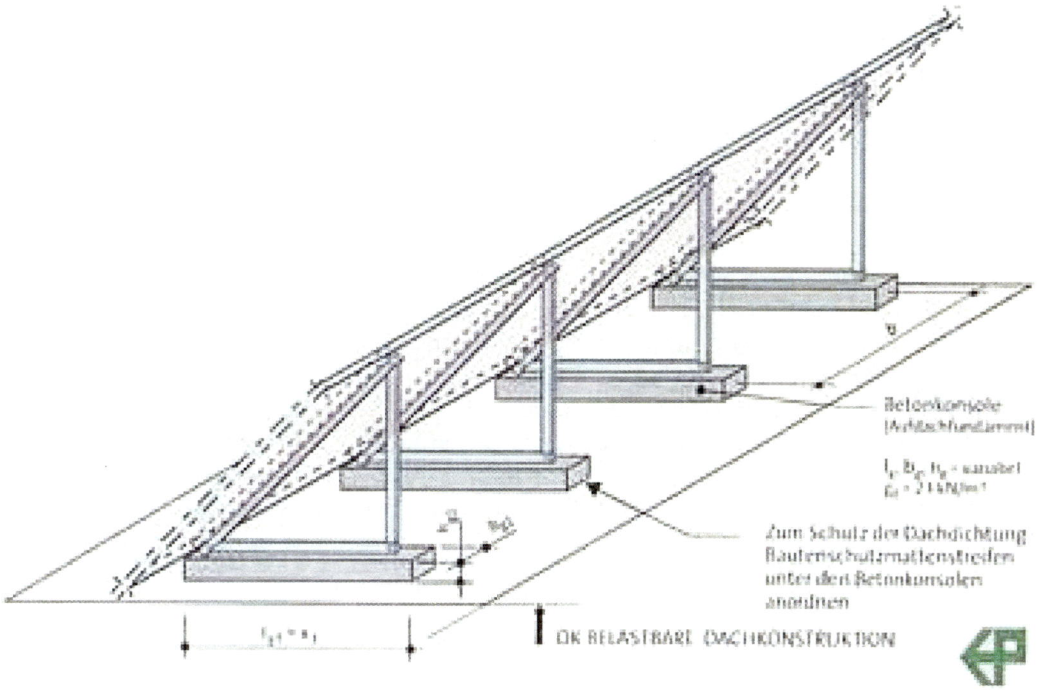

Figure 5.7. Collectors mounted on concrete slab.

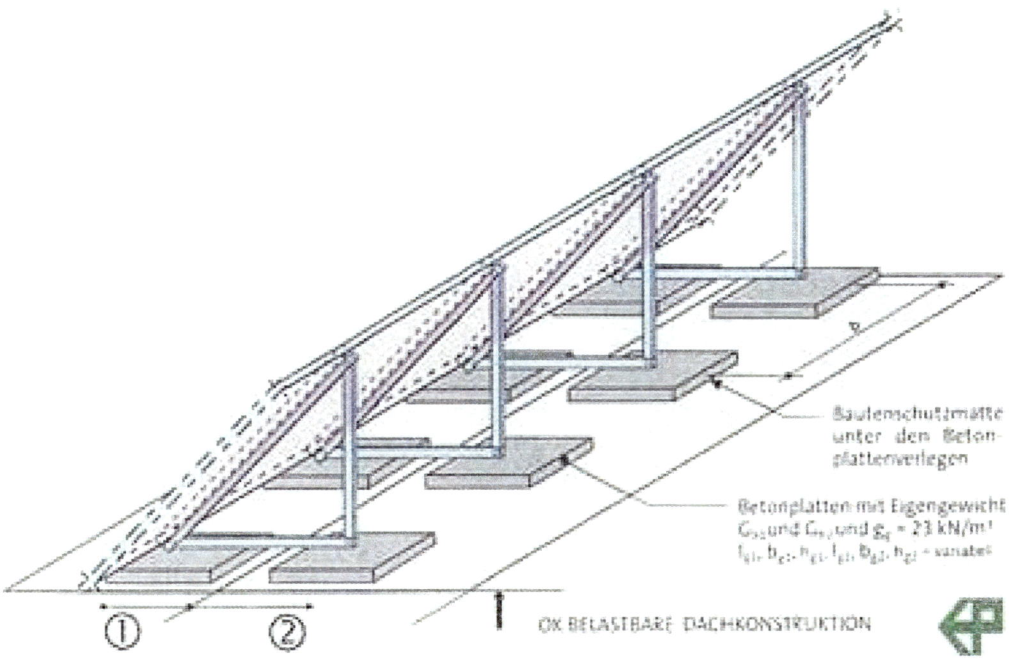

Figure 5.8. Roof support construction example.

5.3 Mechanical

5.3.1 Piping

Generally Type L copper, black steel and stainless steel are appropriate materials for the piping system. When using copper tubing, hard soldering is recommended for the collector loop. Care must be taken in using Teflon tape to seal threaded pipe joints when water /glycol is being circulated in the pipes. With proper dielectrics, black steel piping can be used on the collector side for use with glycol systems.

All piping shall be sloped at 1/4 in/ft (6.35 mm/0.30 m) of run back to the drain back tank. High points shall be kept to a minimum and combination automatic air vent valve/vacuum breakers shall be placed at all high points. Discharge of automatic air vents shall be piped back to the drain back tank and be provided with an in-line sight glass.

Piping should be designed for low pressure drop and the shortest routes used. All exposed piping should be well insulated with approved weather resistant insulation. Dielectric unions should be used at connections between dissimilar metals. Rubber or silicone hose used for connections must be of a high temperature type. The pipe ends should have ferrules to provide a good seal with the hose.

Pipe sizing should be in accordance with recognized methods. Figure 5.9 shows an example of a pipe layout with sizes for a collector field. The pipe sizes shown are metric sizes for copper tube. The equivalent nominal US copper pipe sizes are:

- 16u = 1/2 in.
- 19u = 5/8 in.
- 29u = 1 in.
- 35u = 1-1/4 in.
- 41u = 1-1/2 in.
- 54u = 2 in.

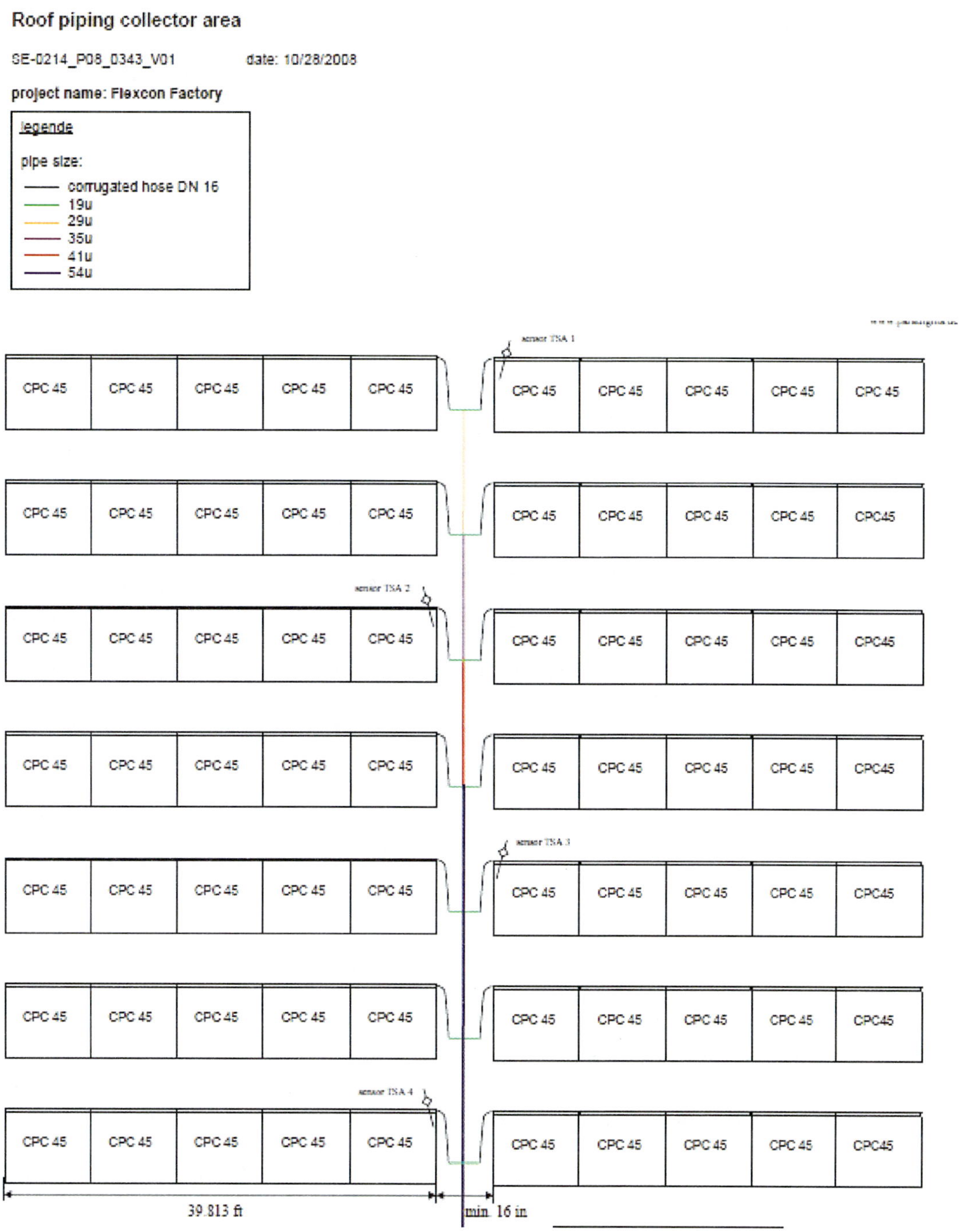

Figure 5.9. Pipe sizes used in a reversed return piping system.

The piping system (valves, pumps, fittings, flanges, connections, and insulation) should be designed to withstand the special conditions caused by the extreme temperatures of stagnation, e.g., 320 °F (160 °C) plus, and frost, e.g., -5 °F (-21 °C); expansion of pipe-length; pressure (e.g., steam) and working fluid (e.g., corrosion). An expansion vessel of sufficient size should also be part of each closed piping network. The system should be designed to operate at a pressure less than 125 psig (861.75 kPa), which will allow the use of standard piping components (class 125). A discussion of system temperatures and pressures is found in Section 3.4.2 "Stagnation" (p 35).

Collector piping must be able to withstand the expansion and contractions of components caused by the changes in temperatures that could be experienced. These temperature changes typically occur daily, which is significantly higher number of cycles than experienced by a normal heating system. The use of offset elbows, high pressure hoses, and expansion couplings should be considered rather than expansion loops unless they are placed horizontally due to drainage difficulties they would create.

Corrosion is a major concern in the solar hot water system. The two types of corrosion that cause the most galvanic damage and pitting corrosion. Solar energy systems generally contain a number of different metals such as aluminum, copper, brass, tin, and steel. This makes the solar system a prime candidate for galvanic corrosion. Heat transfer fluids can contain chemicals and heavy metal ions that would cause local or pit corrosion.

Galvanic corrosion is a type of corrosion caused by an electrochemical reaction between two or more different metals in contact with each other. A chemical reaction between the metals causes a small electrical current that erodes material from one of the metals. If the dissimilar metals are physically joined or if they are contacted by a common storage or heat-transfer fluid, the possibility of galvanic corrosion becomes much greater. Pitting corrosion is a highly localized form of corrosion resulting in deep penetration at only a few spots. This type of corrosion can take years to form, but can be very troublesome since it causes leaks that are difficult to locate.

Pit corrosion occurs when heavy metal ions such as iron or copper plate on a more anodic metal such as aluminum causing a small local galvanic cell can be formed. This corrosion spot or "pit" usually grows downward in the direction of gravity. Pits can occur on vertical surfaces, although this is not as frequent. The corrosion pits may require an extended period (months to years) to form, but once started they may penetrate the metal quite rapidly. Heavy metal ions can either come as a natural impurity in a water mixture heat transfer fluid or from corrosion of other metal parts of the solar system.

Pitting corrosion has the same mechanism (concentration cell) as crevice corrosion. Thus, it can also be aggravated by the presence of chloride or other chemicals that can be part of the water mixture or a contaminant from solder fluxes. Aluminum is very susceptible to pitting corrosion, while copper generally is not.

Several preventive measures will eliminate or at least minimize galvanic and pitting corrosion in collector systems that use an aqueous collector fluid. The best method to prevent galvanic corrosion is to avoid using dissimilar metals. Where this is not possible or practical, the corrosion can be greatly reduced by using nonmetallic connections between the dissimilar metals, thus isolating them. Galvanic protection in the form of a sacrificial anode is another method of protecting the solar system metals. Also, use of similar metals reduces the problems of fatigue failure caused by thermal expansion. Pitting corrosion is essentially eliminated if copper absorber plates are used in the solar collectors. Corrosion inhibitors can minimize pitting corrosion in aluminum absorbers.

When sacrificial anodes are to be used Their placement is important to obtain good protection and it depends on what is being protected, the anode material being used and the electrical conductivity of the heat transfer fluid.

5.3.2 Valves

Valves are used in the solar hot water system for balancing flow, flow adjustment, component isolation, and for temperature and pressure control. These valves should be of the same material as the pipe. At drainage locations of the collector piping ball valves should be used.

For solar collector arrays, the piping design can create nearly equal flow, but a balancing valve may be needed for a final adjustment. Use of these valves should be minimized since their use adds to the pressure drop in the piping system. Each collector array should also have a ball valve at the inlet, and a three-way valve at the outlet, with the third port open to atmosphere for isolation. Use of these valves will allow parts of the system to be taken out service for maintenance, repair, etc. while the remaining parts of the system are operating. These valves can also be used to stop flow as part of a freeze protection plan. Both of these valves are manually operated. The balancing valve should be set during fluid flow system set-up and then fixed at that position.

At heat exchangers, fluid flow adjustment may be necessary to assure the appropriate temperature is being delivered by the solar system components. For example, a mixing valve is used to blend hot water with cooler return water so that a 140 °F (60 °C) domestic hot water is delivered to users. Another example is flow control valves that can be used to direct heated water to the proper height of the storage tank to minimize mixing of different water temperatures. Figures 3.21 through 3.25 show such arrangements.

Valves, other than seasonal or emergency shut-off valves, should be electrically operated and located out of the weather or well protected. A vent must be provided at the high point in liquid systems to eliminate entrapped air and it should also serve as a vacuum breaker to allow draining of the system. To avoid multiple venting, systems should be piped to avoid having more than one high point. Pressure relief by safety relief valves must be provided at some location in each flow circuit that can be isolated by valves. The safety valves must be sized for flow conditions that could occur under stagnation. Check valves can be added to prevent thermally induced gravity circulation. A flow-check valve (used in the hydronic heating industry) will also accomplish the same purpose.

When the solar storage tank contains hot water to be used as DHW, an anti-scald valve is often required on the leaving DHW pipe to limit the outlet temperature, which can reach temperatures of 176 °F (80 °C). Care should be taken to compensate for any pressure drop this valve adds to the DHW circulating piping system.

5.3.3 Strainers and filters

Piping accessories such as fluid strainers and filters are those that would be used with any heating system. They should be place at the inlet of pumps and before control valves.

5.3.4 Fluid pumps

The type pump to be used is a centrifugal type that is used in building heating systems. If the loss of fluid to the environment is a concern a seal-less magnetic drive centrifugal pump should be considered. The pump components (seals, gaskets, bearings, etc.) must be able to withstand hot temperatures that could reach may reach 300 °F (149 °C) for short periods. Basically, three types of pumps are available:

- constant flow pumps
- electronic pressure controlled pumps (variable flow)
- high efficient pressure controlled pumps (variable flow).

Variable speed wet rotor circulators are preferred since they can operate the collectors either at a setpoint, or the most efficient range, while minimizing the electricity used to drive the pump. The use of variable flow is a method for achieving the desired temperature rise of the heat transfer fluid in the solar collectors. If the solar radiation is not very intense, then the fluid flow is reduced, causing the fluid to spend more time in the collector. Figure 5.10 shows the difference in energy use of these different methods of pump operation.

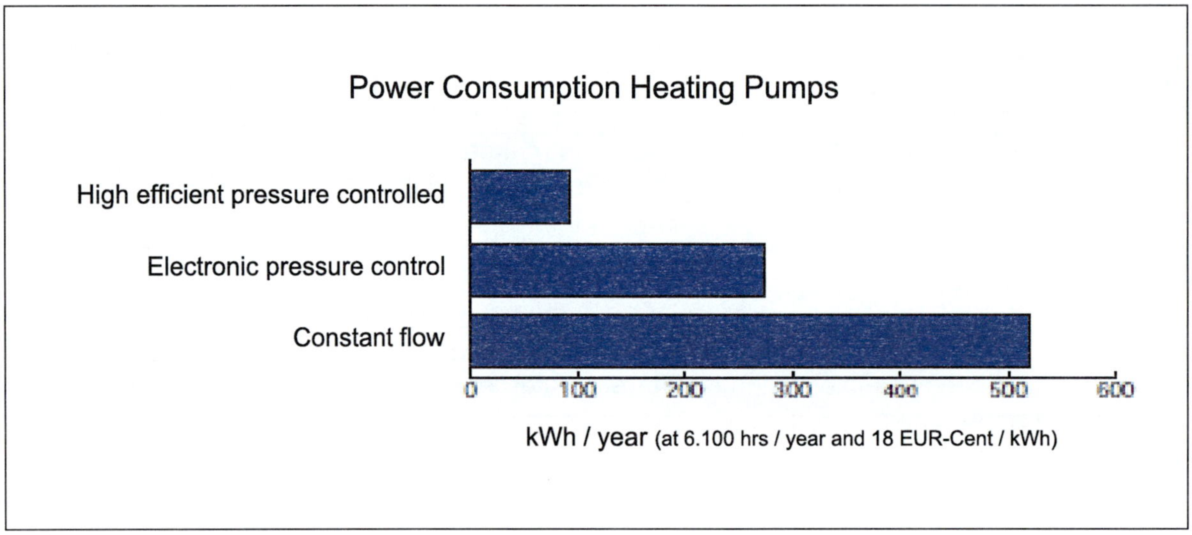

Figure 5.10. Pump motor energy use under various modes of operation.

Sizing of the pump is accomplished in the normal way of heating pump selection. An estimate of the flow rate can be made using the common flow rate for the collector field of 0.020m gpm per sq ft of collector area (when using water as the collector fluid). If a 40% glycol mixture at 140 °F (60 °C) is the collector fluid, then the flow rate would be ~10% more than a water solution. For a more accurate value, use of the computer programs identified in Section 4.3.2 (p 60).

If water is the circulated fluid and the system is open to the atmosphere or the water is potable then the pump wetted components should be made from stainless steel or bronze to minimize corrosion. Solar collector pumps should be placed in locations where leakage would not cause serious damage

5.3.5 Type of heat exchangers used

There are two types of heat exchangers to chose from: plate-and-frame, and tube-and-shell. Plate-and-frame heat exchangers are most economical for commercial systems as they take up little space and do not need insulation. They provide the highest approach temperature (the leaving temperature of the secondary fluid compared to the entering temperature of the hotter primary fluid). Plate-and-frame exchangers need very clean surfaces to obtain this performance and thus require a nightly reverse backflush to break free any accumulated deposit on the water side.

Tube-and-shell heat exchangers are a good option for light commercial applications since they are also practical in this setting (in terms of size), and because they have having good resistance to fouling since the water passages are large. The leaving temperature of the secondary fluid will be lower with this type of heat exchanger.

Conversely, water quality also affects system components; for areas with hard water (hardness > 100ppm), a closed loop or water softener should be used.

Generally stainless steel or copper is selected as a material for heat exchanger construction because of their good heat transfer properties and corrosion resistance.

If the heat transfer fluid is not safe for human consumption or is toxic (like ethylene glycol) then a double wall heat exchanger is required. Figure 5.11 shows an example of this type of heat exchanger where a volumetric space is filled with a nontoxic heat transfer fluid between the two circulating fluids.

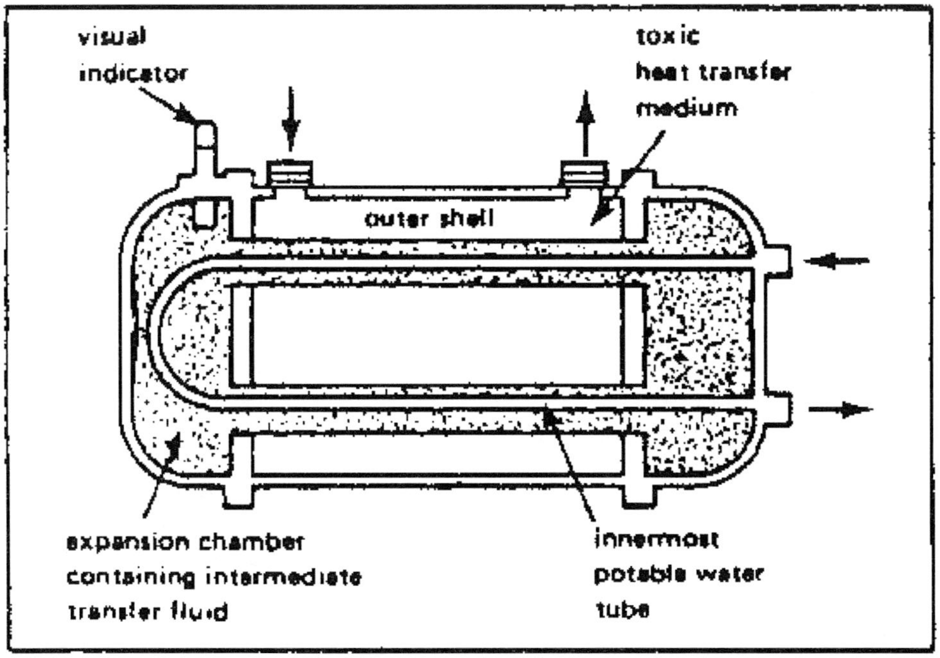

Figure 5.11. Shell and double tube heat exchanger used to protect potable water from harmful heat transfer fluid leaks.

A leak in one side would become visible and the other fluid could not become contaminated. Leak detection would involve noting a change in the fluid level in the interim space or a change in color of the interim fluid. When plate-and-frame heat exchangers are used leak detection can be provided by an additional heat exchanger circulating loop filled with a colored fluid for detection fluid between the glycol used in the collector circulating loop and the domestic water system. A check of the current building code should be made to determine the acceptable method of isolation.

To help minimize pumping energy, the pressure loss of the heat transfer fluids passing through the heat exchanger should be limited to 1 to 2 psi (6.9 to 14 kPa). Heat exchangers used to heat domestic water are exposed to the potable water pressure and thus should be rated for that pressure, typically above 75 psi (517 kPa).

5.3.6 Type storage tank used

The storage tank or tanks for a large hot water heating system should be placed near the solar collectors. These tanks are normally a non-pressurized type, which is open to the atmosphere using a vent in the top cover. The reason for this is the high cost difference of big tanks needed in large solar thermal systems. Pressurized tanks must be ASME rated in accordance with the maximum possible pressures and temperatures. (System designer to reference the requirements of the ASME boiler and pressure vessel Code for determination of requirements.) Non-pressurized systems are allowed and may be necessary to make the system LCCA attractive due to the high cost of ASME rated tanks. In instances where the tanks are non-pressurized, ASME rated tanks are not required. When using a non-pressurized tank, it must be the high point of that circulating fluid system

Domestic hot water storage tanks should be in the building where the hot water is consumed. These tanks would be typically smaller than the system storage tank and should be of a pressurized type. The tank can be piped so its water is warmed by water flowing from the solar heated storage tank and non-solar methods. Here a fossil fuel or electric heater can assure the proper domestic hot water temperature is being maintained.

Storage tanks can be made using steel, fiberglass/plastic, or concrete. Steel tanks make connections with the piping easy, are subject to corrosion; large tanks may require on-site fabrication. Fiberglass/plastic tanks do not corrode, but have a maximum temperature in the range of the temperature that could be expected during stagnation, they cannot be pressurized, and they are more costly. Concrete tanks are low cost, but must be site fabricated, cannot be pressurized, and can make plumbing connections difficult. Steel tanks should be lined with glass, epoxy, or other corrosion resistant material rated for the highest system temperature and working fluid. Alternatively, stainless steel tanks may be provided. Tank life should be at least 15 yrs.

Fiberglass and plastic tanks are corrosion resistant and easily installed, and are available in many shapes and sizes. Although many commonly fabricated tanks will begin to soften at temperatures above the temperature range of 140 to 160 °F (60 to 71 °C). There are more expensive, specially fabricated tanks available that can withstand temperatures up to 250 °F (121 °C). The types of plastics needed to store large quantities of water at high temperatures can be more expensive than steel.

All storage tanks require insulation with a rating at least R-19.It is a good practice to insulate tank supports from the ground if possible. Additional information on tank insulation can be found in the Solar Energy Equipment Chapter of the ASHRAE Handbook - HVAC Systems and Equipment.

All storage tanks for liquids should be located so that if they leak, damage to the building will be prevented. The drain back tank needs a drain line piped to a nearby floor drain. The cost of housing the tank or burying it must be included in the total cost of the solar heating system. Buried tanks must be protected from ground water, and buoyant forces resisted. Underground tanks are not preferred, if other options are available. Tanks must be reasonably accessible for repairs. Tank connections should comply with local codes with regards to backflow preventers, safety relief valves, etc.

5.3.7 Integration of solar collector system into existing hot water distribution system

Solar water heating systems maximize solar heat production when installed in a preheat configuration, i.e., the cold water supply is redirected to the solar storage tank where it can be preheated, and on a draw, the preheated water is directed into the backup or boiler tank for a final bump to temperature before servicing the usage. Of course this requires operation of the boiler equipment adding to the annual fuel use. Also, the firing rates in this mode may not be in the most efficient firing range of boiler operation.

One method to minimize the back-up boiler operation is to heat the DHW entirely by the solar collectors. One method to accomplish is to recirculate the water leaving the collector again through the collectors if it is not hot enough. This set-up requires a three-way valve between the return and either solar storage tank. This will allow the heat transfer fluid warmed by the collector to be directed either to the storage tank or to the line going back to the solar collectors. The control logic is when the solar tank temperature is higher than the return of the recirculation water; the solar collectors can heat it up further. When the solar tank temperature is below the temperature of the heat transfer fluid leaving the collector, it is directed to the tank.

5.3.8 Controls

There should be an automatic control system that senses system performance and properly transfers collected heat to the intended users. Since the amount of heat being collected can vary as does the demand for heat by the users, the control system must tract both of these variables and operate the systems equipment to achieve the maximum performance. The control system accomplishes this by taking information from sensors and, through the controllers, operates pumps, adjusts valves, and activates the auxiliary heating system.

To correctly monitor the solar hot water system, is important to understand system operation and to maintain good efficiencies. The collection of operational data will assist in the tracking of system performance and aid in scheduling system maintenance. This monitored information shall be collected every day of the year on 15-minute interval. All monitor systems should be centrally monitored for consistency; operating personnel must maintain a familiarity with the hardware and software, and the system must be properly managed through trend tracking and effective dispatching of service personnel. The minimum monitoring points are:

- tank temperature
- solar array circulating pump start/stop
- solar array circulating pump status
- solar controller alarm
- solar array inlet temperature
- solar array discharge temperature
- solar array discharge temperature alarm high limit.

The temperature sensors that measure the collector discharge temperature should be monitor the absorber plate temperature near the collector outlet. The storage tank temperature should be at the tank bottom to obtain the coldest temperature. These sensors send this collected information to a controller called a differential temperature thermostat, which then compares it with the adjustable setpoints - usually the high and low values. When the high value is reached (typically 12 °F to 15 °F (-11 to -9.44 °C) an action to withdraw the heat from the collector is initiated; for example starting a pump. As the pump runs the temperature differential drops and after some period of time the low setpoint is reached (typically 4 °F [-15.56 °C]). At this temperature difference, the pump is stopped. The pump is restarted when the high limit of the thermostat is again reached.

When a system has freeze and/or overheating protection, the controller takes the appropriate actions. For example, when freeze setpoint is reached at the collector, the pump could be activated to pump warm water from the storage tank into the collector. The freeze protection setpoint should be set at 40 °F (4 °C) since heat can radiate from the collector to the night sky creating collector freezing conditions above 32 °F (0 °C) outdoor temperatures. The freeze protection sensor must be placed on the collector so that it will sense the coldest water in the collector. This location may be the collector intake or return manifold, the back of the absorber plate near the bottom or center of the collector. The collector center is identified to monitor the irradiation leaving the collector at night. More than one sensor can be used for this function.

The energy produced by solar energy shall be determined by a Btu meter that will measure the domestic hot water flow from the storage tank, the incoming cold make-up water temperature, and the hot water temperature leaving the domestic hot water storage tank. These values should be continually evaluated and compared to previous values to assure proper performance is maintained.

Overheating controls would be initiated when the collector temperature reaches a temperature in the range of 200 to 250 °F (93 to 121 °C) depending on the system. The sensor monitoring this condition would be placed on the back of an absorber plate in the collector. When the collector reaches the high temperature setpoint due to a pump failure or some other event, the controller can take actions to relieve the heat and protect the system. See section 2.4.2 for more information.

The control components are the same that would be used on a commercial hot water heating system.

A typical sequence of operation for the operation of the solar collector circulation pump that sends water to the storage tank is:

- **System run command:**
  If solar array temperature is above the solar tank temperature by 12 °F (-11 °C), energize solar collector circulation pump P-1. Pump shall be energized during daylight hours only. On command to start, pump must remain energized for an adjustable period of time (initial setting of 20 minutes).
- **System stop command (high temperature):**
  If solar collector circulation pump P-1 is energized and if solar storage tank temperature rises above 185 °F (85 °C) (adjustable), activate a high temperature alarm. If solar storage tank temperature rises above 195 °F (91 °C) (adjustable), de-energize solar circulation pump P-1.
- **System stop command (low temperature):**
  If solar collector circulation pump P-1 is energized and solar array discharge temperature falls below the solar tank temperature for a period of 10 minutes (adjustable), solar collector circulation pump shall de-energize.
- **Thermal shock prevention:**
  If solar circulation pump P-1 is de-energized during daylight hours and if solar array temperature is below 180 °F (82 °C) (adjustable), solar circulation pump shall be enabled.

If solar circulation pump P-1 is de-energized and if solar array temperature is 180 °F (82 °C) or above (adjustable), solar circulation pump P-1 shall be disabled for an adjustable time period (8 hr initial setting).

The sequence of control assumes a constant speed pump circulating the heat transfer fluid through the collectors. If a variable speed pump is used, then the pump speed could be altered instead of stopping and starting the pump. Of course the pump operation would be stopped when night begins and when excessive temperatures are reached in the collector. Another option would have a three-way valve that would allow the heat transfer fluid to circulate through the collector again if it is not hot enough to be placed in the storage tank.

5.4 System startup considerations

All pumps, valves and sensors need to be checked for proper operation before starting of fluid flow systems.

5.4.1 Method of filling system and removal of air

Before filling the piping systems with heat transfer fluids the pipes need to be flushed to remove any foreign material and debris. A high head, low flow filling pump will produce the best results. For commercial systems, the arrays must be staggered so air can be methodically purged from the system.

5.4.2 Method of leak detection

On completion of the piping system, it should be pressure tested for leaks. All piping systems should always include a pressure gauge (digital or dial) for monitoring the fluid pressure. The system should be filled first with water and put under pressure of 1½ times the operating pressure for a minimum of 2 hrs, while the pressure is monitored and the distribution is inspected for leaks. The leak inspection should be completed before installing any piping insulation.

6. System Maintenance

6.1 General maintenance

Operation and maintenance (O&M) costs of each solar water heating system is estimated at ½ of 1% of initial cost per year. O&M is similar to that required of any hydronic heating loop and may be provided by site staff, with experts called in if something should fail. Regularly scheduled maintenance includes:

- Check the solar collectors and structure components for any damage. Note location of panel glazing or broken evacuated tubes needing replacement. Note any surface damage on absorber panel and that tubing containing heat transfer fluid is in good condition.
- Check tightness of mounting connectors. Repair any bent or corroded mounting components.
- Drain energy storage tanks for sediment removal.
- Check condition of heat transfer fluids.
- Determine if any new objects, such as vegetation growth, are causing shading of the array and remove them if possible.
- Clean outer surface of collector array annually with plain water or mild dishwashing detergent. Do not use brushes, any types of solvents, abrasives, or harsh detergents.
- Check all connecting piping for leaks. Repair any damaged components.
- Check plumbing for signs of corrosion.
- Check condition of corrosion inhibitors in heat transfer fluids and the state of the sacrificial anodes in the system.
- Observe operational indicators of temperature and pressure to ensure proper operation of pumps and controls.
- Observe that the collector heat transfer fluid pump is running on a sunny day and not at night.
- Use insolation meter to measure incident sunlight and simultaneously observe temperature and energy output values given by system controller. Compare the values with original efficiency of system.
- Check status indicators provided by system controller. Compare indicators with measured values.
- Document all operation and maintenance activities in a workbook available to all service personnel.
- Check proper position of all valves.
- Flush entire piping system to remove mineral deposits every 10 yrs.

A more through checklist can be found in the Solar Energy Use Chapter of the ASHRAE Handbook — HVAC Applications.

6.2 Glycol fluid care

The decomposition rate of glycol varies according to the degree of aeration, high temperature exposure and the service life of the solution. Most water/glycol solutions require periodic monitoring of the pH level and the corrosion inhibitors. The pH should be maintained between 6.5 and 8.0. Replacement of the water/glycol solution may be as often as every 12–24 months or even sooner in high temperature systems. If these solutions are used in the collector loop, the installer should specify the expected life of the solution and the amount of monitoring required. The cost of periodic fluid replacement and monitoring should be considered in the economic analysis. For best performance, the glycol should be replaced every few years. Since glycol-water mixtures do require a lot of maintenance (and since users can be quite negligent) it is recommended that glycols not be used in family housing solar heating and DHW systems, and that glycol-water solutions be reserved for use in large-scale installations, which have regular maintenance schedules.

References

L.D. Danny Harvey. 2007. A handbook on low-energy buildings and district energy systems. Fundamentals, techniques and examples. Earthscan, Sterling, VA.

Solar Thermal Systems, Peuser, Felix A.; Remmers, Karl-Heinz; Schnauss, Martin; Solarpraxis AG, Germany, 2002

Florida Standards for Design and Installation of Solar Thermal Systems, Florida Solar Energy Center; FSEC Standard 104-10; Jan. 2010

Solar Water Heating; Federal Technology Alert; FEMP, DOE; DOE/GO-10098-570; May 1996

Planning and Installing Solar Thermal Systems. A Guide for Installers, architects and engineers. Second Edition. Earthscan. London, Washington, DC. 2010.

Martin Kaltschmitt, Wolfgang Streicger and Andreas Wiese. 2007. Renewable Energy. Technology, Economics and Environment. Springer Verlag. Berlin, Heidelberg, NY.

Les Nelson. 2010. The Importance of Certification for Growing US Markets. "Solar thermal hot water and combisystems, Testing and Certification" – Industry Workshop, ITW, University of Stuttgart, Germany, Monday, February 8th, 2010

Werner Weiss, Renewable Energy in Central and Eastern Europe. MSc Program. AEE INTEC.

Duffie J. A., Beckman William A. 1991. Solar Engineering of Thermal Processes, Second Edition, NY.

Suter J.M., Letz T., Weiss W., Inäbnit J. Solar Combisystems in Austria, Denmark, Germany, Sweden, Switzerland, the Netherlands and the United States, Overview 2000, Bern 2000

Christian Fink and Richard Riva. 2004. Solar-supported heating networks in multi-storey residential buildings. A planning handbook with a holistic approach. Arbeitsgemeinschaft ERNEUERBARE ENERGIE GmbH – Institute fur Nachhaltige Technologien. Gleisdorf, Austria. 1st Edition.

Neue Perspektiven für Luftkollektorsysteme, Gerhard Stryi-Hipp, Korbinian Kramer, Michael Hermann, Jens Richter, Philipp Hofman, Paolo di Lauro, Christoph Thoma, Fraunhofer ISE, 20. Symposium Thermische Solarenergie

Process Heat Collectors," State of the Art within Task 33/IV, Matthias Rommel, Fraunhofer ISE, Werner Weiss, AEE INTEC, Booklet prepared as an account of work done within Task 33 "Solar Heat for Industrial Processes" of the IEA Solar Heating and Cooling Programme, AEE INTEC, Gleisdorf, Feldgasse 19, Austria, 2008

Karl-Heinz Remmers. Große Solaranlagen, Einstieg in Planung und Praxis, Solarpraxis AG, Berlin 2001

Heizungsanlagenverordnung, Sonderdruck "Die wichtigsten Anforderungen der Heizungsanlagenverordnung für den Praktiker," Fachverband Sanitär, Heizung NRW, Düsseldorf

A Guide to Glycols, Dow Chemical Co.

American Society of Heating, Refrigeration and Air-conditioning Engineers: ASHRAE Handbook - HVAC Systems and Equipment

American Society of Heating, Refrigeration and Air-conditioning Engineers: ASHRAE Handbook - HVAC Applications

VDI-Richtlinien 6002, Part 1 Solar heating for domestic water. General principles, system technology and use in residential building. VDI, Düsseldorf, 2004.

VDI-Richtlinien 6002, Part 2, "Solar Heating for domestic water. Application in student housing, senior citizens' homes, hospitals, swimming baths and campgrounds," VDI, Düsseldorf, 2007

ASHRAE Standard 93-2010 -- Methods of Testing to Determine the Thermal Performance of Solar Collectors. American Society of Heating, Refrigerating and Air-Conditioning Engineers, Atlanta, GA. 2010.

ASHRAE Standard 95-1987. Methods of Testing to Determine the Thermal Performance of Solar Domestic Water Heating Systems. American Society of Heating, Refrigerating and Air-Conditioning Engineers, Atlanta, GA. 1987

Related Government Design Specifications

US Department of Defense (DOD). Solar Heating of Buildings and Domestic Hot Water; Unified Facilities Criteria; Department of Defense; UFC 3-440-04N; Jan. 16, 2004

American Society of Sanitary Engineering (ASSE). ASSE 1003 (2001; Errata, 2003) Performance Requirements for Water Pressure Reducing Valves

American Welding Society (AWS). AWS A5.8/A5.8M (2004; Errata 2004) Specification for Filler Metals for Brazing and Braze Welding

ASME INTERNATIONAL (ASME). ASME B16.22 (2001; R 2005) Standard for Wrought Copper and Copper Alloy Solder Joint Pressure Fittings

ASME B16.24 (2006) Cast Copper Alloy Pipe Flanges and Flanged Fittings: Classes 150, 300, 400, 600, 900, 1500, and 2500

ASME B16.39 (1998; R 2006) Standard for Malleable Iron Threaded Pipe Unions; Classes 150, 250, and 300

ASME B31.1 (2007; Addenda 2008) Power Piping

ASME B40.100 (2005) Pressure Gauges and Gauge Attachments

ASME BPVC SEC VIII (2007; Addenda 2008) Boiler and Pressure Vessel Codes: Section VIII Rules for Construction of Pressure Vessels, Division 1

ASTM INTERNATIONAL (ASTM)

ASTM A 193/A 193M (2008b) Standard Specification for Alloy-Steel and Stainless Steel Bolting Materials for High-Temperature Service

ASTM A 194/A 194M (2009) Standard Specification for Carbon and Alloy Steel Nuts for Bolts for High-Pressure or High-Temperature Service, or Both

ASTM B 168 (2008) Standard Specification for Nickel-Chromium-Iron Alloys (UNS N06600, N06601, N06603, N06690, N06693, N06025, and N06045) and Nickel-Chromium-Cobalt-Molybdenum Alloy (UNS N06617) Plate, Sheet, and Strip

ASTM B 209 (2007) Standard Specification for Aluminum and Aluminum-Alloy Sheet and Plate

ASTM B 209M (2007) Standard Specification for Aluminum and Aluminum-Alloy Sheet and Plate (Metric)

ASTM B 32(2008) Standard Specification for solder metal

ASTM B 88(2003) Standard Specification for Seamless Copper Water Tube

ASTM B 88M (2005) Standard Specification for Seamless Copper Water Tube (Metric)

ASTM C 1048(2004) Standard Specification for Heat-Treated Flat Glass - Kind HS, Kind FT Coated and Uncoated Glass

ASTM D 3667(2005) Rubber Seals Used in Flat-Plate Solar Collectors

ASTM E 1(2007) Standard Specification for ASTM Liquid-in-Glass Thermometers

Copper Development Association (CDA)

CDA A4015. (1994; R 1995) Copper Tube Handbook

Manufacturers Standardization Society of the Valve And Fittings Industry (MSS) MSS SP-110 (1996) Ball Valves Threaded, Socket-Welding, Solder Joint, Grooved and Flared Ends

———. MSS SP-25 (2008) Standard Marking System for Valves, Fittings, Flanges and Unions

———. MSS SP-58 (2002) Standard for Pipe Hangers and Supports - Materials, Design and Manufacture

———. MSS SP-69 (2003; R 2004) Standard for Pipe Hangers and Supports - Selection and Application

———. MSS SP-72 (1999) Standard for Ball Valves with Flanged or Butt-Welding Ends for General Service

———. MSS SP-80 (2008) Bronze Gate, Globe, Angle and Check Valves

———. MSS SP-89 (2003) Pipe Hangers and Supports -Fabrication and Installation Practices

Sheet Metal and Air-conditioning Contractors' National Association (SMACNA). SMACNA 1650 (2008) Seismic Restraint Manual Guidelines for Mechanical Systems - Second Edition

Solar Rating and Certification Corporation (SRCC). SRCC CSCWHSR (2004) Summary of SRCC Certified Solar Collector and Water Heating System Ratings

———. SRCC OG-100 (1995) Operating Guidelines for Certifying Solar Collectors

———. SRCC OG-300 (2009) Certification of Solar Water Heating Systems

North American Board of Certified Energy Practitioners (NABCEP).

NABCEP (2009) Solar Thermal Task Analysis

US General Services Administration (GSA)

CID A-A-59617 (Basic) Unions, Brass or Bronze, Threaded Pipe Connections and Solder-Joint Tube Connections

———. FS A-A-50560 (Basic) Pumps, Centrifugal, Water Circulating, Electric-Motor-Driven

———. FS A-A-50561 (Basic) Pumps, Rotary, Power-Driven, Viscous Liquids

———. FS A-A-50562 (Basic) Pump Units, Centrifugal, Water, Horizontal; General Service and Boiler-Feed: Electric-Motor- or Steam-Turbine-Driven

———. FS A-A-50568 (Basic) Gages, Liquid Level Measuring, Tank

———. FS A-A-60001 (Basic) Traps, Steam

———. FS F-T-2907 (Basic) Tanks, Portable Hot Water Storage

———. FS WW-S-2739 (Basic) Strainers, Sediment: Pipeline, Water, Air, Gas, Oil, or Steam

Acronyms and Abbreviations

Term	Definition
AAFES	Army and Air Force Exchange Service
ASHRAE	American Society of Heating, Refrigerating, and Air-Conditioning Engineers
ASME	American Society of Mechanical Engineers
ASSE	American Society of Sanitary Engineering
ASTM	American Society for Testing and Materials
AWS	American Welding Society
BCT	Brigade Combat Team
CDA	Copper Development Association
CDC	Child Development Center
CEP	Central Energy Plant
CERL	Construction Engineering Research Laboratory
CID	Commercial Item Description
COF	Company Operations Facility
COSCOM	Corps Support Command
CPC	compound parabolic concentrating
DH	District Heating
DHW	domestic hot water
DOD	US Department of Defense
DOE	US Department of Energy
EISA	US Energy Independence and Security Act of 2007
EPA	Environmental Protection Agency
EPDM	ethylene propylene diene M-class [rubber]
ERDC	Engineer Research and Development Center
ERDC-CERL	Engineer Research and Development Center, Construction Engineering Research Laboratory
ETC	evacuated tube collector
FEMP	Federal Energy Management Program
FPC	flat plate collector
FRESA	Federal Renewable Energy Screening Assistant
FSEC	Florida Solar Energy Center
GSA	General Services Administration
HT	Heat treated
HVAC	heating, ventilating, and air-conditioning
IAM	Incidence Angle Modifier
IEA	International Energy Agency
IMCOM	Installation Management Command
LCCA	life-cycle cost analysis
MSS	Manufacturers Standardization Society of the Valve And Fittings Industry
NABCEP	North American Board of Certified Energy Practitioners
NIST	National Institute of Standards and Technology
NREL	National Renewable Energy Laboratory
PSP	Pressure Stagnation Protection
PV	photovoltaic
PX	Post Exchange
SEF	Solar Energy Factor
SF	solar fraction

Term	Definition
SHW	solar hot water
SIR	savings to investment ratio
SMACNA	Sheet Metal and Air-conditioning Contractors' National Association, Inc.
SPH	Solar Pool Heating
SPP	steam producing power
SRCC	Solar Rating and Certification Corporation
SSA	Social Security Administration
SWH	solar water heating
TEMF	tactical equipment maintenance facilities
UFC	Unified Facilities Criteria
US	United States
USA	United States of America
USACE	US Army Corps of Engineers
USDOE	US Department of Energy
UV	Ultraviolet

Appendix A: Solar Hot Water Case Studies

Flat plate collectors

FPC – 1

Title: Gneis Moos - solar supported local heating grid with weekly storage and two-pipe network

Site

Gneis-Moos is situated on the outskirts of Salzburg, Austria. From 1998 to 2000, six low-energy terraced houses were built at this site with a total of 61 residential units and a total floor space of 4654 m^2. Some general site information is:

Location:	Gneis-Moos, Salzburg
Latitude:	47.8°
Longitude:	-13.0°
Solar Irradiation	1,113.6 $kWh/(m^2 \cdot a)$
Application:	Solar supported local heating network
	Domestic hot water and space heating
	Medium-term (weekly) energy storage
	Two-pipe network with decentralized heat transfer units
Year of operation start:	2000

Source: Architekturbüro Reinberg ZT GmbH.

Figure A-1. Aerial view: residential terraced house complex Gneis-Moos, Salzburg.

The aim of this project was to realize a highly economic and efficient energy supply system using a functional design and a high building standard (projected specific space heating demand of the building accounts for less than 15,860 Btu/sq ft a [50 kWh/m^2]).

Therefore, the residential complex was equipped with a 4,412 sq ft (410 m^2) (equivalent to 16,336 Btu/min [287 kWth] nominal power) roof-mounted flat plate collector area combined with a 26,420 gal (100 m^3) weekly storage. A condensing gas boiler, acting as auxiliary heating device for the supply of space heating and domestic hot water, is connected to the upper part of the stratified energy storage.

The energy for both space heating and domestic hot water is distributed from the energy storage to the six houses via a two-pipe network with decentralized heat transfer units in the individual flats.

Passive heating for the buildings is achieved through the use of large glazed surfaces on the south side (direct sunrays) and the solar greenhouses. The greenhouses contribute enormously to heating the entire installation, up to 23% of the overall energy needs.

Additionally, ventilation in the homes is automatically controlled using the air heated in the greenhouses and taken to the rooms through heat exchanger.

Solar thermal system characteristics

The solar fraction (*sf*) of the solar thermal system installed was projected to be 34.1%. Monitoring the heating network showed a measured solar fraction of 38.4% in the year 2000 and of 30.6% in the year 2001. The differences between the 2 years are due to variations in the heat demand and the solar irradiation.

The specific annual solar yield *SE* of the system was designed to be111,017 Btu/sq ft (350 kWh/m^2 a). The measured values were even higher and accounted for 119,898 Btu/sq ft (378 kWh/m^2) in the year 2000 and for 120,532 Btu/sq ft (380 kWh/m^2) in the year 2001. The measured solar system efficiency *SN* accounted for 21% in 2001. Figure A-3 shows weather data (heating degree days, collector area irradiation Q_{Coll}) and associated key figures of the solar thermal system (SE, SN, sf) on a monthly basis for the year 2001.

Source: Architekturbüro Reinberg ZT GmbH.

Figure A-2. South oriented glazed façade with greenhouse for passive solar heating.

Due to the reduced heat consumption and the large storage capacity (0.24 m^3 storage volume/m^2 collector area), the mean solar fraction on average accounts for 96% during the summer period (May to August) and 25% during the heating period.

Conversely, the solar system efficiency is lower in summer due to higher solar system losses (Figure A-3).

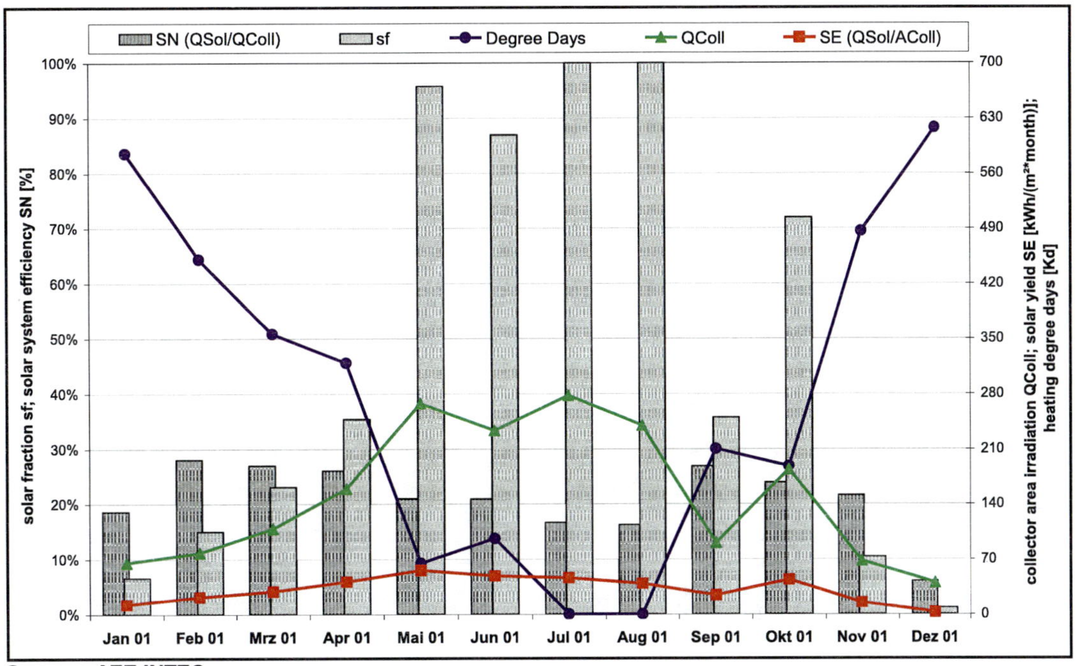

Source: AEE INTEC.

Figure A-3. Solar system performance Gneis-Moos, Salzburg 2001.

Figure A-4 shows the heat balance, where $Q_{Consumer}$ is the final energy consumption for domestic hot water and space heating (storage and net distribution losses considered).

Q_{Sol} is the measured energy input from the solar system into storage and Q_{Aux} is the measured energy input from the condensing gas boiler.

The difference between the input streams Q_{Sol} and Q_{Aux} into storage and the final energy for space heating and domestic hot water ($Q_{Consumer}$) is caused by heat losses of storage as well as of the distribution net. Figure A-5 shows the share of these losses for 2001.

In 2001, the heat losses through storage accounted for 6.1% and through distribution pipes another 25.9% of total heat losses.

The amount of final energy to the customer (DHW and space heating) is reduced by 32%, which means that the annual degree of system utilization $S_{Utilization}$ was 68%.

__Performance of the heat distribution network__

Figure A-6 shows the block diagram of the solar-supported heat supply system, which features heat distribution via a two-pipe network and heat output via so-called decentralized heat transfer units in Gneis-Moos.

The weekly energy storage unit is the central point for all heat flows and also acts as a hydraulic gateway for both the solar thermal system and the conventional heat supply system connected.

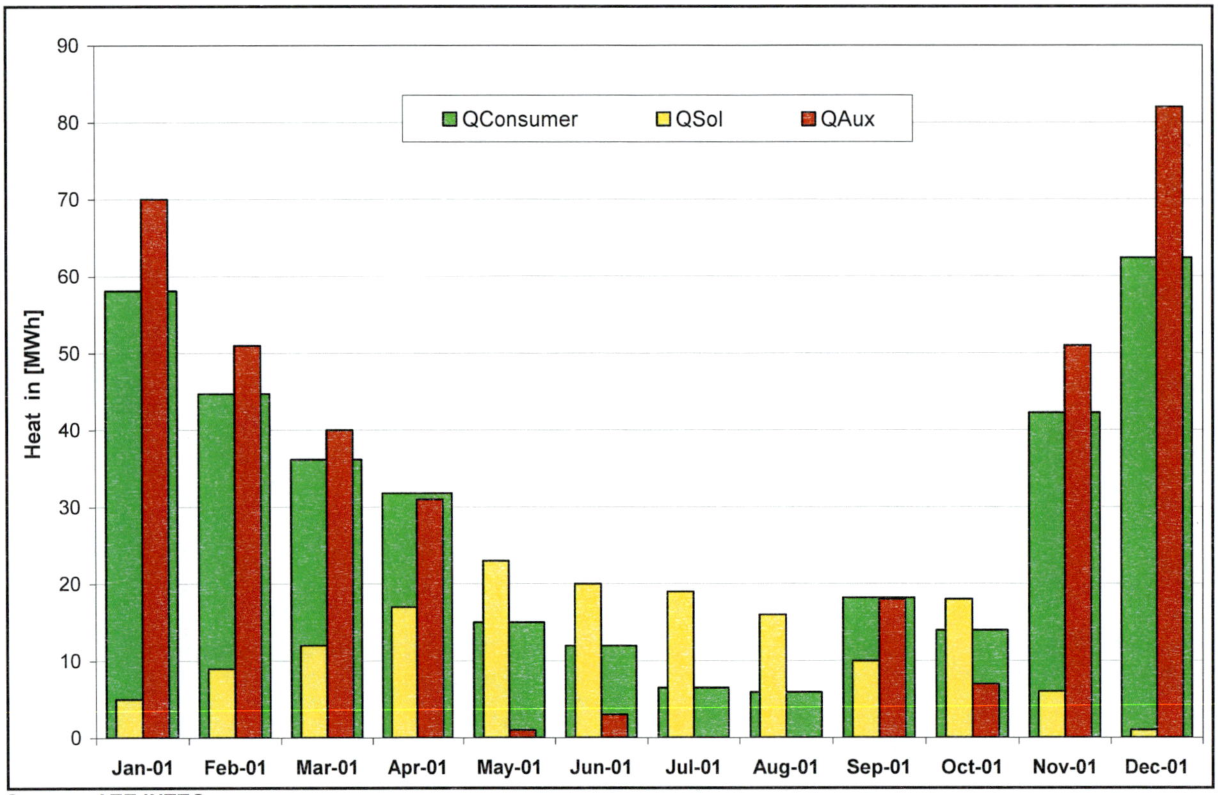

Source: AEE INTEC.

Figure A-4. Heat Balance Gneis-Moos, Salzburg 2001.

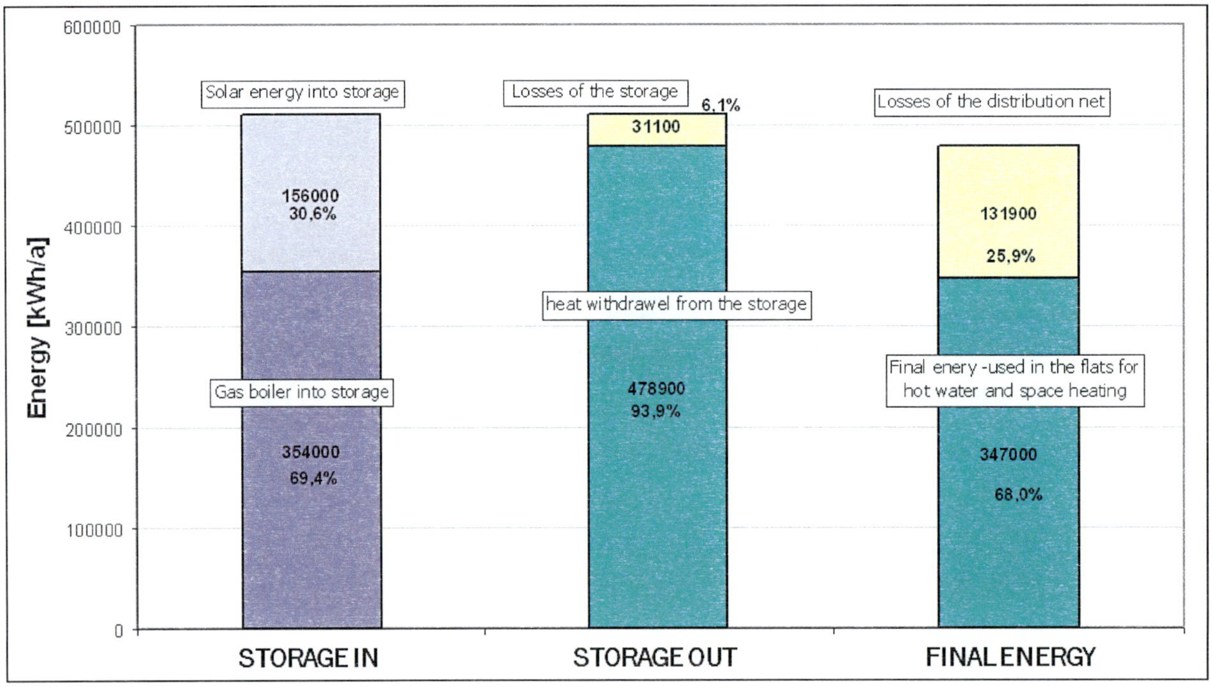

Source: AEE INTEC.

Figure A-5. Energy Balance including system losses Gneis-Moos, Salzburg 2001.

Some details of the solar thermal system are:

		Comments
Collector area installed:	410 m^2	Aperture Area
Type of collector:	*Flat plate collector*	
Brand of collector:	*Gluatmugl GS**	*Manufacturer: Oekotech*
Collector specification:	*Standard flat plate collector*	
c_0 [-]	*0.79*	
c_1 [W/m^{2}K]	*3.979*	
c_2 [W/m^{2}K^2]	*0.014*	
cp [kJ/m^2·K]	*6.53*	
Heat transfer medium:	*Water/propylene-glycol*	*40 vol-%*
Slope from horizontal:	*35 -45°*	
Azimuth:	*South*	
Specific static weight (kg/m^2 collector area)	*24 kg/m^2*	*Collectors only*
Heat storage capacity:	*100 m^3*	
Design peak capacity:	287 kW_{th}	
Calculated (design) annual solar yield:	350 kWh/m^2	
Design solar fraction	*34.1%*	
Measured annual solar yield SE	379 kWh/m^2	*Mean value 2000/2001*
Measured solar fraction sf	34.5%	*Mean value 2000/2001*
Measured solar system efficiency SN	21%	*2001 cf. Figure A-3*
Operation mode:	*Low flow*	
*Details: http://www.solid.at/images/stories/pdf/data%20sheet%20gluatmugl%20gs.pdf		

To guarantee a reliable supply of heat with this design, adequate reserves are permanently stored in the upper part of the energy storage to cover peak demand (Figure A-7).

A network pump and an admixer unit are used to supply the components of the heat transfer unit via a two-pipe network at a constant year-round supply temperature.

It is worth mentioning that the two-pipe network in Gneis-Moos performed well in terms of constant supply temperatures of around 149 °F (65 °C) over the year and almost constant low return temperatures between 86 and 95 °F (30 and 35 °C).

Figures A-8 to A-10 show the supply and return temperatures of the two-pipe network as well as the temperatures of the solar loop for selected days in March, July, and December 2001.

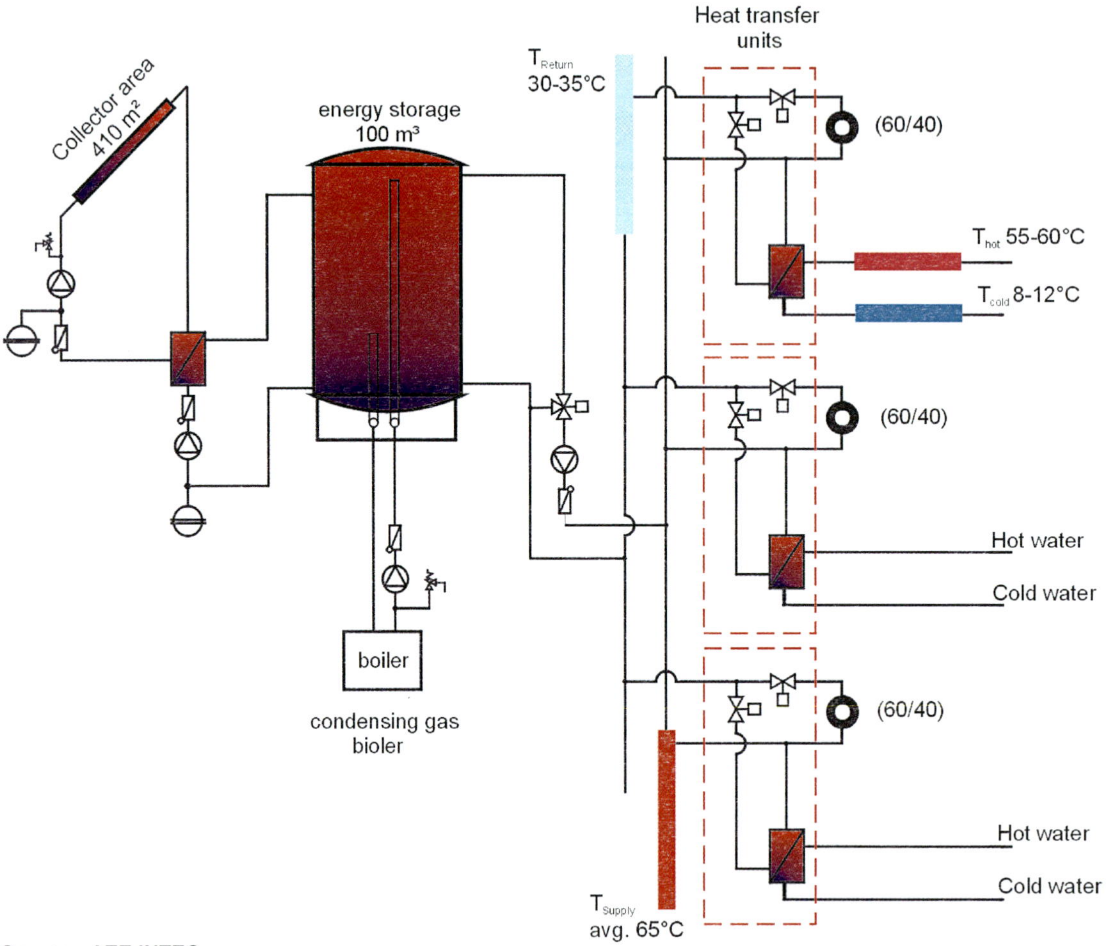

Source: AEE INTEC.

Figure A-6. Solar supported heating grid with weekly storage and two-pipe network connected to decentralized heat transfer units in Gneis-Moos, Salzburg.

Source: AEE INTEC.

Figure A-7. Part of the 26,420 gal (100 m^3) energy storage tank appears from the underground in the middle of the housing estate Gneis-Moos.

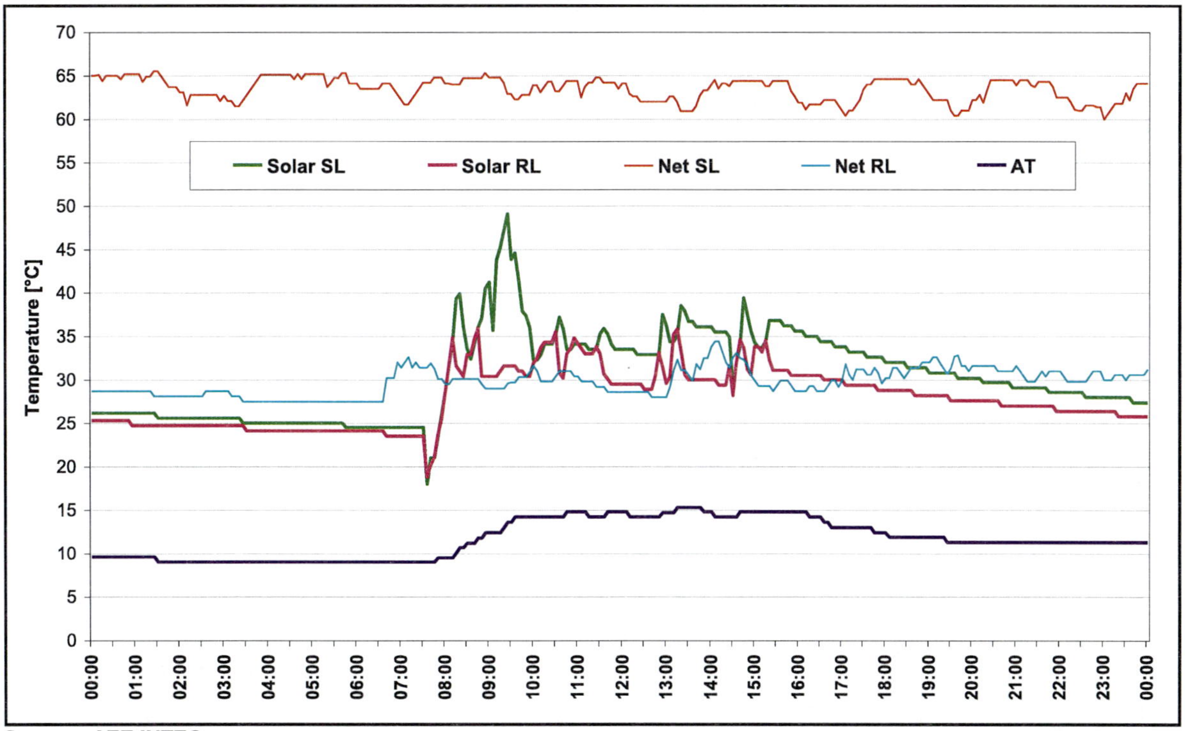

Source: AEE INTEC.

Figure A-8. Temperatures of the solar loop and the distribution grid in March 23rd 2001, Gneis-Moos.

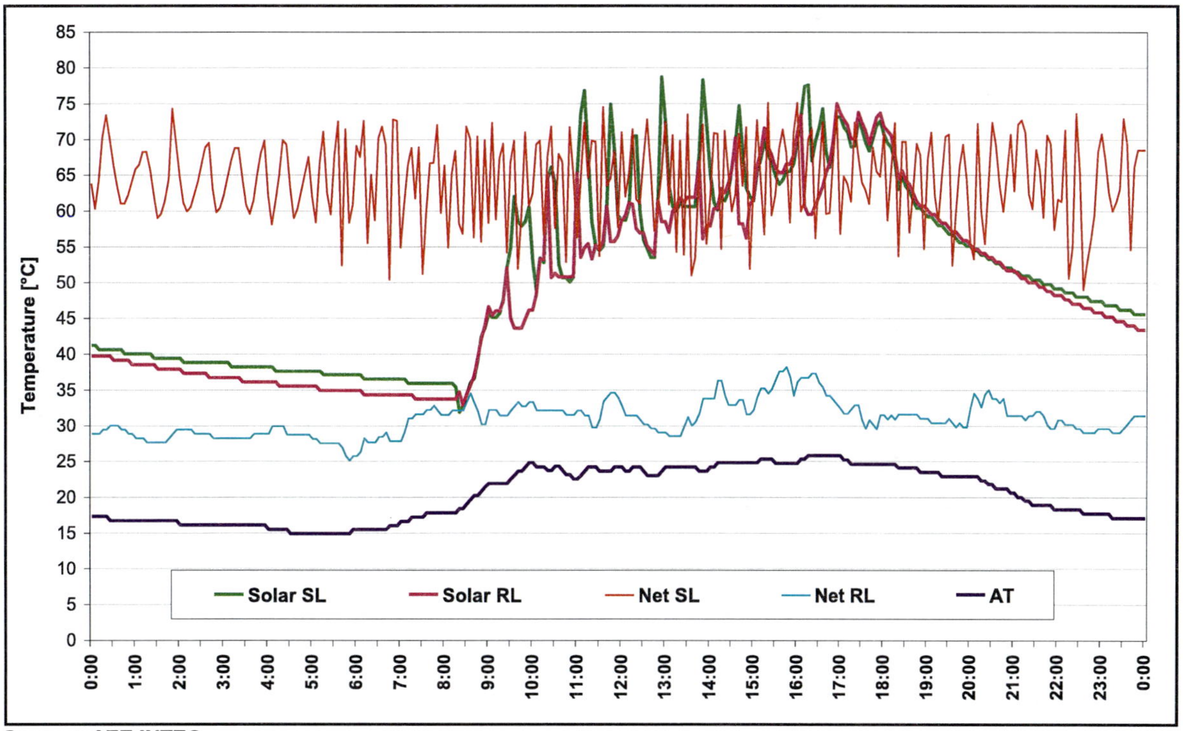

Source: AEE INTEC.

Figure A-9. Temperatures of the solar loop and the distribution grid in July 2nd 2001, Gneis-Moos.

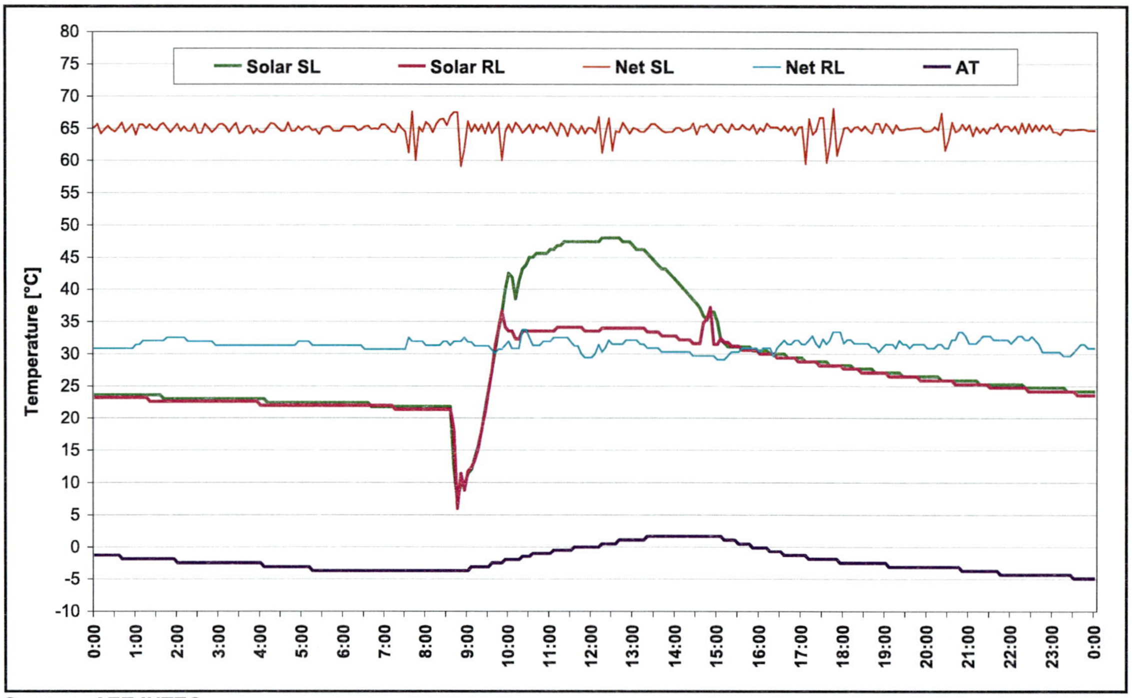

Source: AEE INTEC.

Figure A-10. Temperatures of the solar loop and the distribution grid in December 9th 2001, Gneis-Moos.

Solar SL: Solar loop supply line (from collector to the storage)
Solar RL: Solar loop return line (from storage to the collector)
Net SL: Net supply line (from storage to the apartments)
Net RL: Net return line (from the apartments to the storage)
AT: Ambient temperature

Even in the case of fluctuating consumption rates, as often occurring during the summer period (cf. Figure A-9), the return temperatures of the net remained stable. Temperatures above 95 °F (35 °C) were only measured in individual cases and for short time periods.

Due to the low return temperatures and the use of only two-pipes, the net distribution losses can be considerably reduced compared to a four-pipe network. Furthermore, the efficiency of the solar collectors is positively affected by the low return temperatures (low mean collector temperatures), which results in higher solar yields.

Heating network specification		**comments**
Design peak capacity of the entire grid	250 kW	
Total length of the grid [m]	700	
Auxiliary heating	350 kW	Condensing gas boiler
Design supply and return temperature of the net in summer	65/35	
Design supply and return temperature of the net in winter	65/35	
Measured supply and return temperature of the net in Summer	60-75/<40	*cf. Figure A-9*
Measured supply and return temperature of the net in winter	65/< 35	*cf. Figure A-10*
Total energy supplied by the net per year	478.9 MWh/a	Year 2001: $Q_{Sol} + Q_{Aux} - Q_{Storage}$ (heat losses storage)
Total annual heat transmission losses of the net	131.9 MWh/a	Year 2001: ~28% net losses corresponds to 188 kWh/(m_{grid}·a) cf. Figure A-10

Economics and environmental impacts

		comments
Total investment cost for solar thermal system	166,300 €	410 m^2 (287 kWth)
Auxiliary heating device	101,700 €	350 kWth
Local heating network	268,300 €	700 m trench length
Decentralize heat transfer units	88,200 €	61 units
Specific investment cost for total solar thermal system	~ 410 €/m^2	Reference: Aperture area
operation cost for solar thermal system	~ 500 €/a	
Annual CO_2-savings*	~ 37 tons/a	Corresponds to 920 tons within the entire lifetime of the system (25 years)
Annual natural gas-savings**	~ 183 MWh/a	
*Natural gas emission factor: 0.202 kg_{CO2}/kWh_{gas} **Annual mean efficiency of condensing gas boiler: 85%		

Figure A-11 shows the simple payback time of the investment for the solar thermal system replacing different types of conventional energy sources such as electricity, oil, or natural gas.

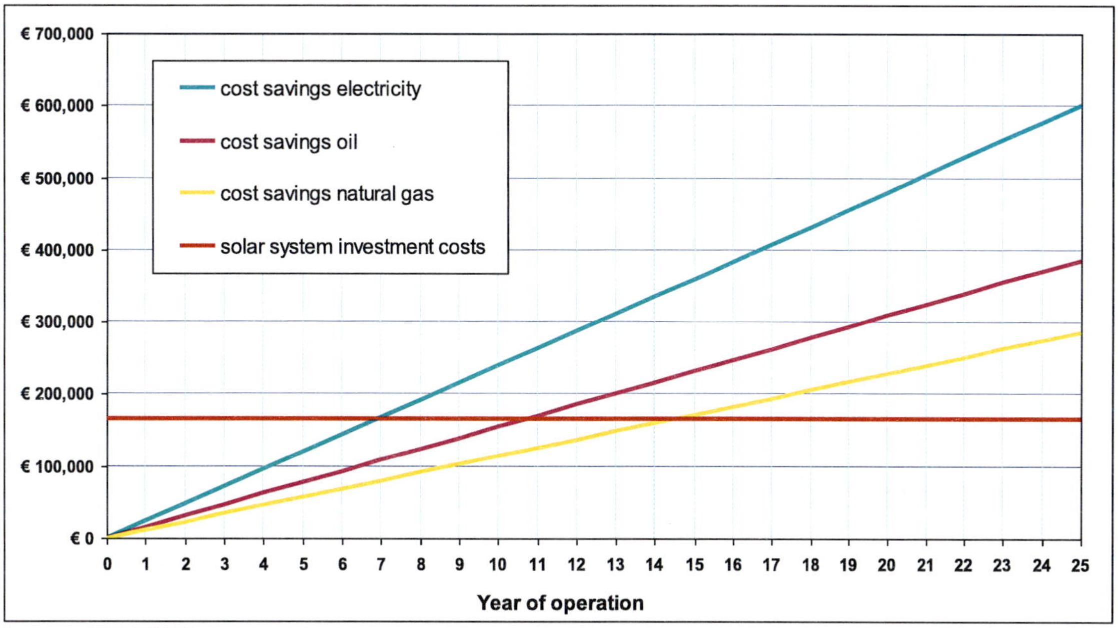

Source: AEE INTEC.

Figure A-11. Simple payback time of the solar thermal system for different energy sources replaced.

Table A-1. Static payback time and cost of electricity, oil, and natural gas.

electricity	oil	natural gas	electricity	oil	natural gas
[years]	[years]	[years]	[€ / kWh]	[€ / liter]	[€ / m^3]
6.9	10.8	14.6	0.16	0.72	0.65

Depending on the energy source replaced, the static payback time varies between 6.9 years in the case of electricity (Table A-1) and 17.6 years in the case of wood-pellets.

The price for solar heat and the dynamic payback time were calculated according to the net present value approach with the energy source natural gas.

Heat source replaced by solar energy	*Natural gas*	*Price: 65 €/MWh*
Discount rate	*4%*	
Annual amelioration natural gas	*6%*	
Technical lifetime of solar system	*25 years*	
Dynamic payback time	***13.2 years***	
Simple payback time	***14.6 years***	
Price for solar heat (€/MWh)	***71.5***	
Internal rate of return (ROI)	***10.5%***	
Net present value (25 years)	***~ $269,800 (190,000 €)***	

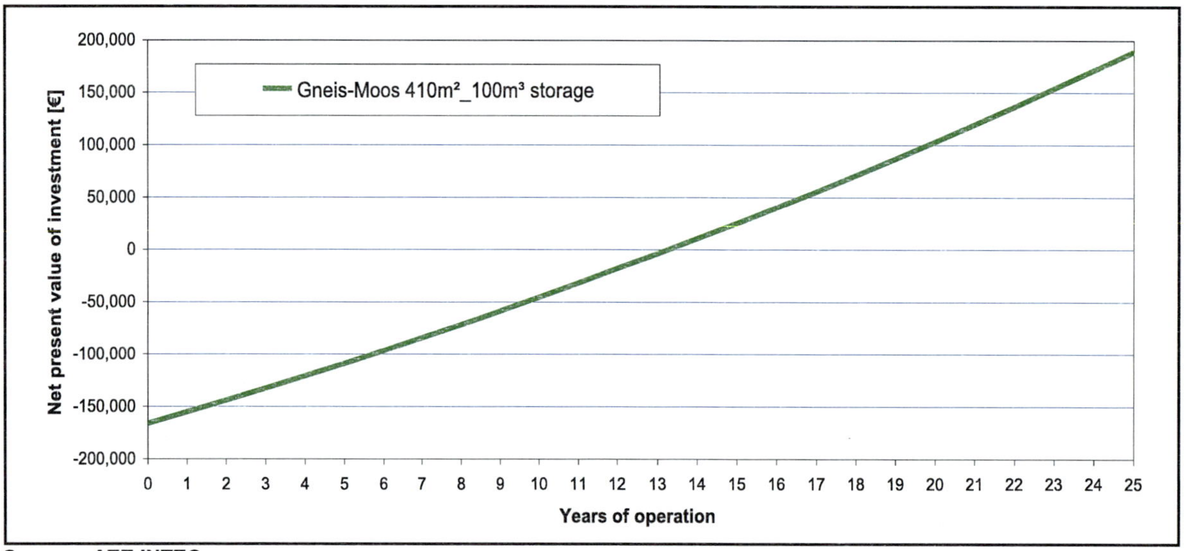

Source: AEE INTEC.

Figure A-12. Net present value of the solar thermal system.

Experiences/lessons learned

Based on the excellent experience gained in this project, the involved housing company (GSWB) erected another 15 multiple family houses with the same building and energy concept. The biggest one is the solar plant at the project Bolaring, which was installed in 2001 with a collector area of 11,363 sq ft (1056 m^2).

POC

S.O.L.I.D. Gesellschaft für Solar installation und Design mbH
Puchstraße 85, A-8020 Graz
Tel: +43 (0) 316 29 28 40 – 0
Fax: +43 (0) 316 29 28 40 – 28
e-mail: office@solid.at

FPC – 2

Title: District heating plant / water works Andritz Graz, Styria

General information

"Water works Andritz" is the largest ground-mounted collector array in Austria (3855.1 m^2), supplying domestic hot water and space heating for a nearby office building and/or feeding into the district heating grid of Graz.

The solar collector installation was built between February and June 2009 on the premises of the water utility Graz AG.

Site

Name	Water works Andritz
Address	Wasserwerksgasse 9-11, A-8045 Graz
Owner	Solar.nahwaerme Energiecontracting GmbH
Operation	From Spring 2009 (3.600 m + 300 m in Spring 2010)
Location:	*Graz, Styria*
Latitude:	47.7°
Longitude:	-15.4°
Solar Irradiation	357,156 Btu/sq ft (1,126 $kWh/m^2\,a$)
Application:	Ground-mounted solar plant (41,481 sq ft [3855.1 m^2]) for domestic hot water and space heating of office building (water utility Graz AG) and for feed-in into district heating grid (Energie Graz GmbH).

Source: S.O.L.I.D.

Figure A-13. Overview of solar supported district heating network Water works, Andritz Graz, Styria.

This project was initiated in response to a strategic decision of the local water supply company (water utility Graz AG) in the course of a rearrangement of the existing energy supply system, which was highly dependent on electricity.

The company was faced with increasing prices for electricity, and therefore decided to adopt alternative sources of energy in an economic and ecological way. Since the existing system had reached the limits of its lifetime, and had become relatively inefficient relative to current technologies, the company decided to increase the system performance to provide the future energy supply by combining solar thermal energy, district heating, and a heat pump.

The project was realized in the framework of a sale of energy contract with a local energy service company (ESCO).

Solar thermal system characteristics

The energy from the solar thermal system is either directly fed into the district heating supply line by heating up the district heating return line (annual mean temperature about 140 °F [60 °C]) or the energy storage connected to the office buildings is charged.

The energy storage has a volume of 17,067 gal (64.6 m^3) and is the central hydraulic gateway between the solar thermal system and the office buildings. At the customer's request, the energy storage was placed in a former water well (partly underground) on the site.

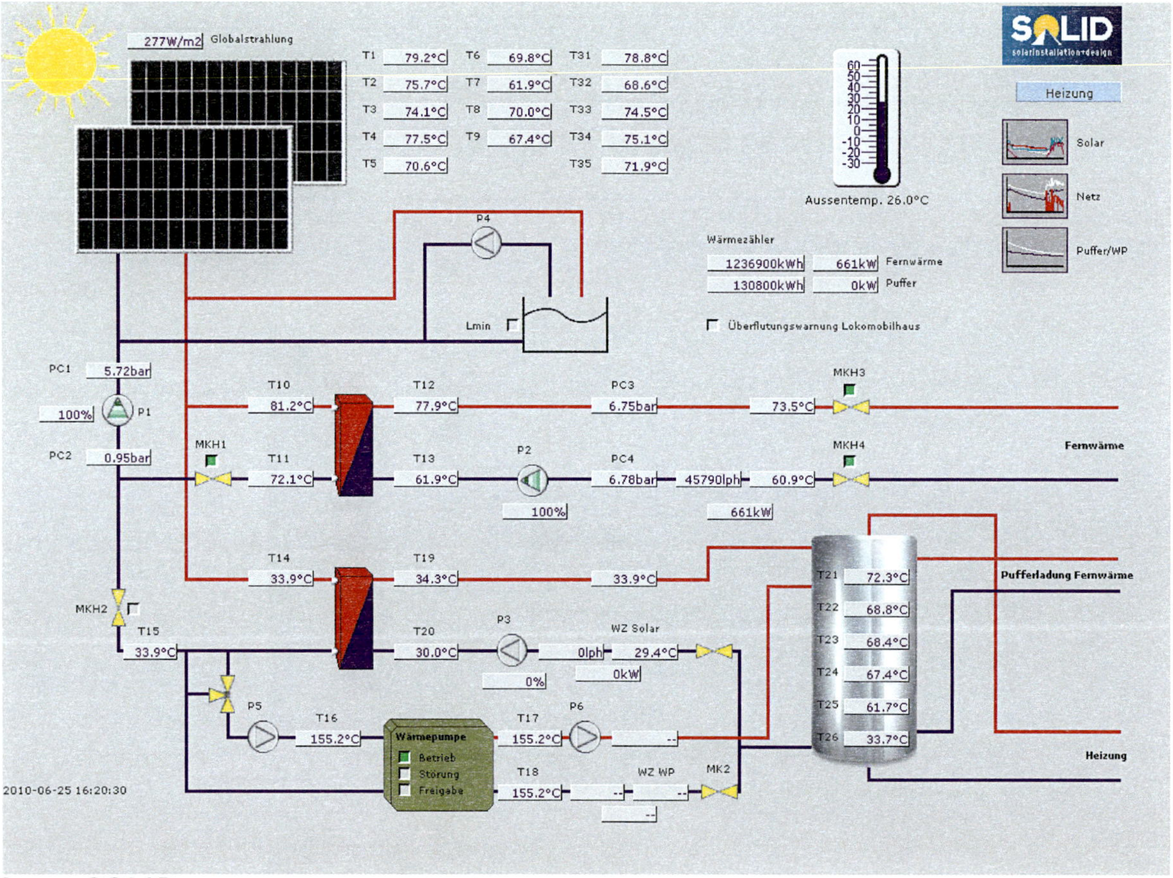

Source: S.O.L.I.D.

Figure A-14. Block diagram of the centralized collector array connected to the district heating network Water works, Andritz Graz, Styria.

Control strategy

The energy storage is charged before the feed-in into the district heating net (economically determined).

If there is a surplus of solar energy (energy storage is fully loaded) and the temperature level is high enough (> 70 °C) the solar thermal system directly heats-up the district heating return line and feeds into the district heating supply line of the Energie Graz GmbH.

Additionally a heat pump (as shown in the block diagram in Figure A-14) is planned to use low-exergetic energy from the solar thermal system during winter.

The heat load of the office buildings connected amounts for 14,230 Btu/min (250 kW_{th}) at the present stage of the project development, but is planned to increase to 28,460 Btu/min (500 kW_{th}) at final stage in 2011.

Here, the upper third of the energy storage also acts as a load compensation utility reducing the peak heating load of the distribution network by 30% (cf. Figure A-15).

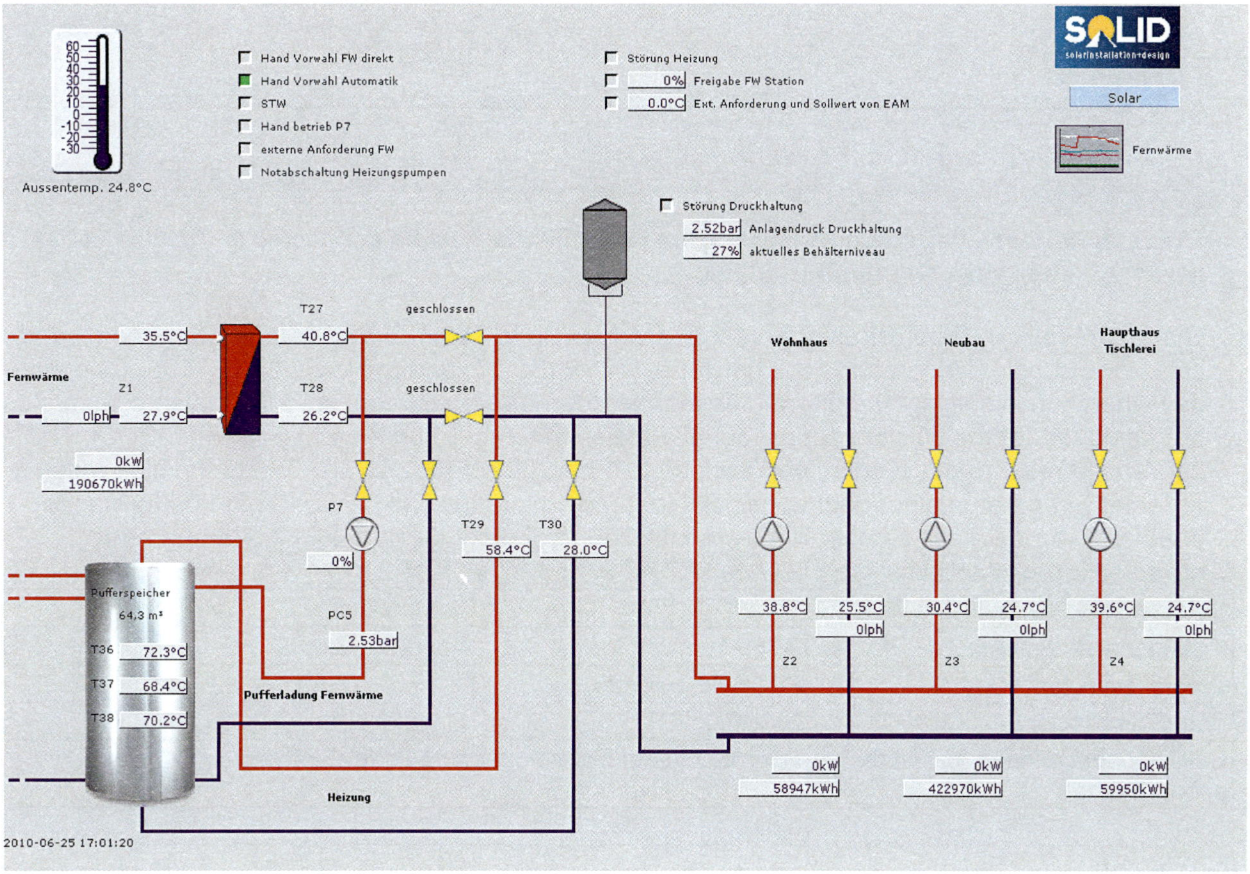

Source: S.O.L.I.D.

Figure A-15. Block diagram of the centralized energy storage connected to the two-pipe building transfer unit Water works Andritz Graz, Styria.

The office buildings receive the energy for space heating and domestic hot water from the central energy storage via a two-pipe network. The district heating grid provides supplementary heating if solar thermal energy cannot meet the demand (cf. Figure A-15).

Figure A-16 shows characteristic temperatures of the solar thermal loop during a warm period in June 2010.

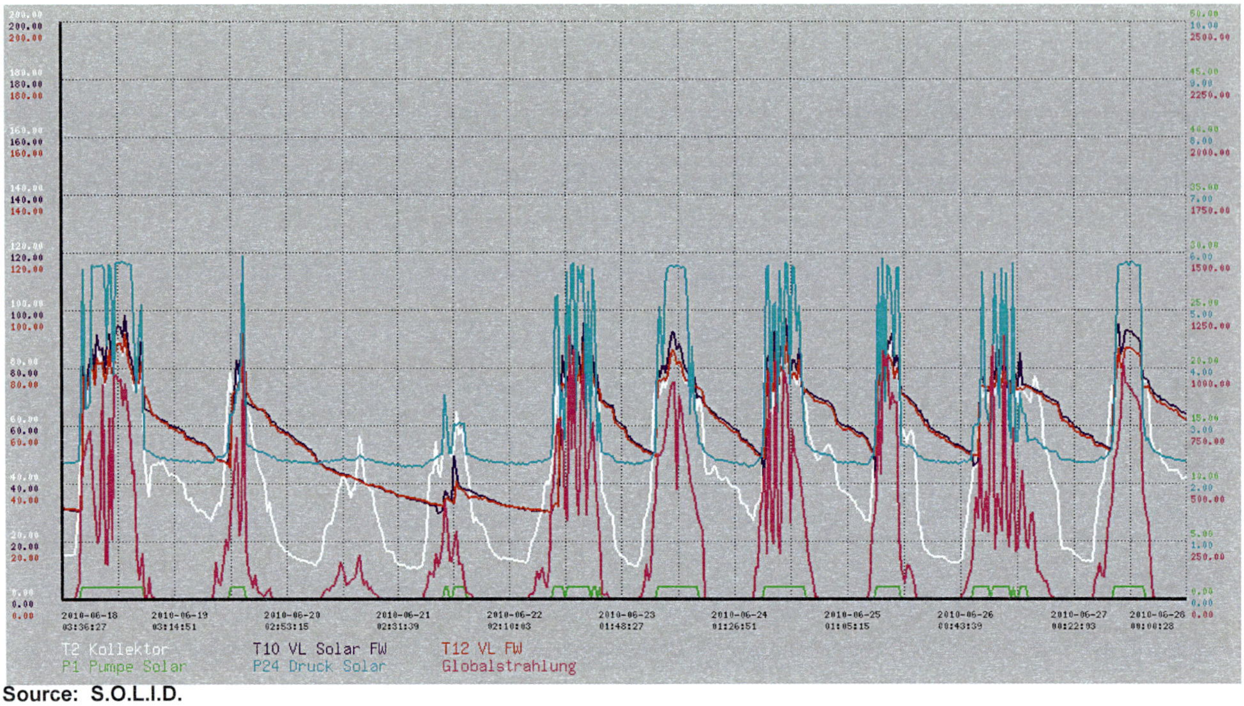

Source: S.O.L.I.D.

Figure A-16. Operating characteristics of the solar thermal system connected to the district heating network Water works Andritz Graz, Styria.

Global Irradiation (Globalstrahlung) reaches values up to 396,488 Btu/sq ft (1250 kWh/m^2) and the solar loop supply temperatures are between 176 °F (80 °C) and (nearly) 212 °F (100 °C) during the day when sun is shining. In general, the high temperature double covered flat plate collectors used are designed for the efficient supply of hot water with temperatures up to 248 °F (120 °C). The operating pressure of the solar loop is close to 6 bar (80 psi [552 kPa]). Due to the high mean collector temperatures needed (especially for feeding into the DH network) high performance double-covered flat plate collectors were installed (Oekotech Gluatmugl high temperature collector→ gross collector area between 77.47 to 153.87 sq ft (7.2 to 14.3 m^2) each).*

Solar thermal system details		comments
Collector area installed:	3,855.1 m^2	Aperture Area
Type of collector:	Flat plate collector	
Brand of collector:	Gluatmugl HT*	Manufacturer: Oekotech
Collector specification (based on aperture area):	Double covered high temperature flat plate collector	
c_o [-]	0.811	
c_1 [W/m^{2}K]	2.710	(15 Btu/hr-sq ft °F)
c_2 [W/m^{2}K^2]	0.010	(0.06 Btu/hr-sq ft °F)
cp [kJ/m^2·K]	7.05	(40.03 Btu/hr-sq ft °F)
Heat transfer medium:	Water / propylene-glycol	30-40 vol-%
Freeze protection strategy:	Water / propylene-glycol	30-40 vol-%
Slope from horizontal:	30°	
Azimuth:	180° (parts 170°, 210°)	Main collector array south oriented, just a few collectors at an azimuth of 210°
Specific weight:	About 0.74 gal/sq ft (30 kg/m^2) collector area	Collectors only

* Detailed datasheet can be downloaded from: http: //www.solid.at/images/stories/pdf/data%20sheet%20gluatmugl%20ht.pdf

Solar thermal system details		**comments**
Heat storage capacity:	17,067.32 gal (64.6 m^3)	
Design peak capacity:	153,604.31 Btu/min (2698.6 kW_{th})	
measured annual solar yield SE	130,048–133,220 Btu/sq ft (410–420 kWh/m^2 a)	Expected specific solar energy yield after 6 months of measurements
*details: http: //www.solid.at/images/stories/pdf/data%20sheet%20gluatmugl%20ht.pdf		

Performance of the heat distribution network

Heating network specification		**comments**
Design peak capacity of the entire District Heating (DH) grid Graz	~1,365 MBtu/hr (~400 MW)	
Design peak capacity of solar thermal system	~ 9.2 MBtu/hr (~2.7 MW)	
Supply and return temperature of the DH grid in summer	167/140 °F (75/60 °C)	
Supply and return temperature of the DH grid in winter	248/140 °F (120/60 °C)	
Total energy supplied to the net per year by the solar thermal system	468,480 Btu/a (1,600 MWh/a)	Expected solar energy yield after 6 months of measurements

Since the heat load connected to the district heating grid in winter is higher than in summer (non-heating season) higher supply temperatures in the main pipes ensure that the demand for space heating *and* hot water can be met. To use solar thermal energy at lower temperatures and during non-heating season the large grid connected collector arrays remain connected to the outer branches of the district heating grid that are operated at lower temperatures. Moreover it is reasonable to additionally realize a direct connection of the solar thermal system to a building close to the site.

Economics

The total investment costs of the entire solar thermal system amount to ~€1.5 million.

The entire system of solar thermal collectors, energy storage, control equipment, piping, pump units etc. was subsidized by the Federal Ministry of Agriculture, Forestry, Environment and Water Management. Kommunalkredit Public Consulting (KPC) managed the funding in charge of the ministry. The funding was 30% percent of the total investment. The solar plant now is operated on a "sale of energy" basis with solar.nahwaerme.Energiecontracting GmbH as the owner and operator of the plant. S.O.L.I.D. GmbH was in charge of design and planning.

Nahwaerme.Energiecontracting GmbH sells the heat, either solar or from district heating grid, to water utility Graz AG for space heating and to DHW at same price as district heating. The rates for district heating comprise an energy tax on fossil fuels of 5 €/MWh ($0.01/Btu). These 5€ are also paid by Graz AG, but go to solar.nahwaerme.Energiecontracting GmbH and not to the treasury. The ground for the solar plant is provided by Graz AG.

Table A-2. Costs.

Element		**comments**
Total investment costs	~ 2,130,000 $ (1,500,000 €)	excl. VAT, excl. subsidies
Specific investment cost	~ $52/sq ft (390 €/m^2)	reference: Aperture area
Operation cost for solar thermal system	Almost negligible	~1-2% of the annual energy gains are Needed for the pumps and the control system. Maintenance is statistically not significant 30%
Subsidies	~ $639,000 (450,000 €)	
Annual CO_2-savings*	~ 186 tons/a (158 m^3)	
Annual primary energy savings**	~ 550 Btu/a (1.880 MWh/a)	

* District heating emission factor (Graz): 0.099 kg_{CO2}/kWh (0.01 lb_{CO2}/Btu)
** Annual mean efficiency of the backup systems: 85%

Lessons learned

Construction

Considerable management efforts were taken as various lines and pipes for water, heating, electricity, glass fiber cables etc. are in the underground of the site. Work on these lines proceeded during construction of the solar plant and the heating system.

Operation

Energy storage management had to be optimized while the system was in operation.

Performance

The heat output of the solar plant met expectations.

General

Exact knowledge about all system parts and partners is essential before planning, e.g., what and when is the exact heat demand, which control systems are used, at which pressure and at which time does the district heating grid operate.

Acknowledgement

DI Harald Blazek
e-mail: h.blazek@solid.at

S.O.L.I.D. Gesellschaft für Solarinstallation und Design mbH
tel: +43 (0) 316 / 29 28 40

Project development
fax: +43 (0) 316 / 29 28 40-28

POC information

S.O.L.I.D. Gesellschaft für Solarinstallation und Design mbH
Puchstraße 85, A-8020 Graz
office@solid.at

Direct further questions to:
DI Harald Blazek
Homepage: http://www.solid.at/index.php?lang=en

FPC – 3

Title: District heating plant/ UPC Arena Graz-Liebenau, Styria

Site

In 2002, Austria's first solar thermal system was built and integrated into a city's district heating grid at the UPC Arena in Graz. This pilot project, which was initiated to show the use of solar energy in district heating systems, demonstrates the commitment of the city of Graz to renewable energies and to the protection of the environment.

General Information	
Name	UPC arena Graz-Liebenau
Address	Liebenauer Hauptstraße 2-6, 8041 Graz
Owner	Solar.nahwaerme Energiecontracting GmbH
Location:	Graz, Styria
Latitude:	47.7°
Longitude:	-15.4°
Solar Irradiation	357,156 Btu/sq ft (1,126 kWh/[m^{2}a]
Application:	Solar supported district heating network Roof-mounted solar thermal system directly connected to the district heating network- removal from return line of the district heating grid and feeding- into the supply line by means of flat plate heat exchangers
Commissioned in	2002

Source: S.O.L.I.D.

Figure A-17. Overview of the solar supported district heating network UPC arena Graz-Liebenau.

The solar plant is operated on a "sale of energy" basis in cooperation with the company nahwaerme.at Energiecontracting GmbH, which was responsible for the funding and the proper operation of the plant, while S.O.L.I.D. was in charge of the design, planning, and erection.

Solar thermal system characteristics

The total collector area is 15,139 sq ft (1407 m^2). The panels are mounted on the roof of the skating hall of the UPC Arena. A transfer station located in the mechanical room of the UPC Area manages the transfer of energy from the solar loop to the district heating grid.

The energy from the solar thermal system is directly fed into the district heating supply line by heating up the district heating return line (annual mean temperature about 140 °F [60 °C]).

Since the solar thermal energy supplied to the net is low compared to the capacity of the entire district heating network, the system is operated without additional energy storage.

Control strategy

The energy from the solar thermal system is used as soon as the solar loop supply temperature is equal to or exceeds than the minimal temperature needed in the district heating supply line.

Speed-controlled pumps are used in the solar loop to reach usable temperature levels and a turbulent flow of the heat transfer medium through the absorber pipe work.

Figure A-19 shows characteristic temperatures of the solar thermal loop during a warm period in June 2010.

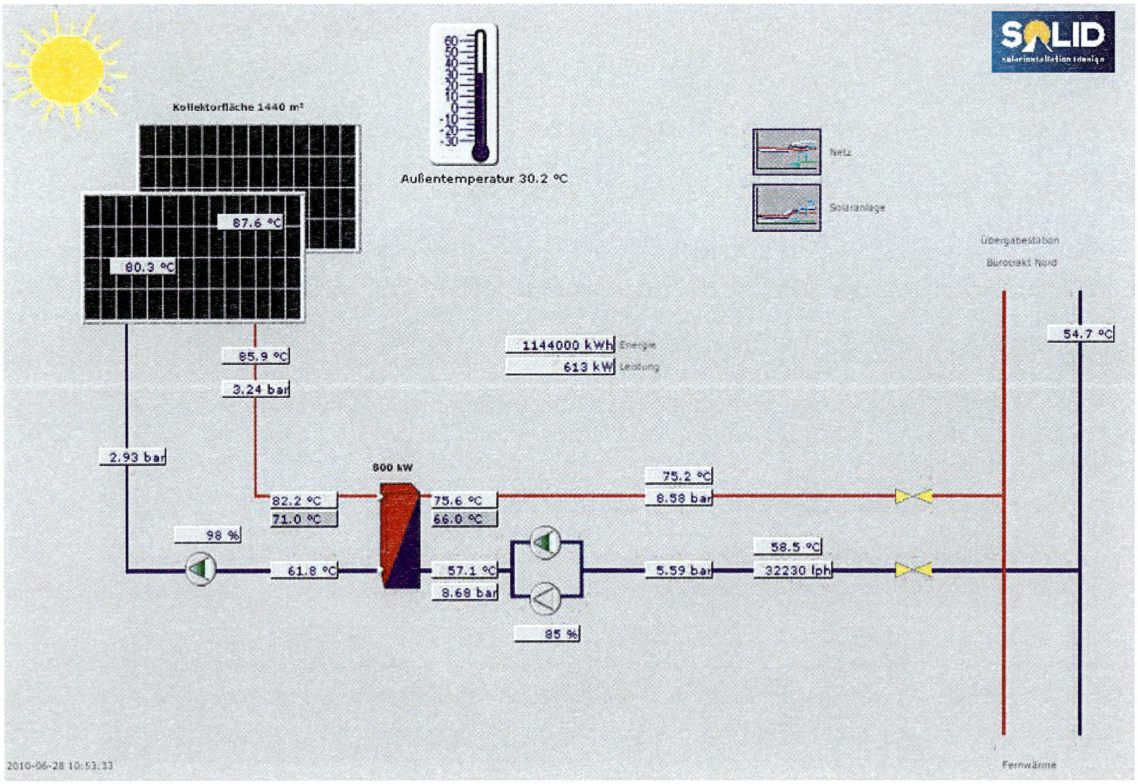

Source: S.O.L.I.D.

Figure A-18. Block diagram of the centralized collector array connected to the district heating network, UPC arena, Graz-Liebenau, Styria.

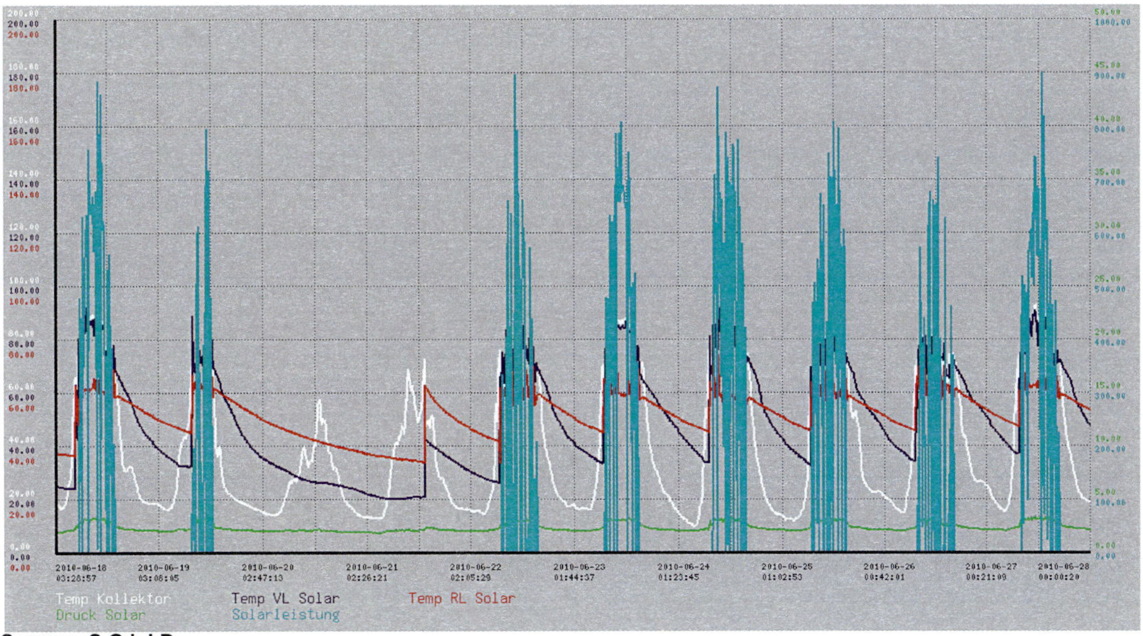

Source: S.O.L.I.D.

Figure A-19. Operating characteristics of the solar thermal system connected to the district heating network, UPC arena, Graz-Liebenau, Styria.

The thermal capacity of the solar system peaks at around 51,228 Btu/min (900 kW_{th}) during sunny periods when global Irradiation amounts for ~396,488 Btu/sq ft (1250 kWh/m^2). The solar loop supply temperatures reach up to 194 °F (90 °C) (cf. Figure A-19), but temperatures up to 230 °F (110 °C) are possible. The operating pressure of the solar loop is kept constant at around 2.5 bar (35 psi [241 kPa]). Some Solar thermal system details are:

		comments
Collector area installed:	15,139 sq ft (1407 m^2)	Aperture Area
Type of collector:	flat plate collector	
Brand of collector:	Gluatmugl GS*	Manufacturer: Oekotech
Collector specification:	standard flat plate collector	
c_o [-]	0.79	
c_1 [W/m^2K]	3.979 (23 Btu/hr-sq ft °F)	
c_2 [W/m^2K]	0.014 (0.08 Btu/hr-sq ft °F)	
cp [kJ/m^2K]	6.53 (0.32 Btu/sq ft)	
Heat transfer medium:	Water / propylene-glycol	30-40 vol-%
Freeze protection strategy:	Water / propylene-glycol	30-40 vol-%
Slope from horizontal:	30°	
Azimuth:	160°	
Specific weight (kg/m^2 collector area)	~ 1 gal/sq ft (24 kg/m^2)	collectors only
Heat storage capacity:	no storage	
Design peak capacity:	56,061 Btu/min (984.9 kW_{th})	
Measured specific annual solar yield SE	114,188 Btu/sq ft (360 $kWh/[m^2 \cdot a]$)	mean annual value
Operation mode:	Low flow speed controlled 30-100%	15 $L/(m^2 \cdot h)$
*details: http: //www.solid.at/images/stories/pdf/data%20sheet%20gluatmugl%20gs.pdf		

Performance of the heat distribution network

Design peak capacity of the district heating network in Graz for winter case (DHW + space heating) is around 1,364.9 MBtu/hr (400 MW_{th}) whereas the main distribution pipes are operated at mean supply temperatures of 167 °F (75 °C) in summer (only DHW) and 248 °F (120 °C) in winter (DHW + space heating). Heating network specifications are:

		comments
Design peak capacity of the entire DH grid	~1,364.9 MBtu/hr (~400 MW)	
Design peak capacity of solar thermal system "UPC arena"	~3.3 MBtu/hr (~0.98 MW)	
Supply and return temperature of the DH grid in summer	167/140 °F (75/60 °C)	
Supply and return temperature of the DH grid in winter	248/140 °F (120/60 °C)	
Total energy supplied to the net per year by the solar thermal system (MWh/yr)	146,400 Btu/a (500 MWh/a)	mean annual value

Since the heat load connected to the district heating grid in winter is higher than in summer (non-heating season) higher supply temperatures in the main pipes ensure that the demand for space heating *and* hot water can be met.

Solar thermal energy may be used at lower temperatures; during the non-heating season, the large grid connected collector arrays remain connected to the outer branches of the district heating grid that are operated at lower temperatures. Moreover it is reasonable to directly connect the solar thermal system to a building close to the site.

Economics and environmental impacts

At the time of its installation, this site was Austria's largest solar thermal system and its first large solar thermal energy contributor to a district heating system; hence, the installation received subsidies from Federal sources as well as from the state and the city to a total level of ~40% of the investment.

The system also was the first large solar thermal "sale of energy" installation, which leads to a significant number of similar projects in the following years and the further development of energy service companies (ESCOs). Some economic parameters are:

		comments
Total investment costs		
Specific investment cost	~Typically $53–$58/sq ft (400–435 €/m^2)	
Operation cost for solar thermal system	Almost negligible	~1-2% of the annual energy gains are needed for pumps and control equipment. Maintenance is statistically not significant
Subsidies	~€	30%
Annual CO_2-savings*	~58 tons/a	
Annual primary energy savings**	~172,752 Btu/a (~590 MWh/a)	
*District heating emission factor (Graz): 0.099 kg_{CO2}/kWh (0.01 lb_{CO2}/Btu)		
**Annual mean efficiency of the backup systems: 85%		

Experiences/Lessons Learned

The system, now in its 8th year of operation, has run very well right from its inception.

The additional local hot water demand in the neighbored offices as well as in the stadium itself could even have been a further advantage for the project in terms of a higher system efficiency and better economics, but negotiations regarding this matter failed. Hence the system solely delivers energy to the city's district heating grid. Nevertheless, it is the oldest Austrian district heat installation larger than 10,760 sq ft (1000 m^2), and still performs "like new."

S.O.L.I.D. Gesellschaft für Solarinstallation und Design mbH
tel: +43 (0) 316 / 29 28 40

Project development
fax: +43 (0) 316 / 29 28 40-28

POC information

S.O.L.I.D. Gesellschaft für Solarinstallation und Design mbH
Puchstraße 85, A-8020 Graz
office@solid.at

Direct further questions to:
DI Harald Blazek
Homepage: http: //www.solid.at/index.php?lang=en

FPC – 4

Title: Solar heating district heating in Ulsted, Denmark

Location: Ulsted, Denmark

Photo

Figure A-20. Aerial picture of the district heating plant of Ulsted and the solar collector field.

Project summary

This project was undertaken to supplement the heat for the district heating grid in the town of Ulsted in Denmark, which supplies all domestic hot water and space heating for the ~475 households and ~1000 inhabitants.

The district heating in Ulsted is supplied by a central boiler fueled by wood pellets.

The solar system is an 53,800 sq ft (5000 m^2) field-based collector on ground-mounted flat plate collectors, installed in 2006.

The solar field covers 20 to 25% of the annual heat demand of the district heating grid and supplied ~2500 MWh of heat in 2009.

Site

Location:	Ulsted, Denmark
Address:	Stadionvej 11, DK- 9370 Hals
Latitude:	55 57.58 N
Longitude:	9.36.26 E
Annual solar radiation:	377 Btu/sq ft (1.190 kWh/ m$^{2)}$ (2009)

Project description

The solar system is constructed of 400 high performance ground-mounted flat plate collectors each with an aperture area of 136 sq ft (12.6 m^2). The collector field totaling a collector area of 53,929 sq ft (5012 m^2) (58,104 sq ft [5400 m^2] gross).

Type/age (when installed)

The system was installed and put into operation in 2006.

Design peak capacity

The solar system is designed for a peak capacity of 12, MBtu/hr (3.5 MW).

Design solar peak capacity

The actual thermal production from the solar field in the first 2 full years of operation was:

	Thermal Production		Solar Radiation	Efficiency in
Year	Btu (MWh)	Btu/sq ft (kWh/m^2)	Btu/sq ft (kWh/m^2)	Aperture Area (%)
2008	678 (2.314)	146,542 (462)	Na	na
2009	660 (2.255)	142,736 (450)	377 (1.189)	38%

Calculated/measured annual solar thermal production

In 2008, the measured share of solar thermal production of heat consumption was 23%.

System details

Description of DH system

The district heating grid provides heating and domestic hot water for the ~750 household in the town of Ulsted.

Solar system

The solar system is constructed of 400 flat plate collectors with a total collector aperture area of 53,929 sq ft (5012 m^2) (58,104 sq ft (5400 m^2) gross area).

Collector type, number, net area

The collectors special high performance flat plate collector designed for large scale systems from ARCON (Figure A-21).

Figure A-21. Cross-section of the ARCON HT collector.

The collectors are designed to accommodate a low pressure loss in the collector, thereby allowing up to 15 collectors to be mounted in series (2,152 sq ft [200 m^2] gross area). This ensures lower costs for installation, piping, pump etc. Collectors In the project are installed in series of up to 12 collectors (1,743 sq ft [162 m^2] gross area). The collectors are also designed to accommodate the highest possible performance of flat plate collectors at the high temperature need for district heating (up to 212 °F [100 °C]) through several features:

- Use of low iron solar glass with a solar transmittance of over 91%
- Use of antireflective coating on the glass, which further increased the solar transmittance.
- Thicker backside and side insulation
- An FEP foil between the glass and the absorber. This foil gives an extra air gab between the absorber and glass (as in doublet glazed windows), which reduced the heat loss through the front of the collector.

The collector field takes up 193,680 sq ft (18,000 m^2) of land.

The solar plant is built as an "island" system with collectors, tank, and control system built as one separate unit. The system is located a few hundred meters from the district heating plant and is connect through a transmission line.

The tank is of 264,200 gal (1000 m^3), equivalent to ~53 gal (200 L) storage capacity per 10.76 sq ft (1 m^2) of collector area. The collector field is built as a separate circuit, which is connected to the tank and district heating grid via an heat exchanger with circulation pumps on each side of the heat exchanger.

The solar system is operated with variable pump speed so that the field supplies flow at a constant temperature, which varies depending on different operating modes. Solar irradiation sensors in the collector field start and stop the pumps and control their speed.

The solar system is configured so that the heat from the solar field can be used in different ways depending on the operating situation at district heating grid and the district heating. The main operating modes are:

- The solar field supplies heat directly in the district heating grid at 158 °F (70 °C).
- The heat is feed to the storage tank at different temperatures depending on the temperature of the storage tank.
- The solar field supplies directly in the transmission line and preheats the return to the wood boilers.

Slope from horizontal

The collectors are mounted at a 33 degree angle.

Economics

System first cost (new or retrofit)

The investment of the solar systems totaled ~$1.7 million (US) (based on an exchange rate of DKK/USD of 5) (8.6 million DKK) including collector field, piping in field, heat exchanger unit, tank and transmission line to heating plan.

The project obtained a $0.3 million (US) (1.5 million DKK) subsidy giving the district heating company a net investment of $1.4 million (US).

The energy costs produced by the solar field-based on an interest rate of 4% over a 25-year lifetime of the system is calculated to ~40 USD/MWh (200 DKK/MWh) (48 USD/MWh or 240 DKK/MWh excluding subsidy).

Savings

The expected simple payback of the solar system is calculated to ~10 years.

User evaluation

General data

POC information

US Army Corps of Engineers
Alexander Zhivov
T +1 217-373-4519
M +1 217-417-6928
Alexander.M.Zhivov@usace.army.mil
Alfred Woody
T +1 248-891-519
awoody@comcast.net

ARCON Solvarme A/S
Anders Otte Jørgensen, CEO
T +45 98 39 14 77
M +45 21 44 75 28
aoj@arcon.dk
Rene Rubak, Project Manager
T +45 98 39 14 77
M +45 61 67 40 91
rr@arcon.dk

Date of the report

05 April 2010

References

www.solvarmedata.dk. Webpage with online production data on solar district heating plants in Denmark and key figures on of the systems.

www.arcon.dk. Web page of ARCON with data sheets on the collectors and key figures of this project and other similar project installed by ARCON.

FPC – 5

Title: Solar water heating connected to combined heat & power (CHP) district heating system of Strandby, Denmark

Location: Strandby, Denmark

Photo of the installation

Figure A-22. Aerial image of the district heating plant of Strandby and the solar collector field.

Project summary

This project was undertaken to supplement the heat for the district heating grid in the town of Strandby in Denmark, which supplies all domestic hot water and space heating for the ~830 households and ~1750 inhabitants.

The district heating in Strandby is supplied by a combination of natural gas boilers and a gas-driven CHP (Combined Heat and Power) plant.

The solar system is an 86,080 sq ft (8000 m^2) collector field-based on ground-mounted flat plate collectors, which was installed in 2008.

The solar field covers ~18% of the annual heat demand of the district heating grid and supplied ~995,519.96 Btu (3400 MWh) of heat in 2009.

Site

Location:	Strandby
Denmark Address:	Ravnmarken 8, DK-9970 Stranby
Annual Solar radiation:	314,018 Btu/sq ft (990 kWh/m^2) (2009)

Project description

The solar system is constructed of 641 high performance ground-mounted flat plate collectors each with a aperture area of 136 sq ft (12.6 m^2). The collector field totaling a collector area of 86,769 sq ft (8064 m^2) (93,117 sq ft [8654 m^2] gross).

Type/age (when installed)

The system was installed and put into operation in 2008.

Design peak capacity

The solar system is designed for a peak capacity of 20.5 MBtu/hr (6.0 MW)

Calculated/measured annual solar thermal production

The actual thermal production from the solar field in the first full year of operation was:

	Thermal Production		Solar Radiation	Efficiency in %
Year	**Btu (MWh)**	**Btu/sq ft (kWh/m²)**	**Btu/sq ft (kWh/m²)**	**Aperture Area**
2009	995.5 (3.400)	133,854 (422)	314,018 (990)	43

In 2009, the share of solar thermal production of the total heat consumption was ~18%.

System details

Description of DH system

The district heating grid (Figure A-23) provides heating and domestic hot water for the ~830 household in the town of Strandby and supplies ~4.4 MBtu (15,000 MWh) of heat annually.

The district heating system is a traditional district heating grid with a forward and return lines.

The grid covers 25 km of main line and 26 km of connection line.

The heat for the grid is produced and distributed to the grid from the district heating plant, which uses natural gas.

The plant consists of one Rolls Royce gas engine (12.5 Btu/hr [3.66 MW] power and 14 MBtu/hr [4.1 MW] heat) and a back-up gas boiler of 34 MBtu/hr (10 MW) for heating only.

The gas engines works as part of the back-up source for the power grid in the western part of Denmark and is in operation at peak power demand or when the production from wind turbines are low. The engines produce approx 3.2 MBtu (11,000 MWh) of power annually.

The district heating grid supplied by the gas boilers when the engines are not running. The solar system supplement the system.

The gas boiler is fitted with a 13,660 Btu/min (240 kW) absorption chiller, which extracts heat from the boiler exhaust gas. The absorption chiller is collected to the storage tanks, which works as the "cooling tower" of the absorption chiller.

Solar system

The collectors are designed to accommodate a low pressure loss in the collector allowing up to 15 collectors to be mounted in series (2,152 sq ft (200 m^2) gross area). This ensures lower costs for installation, piping, pump etc. Collectors in the project are installed in series of up to 1,743 sq ft (162 m^2) gross area).

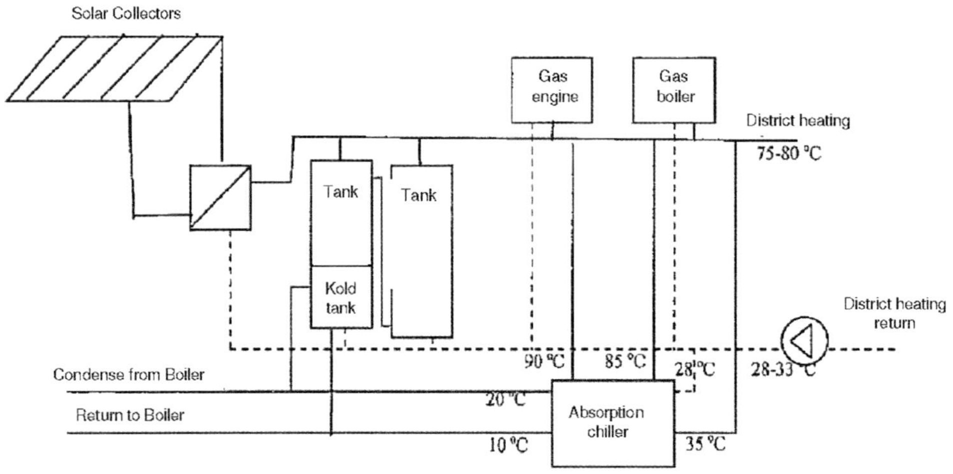

Figure A-23. Diagram of district heating system.

The collectors are also designed to accommodate the highest possible performance of flat plate collectors at the high temperature need for district heating (up to 212 °F [100 °C]) through several features:

- Use of low iron solar glass with a solar transmittance of over 91%
- Use of antireflective coating on the glass, which further increased the solar transmittance
- Thicker backside and side insulation
- An FEP foil between the glass and the absorber. This foil gives an extra air gab between the absorber and glass (as in doublet glazed windows), which reduced the heat loss through the front of the collector.

Collector type, number, net area

The special high performance flat plate collectors were designed for large scale systems by ARCON (Figure A-24). The solar system is constructed of 641 flat plate collectors with a total collector aperture area of 86,768 sq ft (8064 m^2) (93,117 sq ft [8654 m^2] gross area). Each collector has an aperture area of 135 sq ft (12.58 m^2) (145 sq ft [13.5 m^2] gross area). The collector field covers 269,000 sq ft (25,000 m^2) of land.

Figure A-24. Cross section of the ARCON HT collector.

Slope from horizontal

The collectors are mounted at a 35 degree angle.

Storage: Yes/no, capacity

Two water storage of each 396,300 gal (1500 m^3) secures the necessary flexibility in the heating plant and stores the heat from the gas engines, solar system and gas boilers at 185–194 °F (85–90 °C). One of the 396,300 gal (1500 m^3) storage tanks were added in combination with the installation of the solar system.

The use of the absorption chiller to extract heat from the exhaust of the gas boilers secures a low temperature in the storage tanks. The solar system heats from the bottom of the storage tank and the absorption chiller secures lower inlet temperatures in the collectors, which increases the energy output from the collector field.

Freeze protection strategy

The solar system runs on a glycol mixture to avoid freezing in winter.

Operation modes

The solar field is built as a separate circuit, which is connected to the district heating plant via an heat exchanger with circulation pumps on each side of the heat exchanger.

The solar system is operated with variable pump speed so that the field supplies flow at a constant temperature, which varies depending on different operating modes. The speed of the pump and the start and stop of the pumps are controlled by solar irradiation sensors in the collector field.

The solar system is configured so heat from the solar field can be used in different ways in the district heating plant depending on the plant's operating situation. The main operating modes are:

- The solar field supplies heat directly in the district heating grid at 167 °F (75 °C).
- The heat is feed to the storage tank at different temperatures depending on the temperature of the storage tank.
- The solar field heats the return temperatures of the district heating network before it is heated by the gas motors. This operating mode is used if the solar radiation is insufficient to run in the other operating modes.

Economics

System first cost (new or retrofit)

The investment of the solar systems totaled ~$2.9 million (US) (based on an exchange rate of DKK/USD of 5) (14.5 million DKK) including collector field, piping in field and heat exchanger unit. This is excluding the additional storage tank of 1500 m^3.

The project obtained a $0.7 million (US) subsidy giving the district heating company a net investment of $2.2 million (US).

Operation Costs

The energy costs produced by the solar field-based on an interest rate of 4% over a 25-year lifetime of the system is calculated to ~$55 (US)/MWh (273 DKK/MWh) excluding subsidy ($42 (US)/MWh or 209 DKK/MWh).

Savings

The estimated simple payback of the solar system (including subsidy) is calculated at ~8 years.

References

www.solvarmedata.dk. Webpage with online production data on solar district heating plants in Denmark and key figures on of the systems.

www.strandbyvarmeveark.dk. Webpage of the District Heating company of Strandby, which included key figures of the company.

www.arcon.dk. Web page of ARCON with data sheets on the collectors and key figures of this project and other similar project installed by ARCON.

FPC – 6

Title: Individual solar hot water system for a residential apartment complex in Frederikshavn, Denmark

Location: Frederikshavn, Denmark

Site

Rønneparken, Fælledvej
DK-9900 Frederikshavn

Photos of the installation

Figure A-25. Different types of buildings in the complex.

Figure A-26. Service building with collectors.

Project summary

The system is a 1,614 sq ft (150 m^2) solar system project for a domestic hot water residential complex with 126 apartments in mixed buildings with attached houses and apartment buildings.

The back energy for the domestic hot water system is based on district heating. The system is constructed of 12 large collector installed on the sloped roof of a service building.

Site

Location:	Frederikshavn
Denmark Address:	Rønneparken, Fælledvej DK-9900 Frederikshavn
Annual Solar radiation:	~1000 kWh/ m^2

Expected/measured share of solar thermal energy production

The system has been in operation for 10 years and has in this period supplied ~204,959 Btu (700 MWh) of heat. This corresponds to 20,496 Btu/yr (70 MWh/yr) and ~149,079 Btu/sq ft (470 kWh/m^2) annually, which is a solar efficiency of ~45%.

System details

Description of DH system

The system (Figure A-27) was installed in 1999 to supplement the domestic hot water supply for the complex of 126 apartments.

The system is controlled by a simple temperature difference solar controller.

The domestic hot water tanks supplies the complex with domestic hot water directly through a pipe grid in the complex.

A back-up for the solar system are to identical domestic hot water tanks, which are heated by the district heating grid. These are installed in series with the solar tanks.

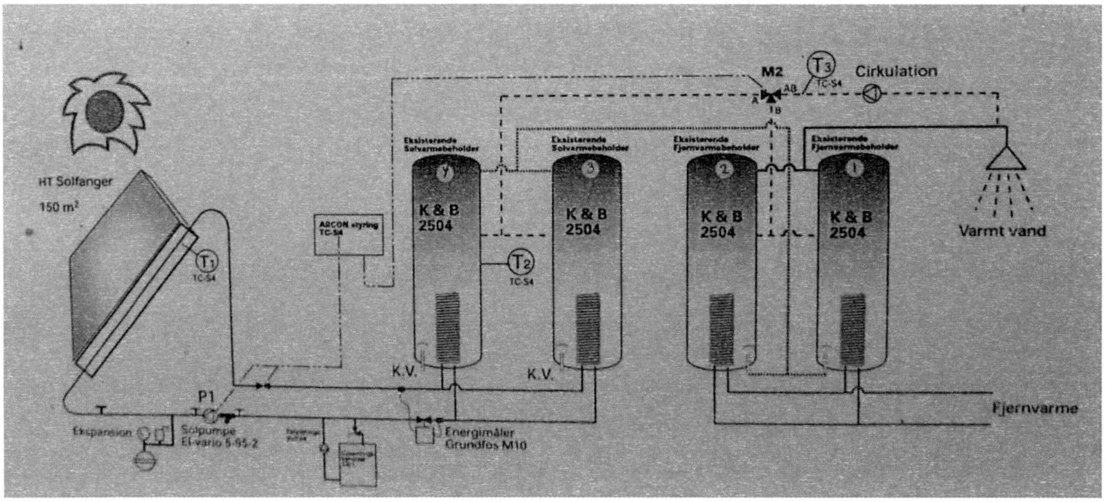

Figure A-27. Diagram of the heating system.

Connection of SWH to DH

The 12 solar collectors are connected to two parallel 2500 L domestic hot water tanks.

Solar system

The solar system is constructed of 12 high performance roof-mounted flat plate collectors each with a aperture area of 135 sq ft (12.5 m^2).

Collector type, number, net area

The collector field totals a collector area of 1,614 sq ft (150 m^2) (1,743 sq ft [162 m^2] gross area).

Freeze protection strategy

The hot water tanks are heated by the collector circuit through internal coils in the tanks. The solar collector circuit runs on a glycol mixture to avoid freezing in winter.

Economics

The investment of the solar systems totaled ~$50,000 (US) (based on an exchange rate of DKK/USD of 5) (250,000 DKK) including collector field and installation, but excluding tanks and heat exchangers.

FPC – 7

Title: Solar Water Heating for small residential community in the city of Skørping, Denmark

Location: Skørping, Denmark

Photo of the installation

Figure A-28. The collector field.

Project summary

The project was undertaken as demonstration system to show how a 100% renewable heating system can be constructed for a residential community, in this case with 22 houses.

The heating system for space heating and hot water for the 22 houses and is constructed as a central heating system consisting of a solar thermal plan and a wood boiler.

The system was built in 1994 when the 22 houses where build.

Later, the wood boiler was taken out of operation, The community is now connected to the local district heating grid in the town of Skørping. The solar system is still in operation and is supplemented by the district heating grid.

The solar system is an 5,918 sq ft (550 m^2) collector field-based on ground-mounted flat plate collectors, which is connected to a 16,140 sq ft (1500 m^2) seasonal storage system.

The solar field produces ~58,560 Btu (200 MWh) annually, which fills ~50% of the annual heat demand of the 22 houses, which house ~70 people.

Site

Location:	Skørping, Denmark
Address:	Ottrupgård DK-9520 SkørpingStranby
Annual solar radiation:	306,722 Btu/sq ft (967 kWh/m^2) (2001)

Project description

Type/age (when installed)

The system was installed in 1994 as a part of the heating system for the community of 22 houses. The solar system is constructed of 45 high performance, ground-mounted flat plate collectors each with a aperture area of 135 sq ft (12.5 m^2). The collector field totaling a collector area of 6,047 sq ft (562 m^2) (6,531 sq ft [607 m^2] gross area).

Design peak capacity

The solar system is designed for a peak capacity of 1.4 MBtu/hr (0.4 MW).

Calculated/measured annual solar thermal production

The actual thermal production from the solar field for the period of 1996 to 2001 has been 61,780 Btu (211 MWh) annually, on average:

	Thermal Production				Solar Radiation		Efficiency
Year	Btu	MWh	Btu/sq ft	kWh/m^2	Btu/sq ft	kWh/m^2	% Aperture Area
1996	64,709	221	124,656	393	306,406	966	41%
1997	74,078	253	142,736	450	330,512	1042	43%
1998	53,582	183	103,404	326	285,788	901	36%
1999	65,294	223	125,924	397	301,648	951	42%
2000	55,046	188	106,259	335	288,009	908	37%
2001	581,793	1987	111,651	352	306,723	967	36%

System details

Description of DH system

The heating system was originally constructed of two wood-pellet boilers and the solar system, which was connected to a 396,300 gal (1500 m^3) seasonal storage tank. Later the wood pellet boilers ("Kedel") were taken out of operation and replaced with a connection the district heating grid in the town of Skørping.

The solar system and wood boilers are connected to the 22 houses via a local district heating grid. The forward temperature to the houses is typically 149 °F (65 °C) and the return temperature 77 °F (25 °C). Each house is directly connected to the heating grid and heated by floor heating. For the domestic hot water system, each house is fitted with a 79 gal (300 L) domestic hot water boiler, which is heated by heating grid via a coil in the tank.

Solar system

Each collector has an aperture area of 135 sq ft (12.58 m^2) (145 sq ft [13.5 m^2] gross area). The solar field is built as a separate circuit, which is connected the storage system and the heating system via a heat exchanger with circulation pumps on each side of the heat exchanger.

Collector type, number, net area

The solar system is constructed of 45 flat plate collectors with a total collector aperture area of 6,047 sq ft (562 m^2) (6,531 sq ft [607 m^2] gross area). The collectors special high performance flat plate collector designed for large scale systems from ARCON.

Slope from horizontal

The collectors are mounted at a 37 degree angle.

Storage: Yes/no, capacity

The storage system is a insulated pond covered with a insulating floating lid and sealed with a liner.

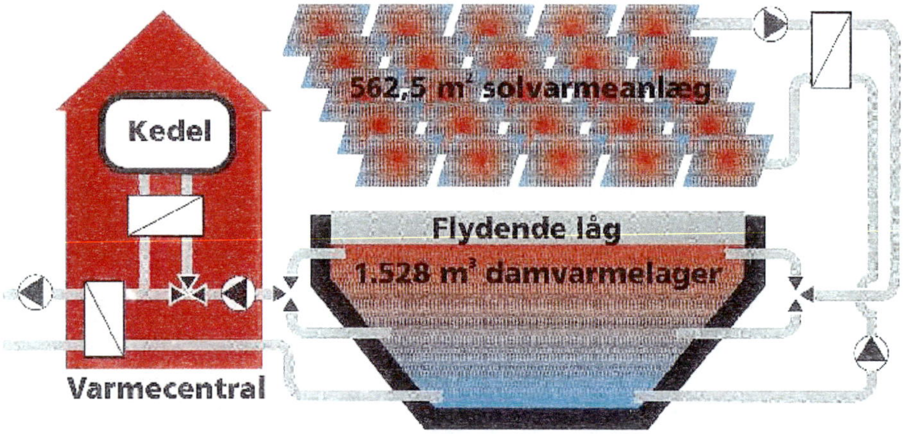

Figure A-29. Diagram of the heating system.

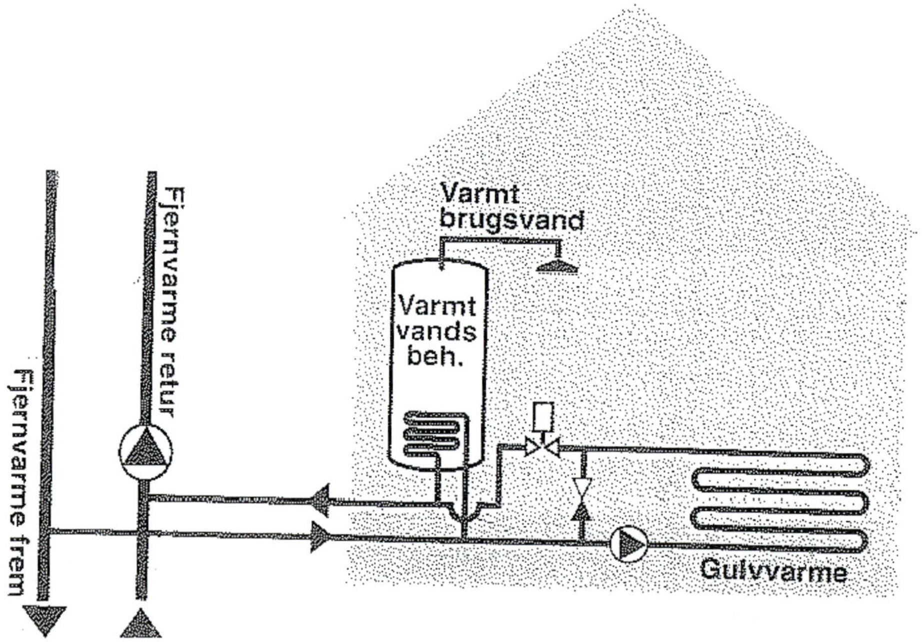

Figure A-30. Diagram of house installation.

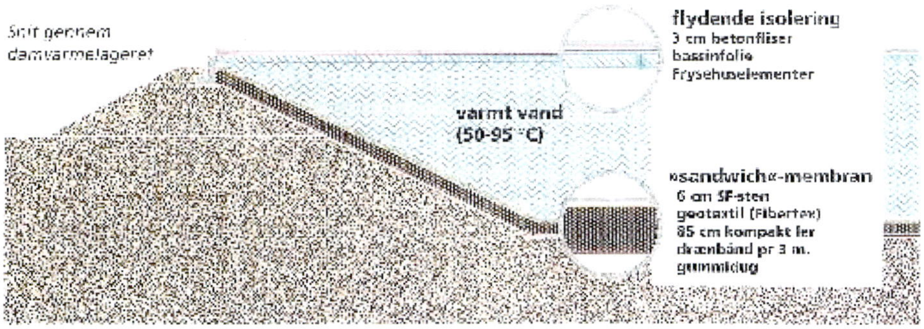

Figure A-31. Cross section of seasonal storage system.

__Freeze protection strategy__

The solar system runs on a glycol mixture to avoid freezing in winter.

__Operation modes__

The solar system is configured so that the heat from the solar field can be used in different ways:

- The solar system can directly feed the local district heating grid at 149 °F (65 °C).
- The solar system feeds the seasonal storage tank.

Economics

The investment of the solar systems totaled ~$190,000 (US) D (Based on an exchange rate of DKK/USD of 5) (0.95 million DKK) including collector field, piping in field and heat exchanger unit.

The investment of the seasonal storage system was ~$300,000 (US) D (1.6 million DKK)

Acknowledgement

are:

with support and input from, per Alex Sørensen (of PlanEnergi), who was the main engineering planning consultant on the project and Anders Otte Jørgensen (CEO of ARCON Solar,) who was the supplier of the system.

References

Internal project report of the project prepared by PlanEnergi and others

www.arcon.dk. Web page of ARCON with data sheets on the collectors and key figures of this project and other similar project installed by ARCON.

FPC – 8

Title: Solar water heating connected to combined heat & power (CHP) district heating system of Brædstrup, Denmark.

Location: Brædstrup, Denmark

Project summary

This project was undertaken to supplement the heat for the district heating grid in the town of Brædstrup in Denmark, which supplies all domestic hot water and space heating for the ~1400 households and ~2950 inhabitants.

The district heating in Brædstrup is supplied by a combination of natural gas boilers and a gas-driven CHP (Combined Heat and Power).

The solar system is an 86,080 sq ft (8000 m^2) collector field-based on ground-mounted flat plate collectors, which was installed in 2007.

The solar field covers ~8% of the annual heat demand of the district heating grid and supplied ~936,960 Btu (3200 MWh) of heat in 2009.

Figure A-32. Aerial image of the district heating plant of Brædstrup and the solar collector field.

Site

Location:	Brædstrup, Denmark
Address:	Fjernvarmevej 2, DK- 8740 Brædstrup
Latitude:	55 57.58 N
Longitude:	9.36.26 E
Annual solar radiation:	366,354 Btu/sq ft (1,155 kWh/m^2) (2009)

Project description

The solar system is constructed of 641 high performance ground-mounted flat plate collectors each with a aperture area of 136 sq ft (12.6 m^2). The collector field totaling a collector area of 86,769 sq ft (8064 m^2) (93,117 sq ft [8654 m^2] gross).

The system was installed during 2007 and put into operation in 2007.

The solar system is designed for a peak capacity of 20.5 MBtu/hr (6.0 MW).

The actual thermal production from the solar field in the first 2 full years of operation was:

	Thermal Production	Solar Radiation	Efficiency in %
Year	Btu (MWh)	Btu/sq ft (kWh/m^2)	Aperture Area
2008	895 (3.055)	373 (1.176)	32
2009	945 (3.229)	366 (1.155)	35

In 2008, the measured share of solar thermal production for heat consumption was 7.6%.

System description

The district heating system

The district heating grid provides heating and domestic hot water for the ~1400 household in the town of Brædstrup and supplies ~9.7 MBtu (33,000 MWh) of heat annually.

Figure A-33. The district heating plant at Brædstrup, Denmark.

The district heating system is a traditional district heating grid with a forward and retuning line.

The grid covers 15.5 mi (25 km) of main line and 16.2 mi (26 km) of connection line.

The line losses totals ~19% of the total heat production of ~12 MBtu (41,000 MWh) annually.

The grid is a direct network where the heating system in each house is connected directly to the grid. Domestic hot water is produced by an exchanger either direct flow or via an water heating tank in each house.

The typical operating temperatures of the grid are a forward temperature of ~158 °F (70 °C) and a return of approx 100 °F (38 °C) at winter and ~95 °F (35 °C) at summer.

The heat for the grid is produced and distributed to the grid from the district heating plant, which uses natural gas.

The plant consists of two gas engines of each 25 MBtu/hr (7.3 MW) power and 28 MBtu/hr (8.2 MW) heat (combined heat and power engines) with two back-up gas boilers of 35.8 MBtu/hr (10.5 MW) and 46 MBtu/hr (13.5 MW) for heating only.

The gas engines works as part of the back-up source for the power grid in the western part of Denmark and is in operation at peak power demand or when the production from wind turbines are low. The engines produce approx 4 MBtu (14,000 MWh) of power annually.

The district heating grid supplied by the gas boilers when the engines are not running. The solar system supplement the system.

A water storage of 528,400 gal (2000 m^3) secures the necessary flexibility in the heating plant and stores the heat from the Gas engines, solar system and gas boilers at 185 to 194 °F (85 to 90 °C).

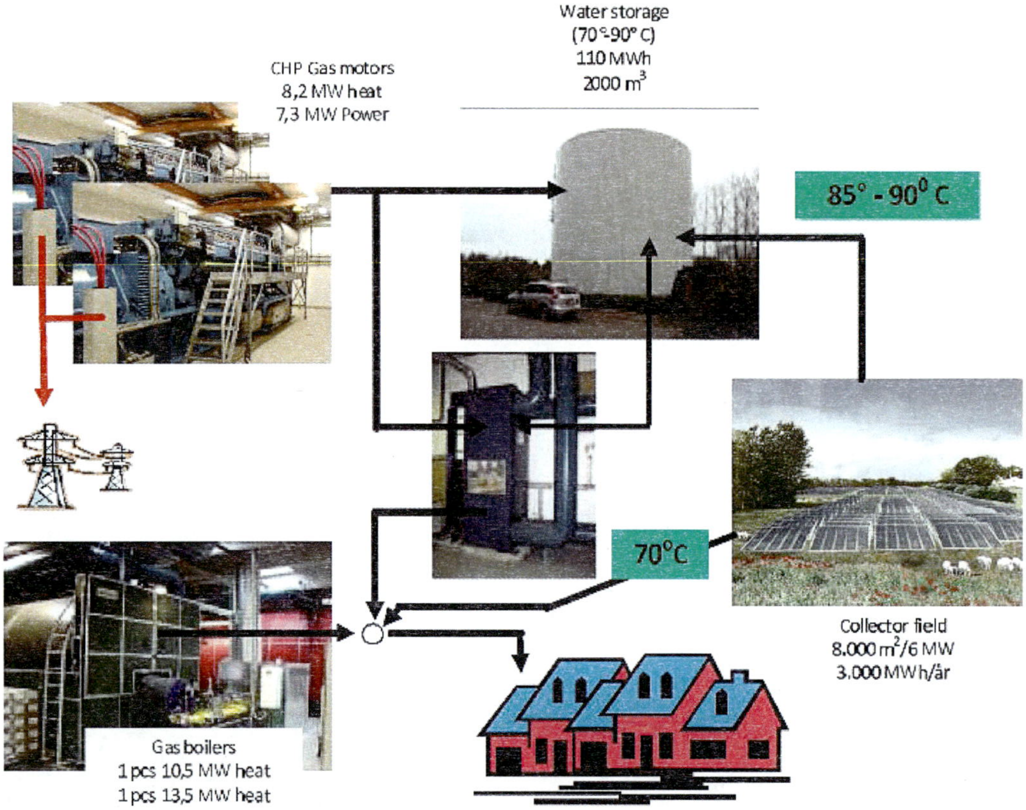

Figure A-34. Overview of the District Heating System of Brædstrup.

<u>The solar system</u>

The solar system is constructed of 641 flat plate collectors with a total collector aperture area of 86,769 sq ft (8064 m^2) (93,117 sq ft [8654 m^2] gross area).

The collectors special high performance flat plate collector designed for large scale systems from ARCON.

The collectors are designed to accommodate a low pressure loss in the collector allowing up to 15 collectors to be mounted in series (2,152 sq ft [200 m^2] gross area). This ensures lower costs for installation, piping, pump etc. Collectors in the project are installed in series of up to 11 collectors (1,603 sq ft [149 m^2] gross area).

The collectors are also designed to accommodate the highest possible performance of flat plate collectors at the high temperature need for district heating (up to 212 °F [100 °C]) through several features:

- Use of low iron solar glass with a solar transmittance of over 91%
- Use of antireflective coating on the glass, which further increased the solar transmittance
- Thicker backside and side insulation
- An FEP foil between the glass and the absorber. This foil gives an extra air gab between the absorber and glass (as in doublet glazed windows), which reduced the heat loss through the front of the collector.

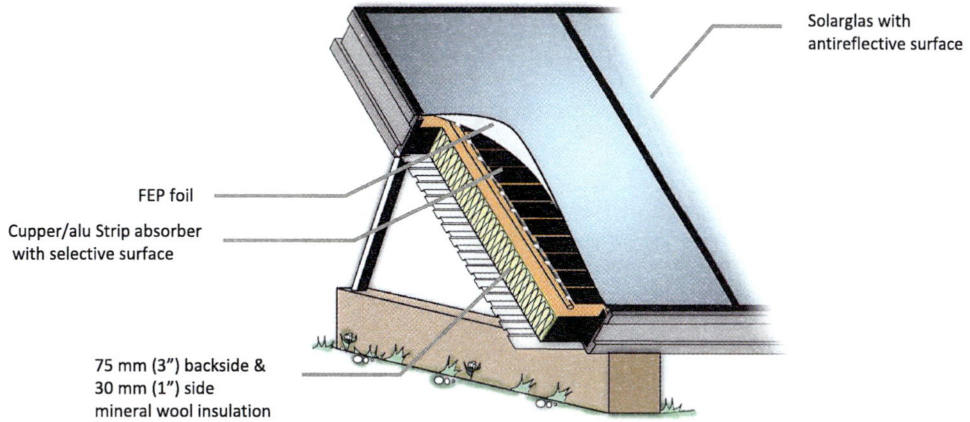

Figure A-35. Cross section of the ARCON HT collector.

Each collector has an aperture area of 135 sq ft [12.58 m^2] (145 sq ft [13.5 m^2] gross area).

The collectors are mounted at a 33 degree angle with a distance of the rows of 14.1 ft (4.3 m) from the front of one collector to the front of the next collector.

The rows are constructed in such a way that the first one or two collectors in the rows, where the temperature are low, are without the FEP foil and the last collectors in the row where the temperatures are high are fitted with the FEP foil. This to reduce costs, as the FEP foil at the low temperature does not improve performance.

The collector field takes up 247,480 sq ft (23,000 m^2) of land.

The solar system runs on a 30% glycol mixture to avoid freezing in winter.

The solar field is built as a separate circuit, which is connected to the district heating plant via an heat exchanger with circulation pumps on each side of the heat exchanger.

The solar system is operated with variable pump speed so that the field supplies flow at a constant temperature, which varies depending on different operating modes. The speed of the pump and the start and stop of the pumps are controlled by solar irradiation sensors in the collector field.

The solar system is configured so that the heat from the solar field can be used in different ways in the district heating plant depending on the operating situation at the plant. The main operating modes are:

- The solar field supplies heat directly in the district heating grid at 158 °F (70 °C)
- The heat is feed to the storage tank at different temperatures depending on the temperature of the storage tank.
- The solar field heats the return temperatures of the district heating network before it is heated by the gas motors. This operating mode is used if the solar radiation is insufficient to run in the other operating modes.

Economics

Brædstrup District Heating provides some of the lowest heating costs for the consumers of district heating in Denmark that uses natural gas.

The average heating costs for the consumer in Brædstrup in 2008 for was ~10 US cent/kWh (~$3/kBtu) including all energy taxes. This as ~50% of the costs of heating with an individual gas or oil boiler.

The investment of the solar systems totaled ~$2.5 million (US)* (12.3 million DKK) including collector field, piping in field and heat exchanger unit. This investment did not include the storage system, which was a part of the district heating plant.

The project obtained a $0.5 million (US) subsidy giving the district heating company a net investment of $2 million (US).

The cost of natural gas varies according to the general oil price development. The average cost of producing heat on the gas boilers in 2008 and 2009 at Brædstrup district heating, which is the alternative to the solar, has been ~$88 (US)/MWh (438 DKK/USD). This includes a general CO_2 tax of $37 (US)/MWh.

The operating costs of the plant have been at a minimum. The electricity costs for the solar system has been ~6,826 Btu (2 kWh) power per produced 293 Btu (1 MWh) heat. The maintenance costs have been at a level of $1000 (US) annually.

The total cost savings in the first 2 years of operation with a thermal production of solar field of 907,679 and 936,960 Btu (3100 and 3200 MWh) in 2008 and 2009 respectively have been ~$550,000 (US). This corresponds to a simple payback of the solar system of ~9 years (7 years including subsidy).

The energy costs produced by the solar field-based on an interest rate of 4% over a 25-year lifetime of the system is calculated to ~5 US cent/kWh ($1.77/kBtu).

Evaluation and lessons learned

The system was the first solar system ever to be installed in combination with CHP. All other large scale systems previously installed have been on boiler driven district heating (biomass oil etc.)

Experience with the system has proven that solar systems work well in combination with CHP plants.

---

* Based on an exchange rate of DKK/USD of 5.

After the completion of the system in Brædstrup, several of other CHP district heating plants in Denmark have initiated similar projects.

Experience of using the low temperature from the solar field to preheat the gas engines has proven to be successful, which increases the efficiency of the solar system.

In general the Brædstrup District Heating has had good experience with the operation of the systems, which has required a minimum of supervision and maintenance.

The system has proven to deliver a performance very close to the performance simulated in advance.

Acknowledgement

These case studies were prepared with support and input from per Kristensen, CEO at Brædstrup District Heating, per Alex Nielsen, PlanEnergi, who was the main engineering planning consultant on the project, and Anders Otte Jørgensen CEO of ARCON Solvarme who was the supplier of the system.

References

www.solvarmedata.dk. Webpage with online production data on solar district heating plants in Denmark and key figures on of the systems.

www.breadstrup.fjernvarme.dk. Webpage of the District Heating company of Brædstrup, which included key figures of the company.

www.arcon.dk. Web page of ARCON with data sheets on the collectors and key figures of this project and other similar project installed by ARCON.

FPC – 9

Title: Wohnhochhaus Frankfurt Peter-Fischer-Alle

Location:

Peter-Fischer-Allee
65929 Frankfurt/Main
Germany

Project summary

Facade collectors in a redeveloped multistory building with apartments.

Project description

Installation date: 2004
Collector area: 2,712 sq ft (252 m^2)
Designs solar system yield: 23,424 Btu (80 MWh) per annum
Measured solar system yield: 19,325 Btu (66 MWh) per annum
Measured spec. system yield: 83,421 Btu/sq ft (263 kWh/m^2) per annum
Meas. electric energy for solar syst.: 451 Btu (1.54 MWh) per annum
Designed hot water consumption: 1,030,380 gal (3900 m^3) per annum
Measured hot water consumption: 598,413 gal (2265 m^3) per annum
Measured hot water energy demand: 35,136 Btu (120 MWh) per annum
Measured circulation losses: 39,235 Btu (134 MWh) per annum
Measured solar share: 55% (tapping), 26% (tapping+circ.)
System efficiency: 32.7% (system energy yield/ radiation input)

Solar system details

DH system: standby storage with circulation
Connection SWH to DH: via buffer tank
Solar system: Schüco
Collectors: standard flat plate collector Schüco GK, purpose-build for façade
Slope: 90 degree
Orientation: 240 degree
Storage volume: 2,774 gal (10.5 m^3)
Specific storage volume: 0.31 gal/cu ft (41.6 L/m^3) (storage volume per square feet absorber area)
Freeze protection: Water-glycol-mixture
System strategy: Domestic hot water heating with solar preheating storage

Economics

Costs solar system: $259,630 (182,838 €) (incl. Tax)
Annual costs for loan(6% rate for 20 years): $22,639.06 (15,943 €)
Costs of solar energy (planned): ~$0.01/Btu (0.20 €/kWh)
Costs of solar energy (measured): ~$0.01/Btu (0.24 €/kWh)
Conventional heating: two gas-fired boiler (2 x 56,920 Btu/min [1000 kW]) one combined heat and power unit (4,611 Btu/min [81 kW] thermal)

Figure A-36. Subprogram 2, domestic hot water, 2006 Apartment House Frankfurt, 2700 sq ft (250 m^2).

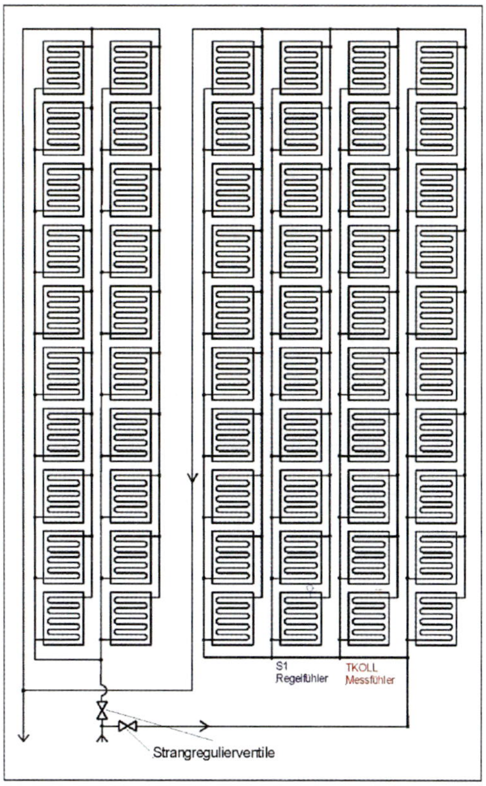

Figure A-37. Piping schematic of the fassade collector field (Apartment House Frankfurt).

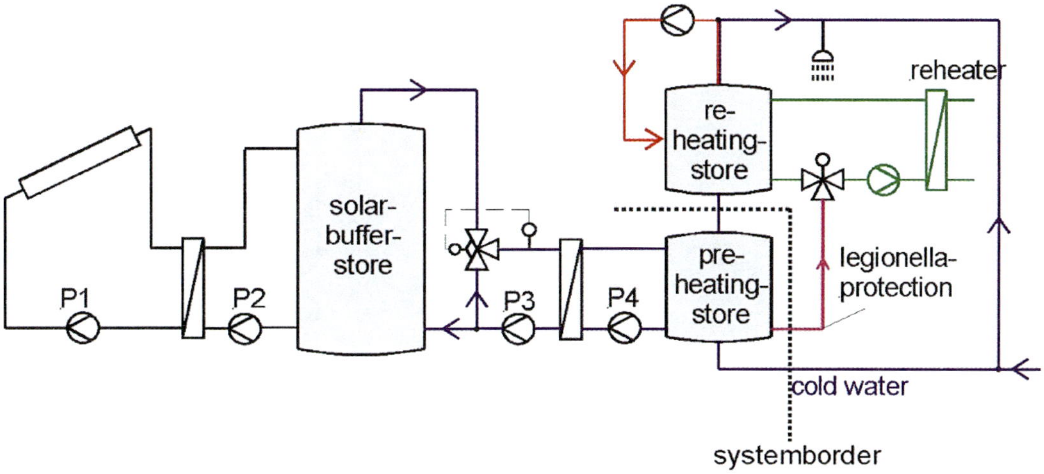

Figure A-38. Solar system for drinking water heating with pre-heating store and thermal disinfection for legionella-protection (no coverage of circulation losses).

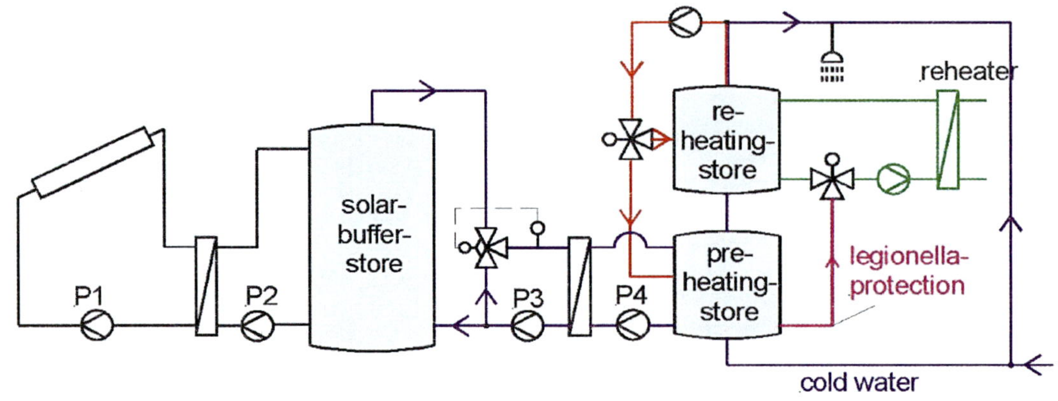

Figure A-39. Solar system for drinking water heating with pre-heating store and thermal disinfection for legionella-protection (with coverage of circulation losses).

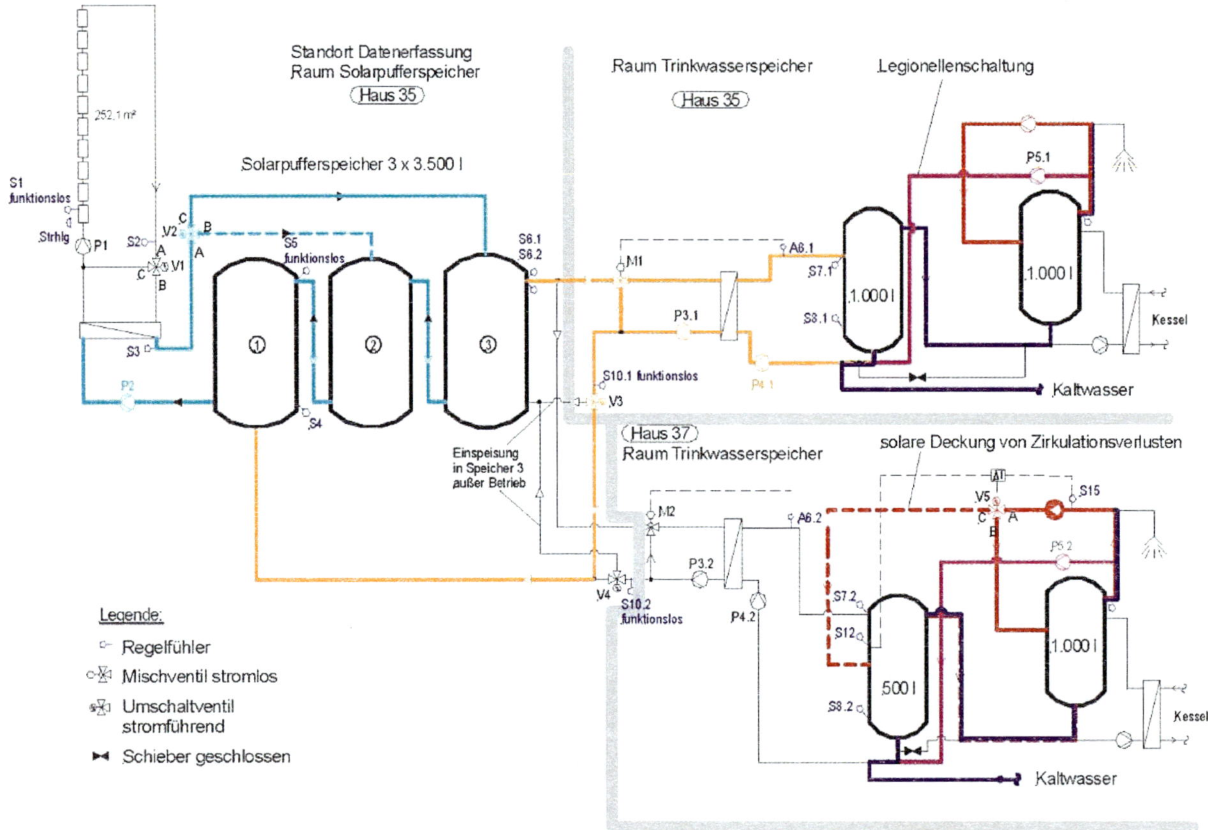

Figure A-40. Piping schematic of the solar system (Apartment House Frankfurt).

FPC – 10

Title: Wohngebiet Ehemaliger Schlachthof Speyer residential area (Old Slaughterhouse Speyer, Germany)

Location: Mausbergweg
67346 Speyer, Germany

Project summary

District heating network with 61 one-family houses

Project description

Installation date: 2005
Collector area: 5,864 sq ft (545 m^2)
Design solar system yield: 52,704 Btu (180 MWh) per annum
Measured solar system yield: 58,853 Btu (201 MWh) per annum
Measured spec. system yield: 117,043 Btu/sq ft (369 kWh/m^2) per annum
Meas. electric energy for solar syst.: 703 Btu (2.4 MWh) per annum
Measured energy demand for DH: 243,317 Btu (831 MWh) per annum
Measured energy losses in DH: 20% from energy demand
Measured solar share: 22.4%
System efficiency: 26.8% (system energy yield / radiation input)

Solar system details

DH system: Four pipes, bivalent storage tank
Connection SWH to DH: Via storage tank
Solar system: Wagner & Co
Collectors: Roof integrated standard flat plate collector: Solar Roof FDK
Slope: 30 degrees
Orientation: 180 degrees
Storage volume: 26,420 gal (100 m^3)
Specific storage volume: 1.37 gal/cu ft (183 L/m^3) (storage volume per square feet absorber area)
Freeze protection: Water-glycol-mixture

System strategy

Four pipes DH with bivalent storage and three collector fields

Economics

Costs solar system: $506,968 (357,020 €) (incl. tax)
Annual costs for loan (6% rate for 20 years): $44,207 (31,132 €)
Costs of solar energy (planned): ~$0.01/Btu (0.176 €/kWh)
Costs of solar energy (measured): ~$0.01/Btu (0.166 €/kWh)
Conventional heating: One gas-fired boiler (34,152 Btu/min [600 kW])

District heating net

Advance temperature of DH: 153 °F (67 °C)
Return temperature of DH: 90 °F (32 °C) (winter), 99 °F (37 °C) (summer)
Planed annual energy demand: 222,528 Btu (760 MWh) per annum
Planed energy losses in DH underground piping: 31,915 Btu (109 MWh) per annum
Heat transfer stations: Radiator and under-floor heating, tankless water heating

Figure A-41. District heating Speyer "Old slaughterhouse" 5866 sq ft/26,400 gal (545 m^2 / 100 m^3).

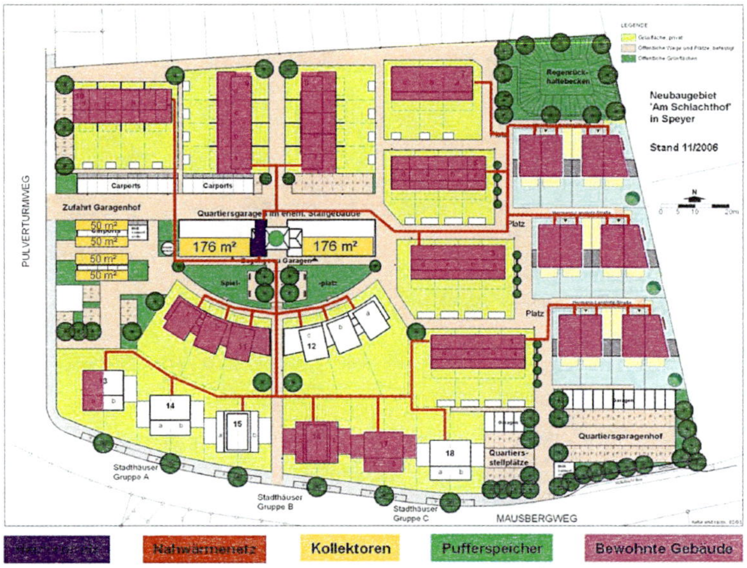

•Quelle: egs/sws

Figure A-42. District heating Speyer "Old Slaughterhouse."

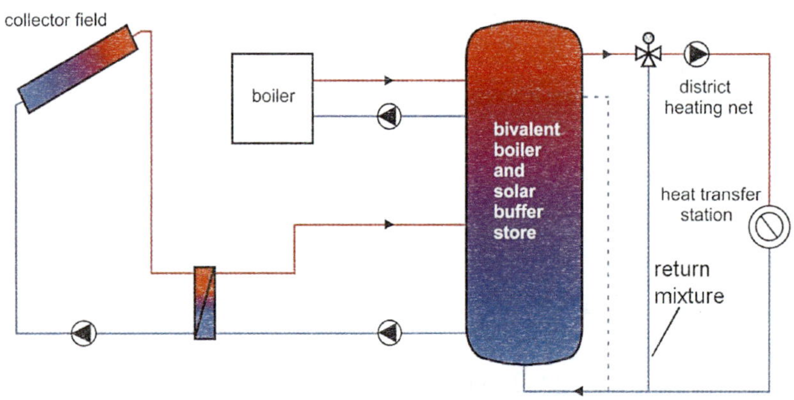

Figure A-43. Heating net with four-pipe boiler reheating inside buffer store.

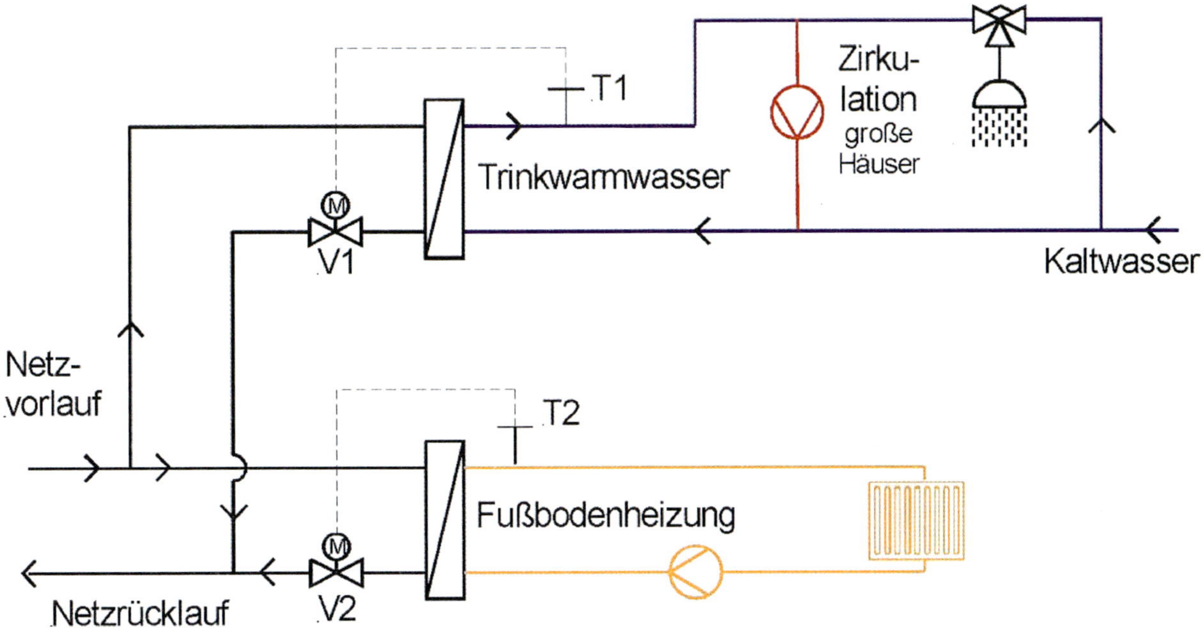

Figure A-44. Piping schematic of heat transfer stations in family houses (District heating Speyer "Old Slaughterhouse").

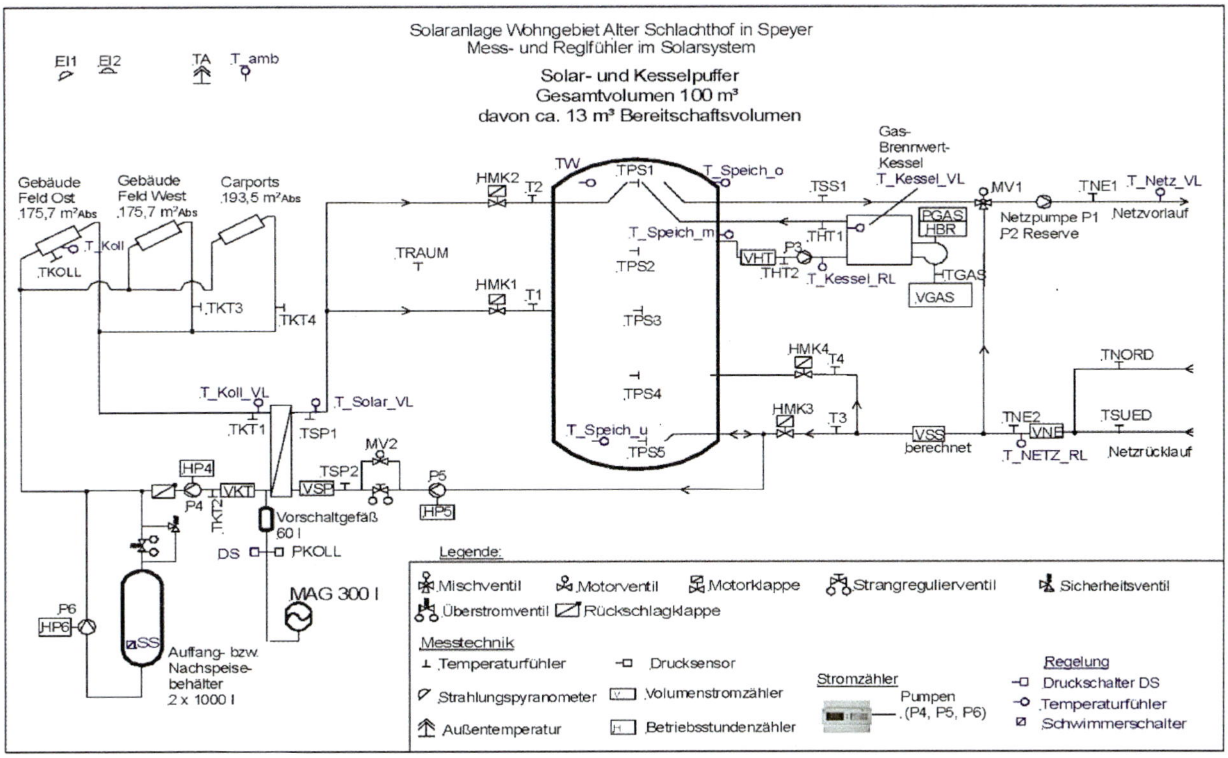

Figure A-45. Piping schematic of the solar system and the connection to district heating net (Speyer "Old Slaughterhouse").

FPC – 11

Title: Wohngebiet ehemalige Kaserne Normand
(Residential area former barracks Normand)

Location

Paul-Egell-Straße
67346 Speyer
Germany

Photo of the installation

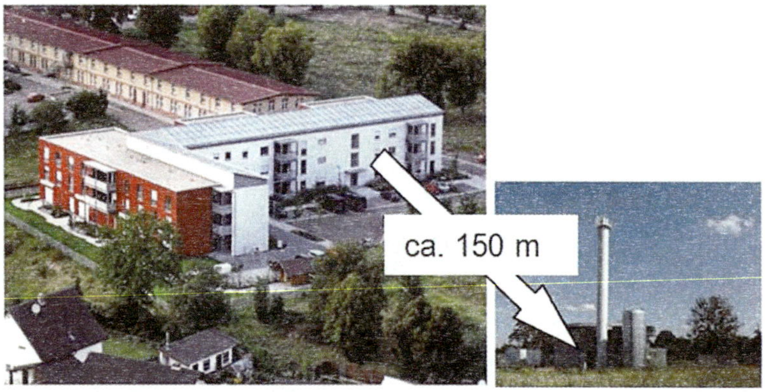

Figure A--46. Residential area former barracks Normand, highlighting CHP.

Project summary

District heating network with former barracks refurbished as apartment houses, some new residential houses, new residential home for handicapped people

Project description

Installation date: 2007
Collector area: 3,077 sq ft (286 m^2) at present, (planned: 7,532 sq ft [700 m^2])
Design solar system yield: 28,314 Btu (96.7 MWh) per annum (3,077 sq ft [286 m^2])
Measured solar system yield: 24,537 Btu (83.8 MWh) per annum
Measured spec. system yield: 92,937 Btu/sq ft (293 kWh/m^2) per annum
Measured electric energy for solar system: 351 Btu (1.2 MWh) per annum
Measured energy demand for DH: 617,222 Btu (2108 MWh) per annum
Measured solar share: 3.4%
System efficiency: 21.2% (system energy yield / radiation input)

Solar system details

DH system: Three pipes, monovalent solar storage tank
Connection SWH to DH: Via storage tank
Solar system: Wagner & Co
Collectors: Roof integrated standard flat plate collector: Solar Roof FDK
Slope: 15 degree
Orientation: 140 degree
Storage volume: 6,605 gal (25 m^3)
Specific storage volume: 0.65 gal/cu ft (87 L/m^3) at present (storage vol./m^2 absorber area) 0.26 gal/cu ft (35 L/m^3) planned

Freeze protection: Water-glycol-mixture.

System strategy

Three pipes DH with solar storage and one Collector field.

Economics

Costs solar system: $268,664 (189,200 €) (incl. tax)
Annual costs for loan (6% rate for 20 years): $23,430 (16,500 €)
Costs of solar energy (planned): ~$0.01/Btu (0.176 €/kWh)
Costs of solar energy (at present): ~$0.01/Btu (0.23 €/kWh)

Conventional heating

One wood chip boiler (36,998 Btu/min [650 kW]) for base load in winter, one gas-fired boiler (50,943 Btu/min [895 kW]).

District heating net

Advance temperature of DH: Min. 160 °F (71 °C), max. 180 °F (82 °C)
Return temperature of DH: 126 °F (52 °C) (winter), 140 °F (60 °C) (summer)
Planned annual energy demand: 1.2 MBtu (4000 MWh) per annum
Planned energy losses in DH underground piping: 98,966 Btu (338 MWh) per annum
Heat transfer stations: Radiator and under-floor heating, mostly water heating with standby storage and circulation.

References

Croy, R.; Wirth, H.P. final report about measurements until 31.12.2009 (in German only), http: //www.zfs-energietechnik.de/data/docs/20100112154110_0.pdf?RND=16036

FPC – 12

Solar water heating (SWH) connected to district heating networks (DH)

Title: District heating network residential area "Gorch-Fock-Weg,"

Location: Norderney, Germany

Photo of the installation

Figure A-47. Front view of the heating central with two collector rows on a flat roof.

Figure A-48. Collector row piping seen from the backside of first collector row (left side) one of two solar storage tanks (right side).

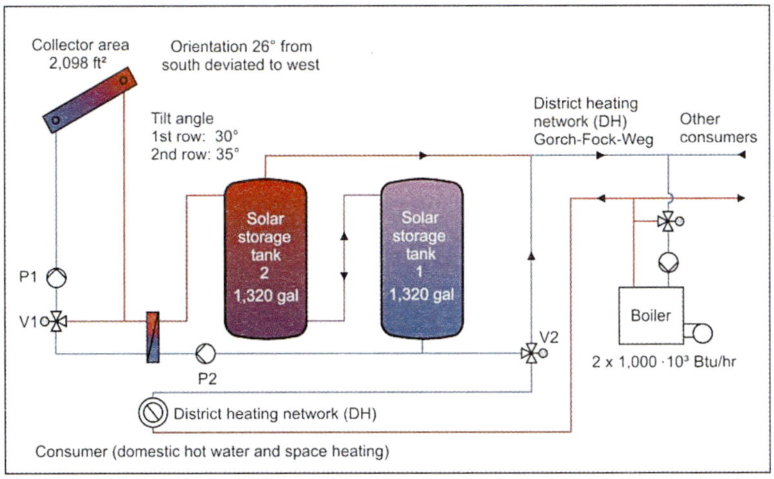

Figure A-49. Highly simplified schematics of the solar system, solar storage tanks arrangement and the integration in the district heating network (DH).

Project summary

Near the harbor of the island of Norderney, well-known in Germany as a seaside vacation spot, a residential area with 23 row houses and one multi-family house is supplied with energy from a district heating network (DH). The central heating plant is located near to the community (just on the other side of the street) so the piping of the DH is very short. The public service of the city of Norderney (Wirtschafts- betriebe Norderney) operates the heating central and the underground DH to supply the area with heat for domestic hot water and space heating. For this purpose, the central is equipped with two natural gas-fired boilers (2 x 1,000 10^3 Btu/hr [2,928,000 kWh]), a solar energy system with a collector area of 2098 sq ft (195 m^2) and a solar storage tank capacity of 2 x 1320 gal (4996 L). The DH was built in 2002, while the solar system was added to the DH in 2007—an important detail, because the DH was neither planned nor designed to meet the strict requirements of the later added solar system.

The heat transfer stations (HTS) in the houses are owned by the homeowners, so the technical influence of the carrier of the DH (Wirtschaftsbetriebe Norderney) is limited.

The power supply and the DH advance temperature (in the range between 149 and 176 °F (65 and 80 °C) depending on ambient temperature) are guaranteed by the Wirtschaftsbetriebe Norderney. The installed return temperature limiters restrict the return temperature of space heating loops in the houses to not more than 122 °F (50 °C). To protect against Legionella bacteria (for which domestic hot water must be at least 140 °F [60 °C]), it is not advised to integrate return temperature limiters in the domestic hot water system.

This arrangement of solar system, boilers, and district heating network works well. No severe problems were found in the concept. One problem of excessive thermal loss in the solar storage tanks is under investigation, and a technical solution, now under discussion, has not yet been implemented.

Site

Location:	Residential Area "Gorch-Fock-Weg"
Town:	Norderney
Country:	Germany
Latitude:	53 42' North
Longitude:	7 08' East

Project description

A collector area of 2098 sq ft (195 m^2) is installed on a flat roof on top of the heating central, which houses the boilers, the solar storage tanks, and the control facilities. The energy from the collector loop is transferred by a flat plate heat exchanger to the charge loop of the solar storage tanks. Tanks 1 and 2 (both with 1320 gal [4,996 L] capacities) were located in a row to maintain the temperature of the charge loop.

Depending on the DH return temperature, the volume flow can be guided by shutoff valves to discharge the solar storage tanks or to bypass them (primary function). A secondary function of this valve is to control the temperature of the discharge flow from the solar storage tanks into the DH return pipe. This can be important in the case the temperature in the solar storage tanks rises higher than is needed in the DH advance. Gas boilers can supplement to guarantee the DH advance temperature in periods with low irradiation and less solar energy availability. The correct DH advance temperature is controlled by a modulated boiler firing. There is no need for a separate control valve, which compensates for fluctuations in temperature generated by boiler stop and go operation because of the modulating firing of the boiler. The solar system was connected to the DH in 2007.

Table A-3. Expected data during planning.

Number of buildings supplied by DH (row house, multi-family house):	3 + 1
Number of people supplied by DH:	124
Maximum energy demand for space heating supplied from DH:	1,000*10^6 Btu/yr (2,928 *10^6 kWh)
Maximum energy demand for domestic hot water supplied from DH:	340*10^6 Btu/yr (996*10^6 kWh)
Maximum energy losses in DH underground piping:	140*10^6 Btu/yr (410*10^6 kWh)
Total energy demand from heating central:	1,500*10^6 Btu/yr (4,392*10^6 kWh)
Total energy supplied from solar system to DH:	249*10^6 Btu/yr (729*10^6 kWh)
Solar fraction of total energy demand DH:	17.0%
Solar system efficiency (energy output from solar storage tanks/irradiation energy):	31.6%
Max. DH advance temperature (depends on ambient temperature):	149 – 176 °F (65–80 °C)
Max. DH return temperature:	122 °F (50 °C)

The data in Table A-3 were calculated using the following equations:

Solar system efficiency = Solar energy output from solar storage tanks/Irradiation in collector area

Solar fraction of total energy demand DH = Solar energy output from solar storage tanks/Total energy demand for DH

Table A-4 lists the measured data from 2008 to 2009. Irradiation in the collector areas differ in a small range. As one would expect, solar output from the collector loop and from the solar storage tanks (which includes the thermal losses of the tanks) depend directly on solar irradiation. However, the difference between solar energy output from the collector loop and the solar output from the solar storage tanks (in 2008: 236.2*10^6 Btu/yr to 190.4*10^6 Btu/yr [692 *106 kWh/y to 557*106 kWh/yr]) is too large. The thermal loss in the storage tanks reaches 20%. The cause for this phenomenon is currently under investigation.

This proportion between irradiation and solar output, termed "solar system efficiency," operates in the range between 22.9 and 24.3%. The rise in solar efficiency from 2008 to 2008 can be attributed to the slightly increase in irradiation and the slightly decline of DH return temperature in that period. Generally the lower DH return temperature is, the higher is the efficiency of the solar system.

Measured data

Table A-4. Measured data from 2008 and 2009.

(IP)		2008	2009
Irradiation in horizontal area	10^3 Btu/(sq ft*yr)	323.2	335.7
Irradiation in collector area	10^6 Btu/yr	831.5	846.6
Irradiation in collector area, specific	10^3 Btu/(sq ft*yr)	396.2	403.5
Solar energy output collector loop	10^6 Btu/yr	236.2	241.8
Solar energy output solar storage tanks	10^6 Btu/yr	190.4	205.3
Total energy demand of DH	10^6 Btu/yr	1,511	1,521
Solar system efficiency	%	22.9	24.3
Solar fraction of total energy demand DH	%	12.6	13.5
DH advance temperature, yearly average	°F	156.4	152.2
DH return temperature, yearly average	°F	134.6	131.7
(SI)		**2008**	**2009**
Irradiation in horizontal area	10^3 kWh/m^2*yr	1.02	1.06
Irradiation in collector area	10^6 Btu/yr	2,434.63	2,478.84
Irradiation in collector area, specific	10^3 kWh/m^2*yr	1.25	1.27
Solar energy output collector loop	10^6 kWh	691.59	707.99
Solar energy output solar storage tanks	10^6 kWh	557.49	601.12
Total energy demand of DH	10^6 kWh	4,424.21	4,453.49
Solar system efficiency	%	22.9	24.3
Solar fraction of total energy demand DH	%	12.6	13.5
DH advance temperature, yearly average	°C	69	67
DH return temperature, yearly average	°C	57	55

The solar fraction of total energy demand DH climbs from 12.6 in 2008 to 13.5% in 2009. This is understandable because the energy output of the solar storage tanks rises from 190.4*10^6 Btu/yr to 205.3*10^6 Btu/yr (557.49*10^6 kWh/yr to 601.12*10^6 kWh/yr) in 2009 while the energy demand of the DH in both years is near constant at circa 1500*10^6 Btu/yr (4,392.00*10^6 kWh/yr).

On a yearly average, the DH advance temperature normal falls between 152.2 and 156.4 °F (66.78 and 69.11 °C). However, in this case, the return temperature between 131.7 and 134.6 °F (55 and 57 °C) on yearly average lies clearly above the planned 122 °F (50 °C), a clearly undesirable situation. All efforts to reduce the DH return temperature to 122 °F (50 °C) in yearly average by adjusting the heat transfer stations in the houses failed. A "lesson learned" from this experience is that one cannot necessarily improve the performance of a solar system designed DH (built in 2002) by simply adding a more modern solar system (built in 2007).

Figure A-50 shows the measured daily data of irradiation, and solar energy output from solar storage tanks in 2008, and the efficiency of the solar system calculated from that data. The irradiation over the year is typical for Germany, with broad differences between summer and winter. The solar energy outputs follows suit; the main harvest of energy is found from February to October, and energy produced from November to January is almost negligible. In more northern regions, this effect is even more striking, while less so in more southern regions. While the summer solar system efficiency can reach up to 40%, poor winter performance reduces the yearly average to only 22.9%.

System details

Solar system

The solar system contains a collector area divided in two rows, situated on the flat roof of the heating central building roof, piping, a flat plate heat exchanger, two solar storage tanks, the connection to the DH, and the controls.

Connection of SWH to DH

The Solar system connects to the DH via a return pipe located in the heating central building (Figure A-51).

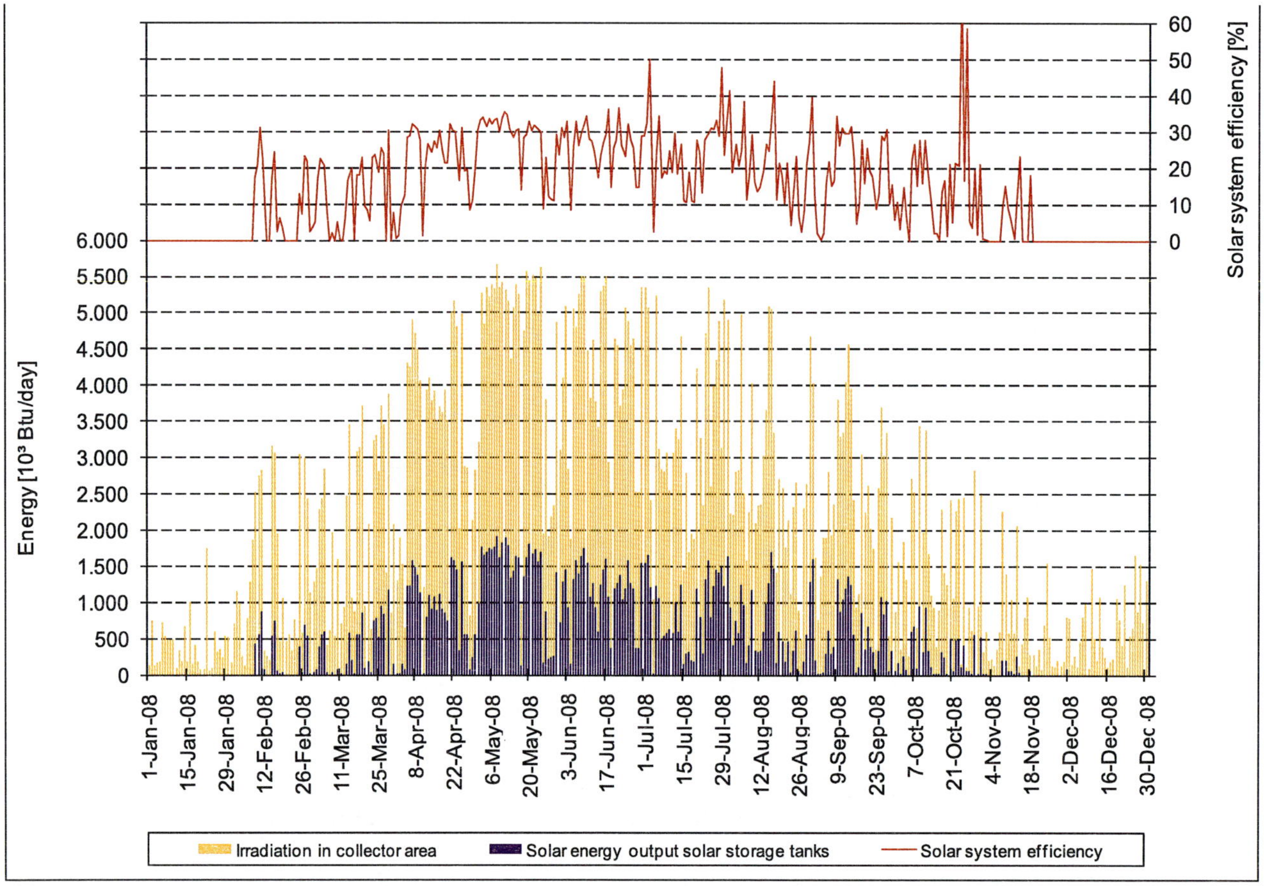

Figure A-50. Irradiation, energy output, and efficiency solar system in 2008, daily data solution.

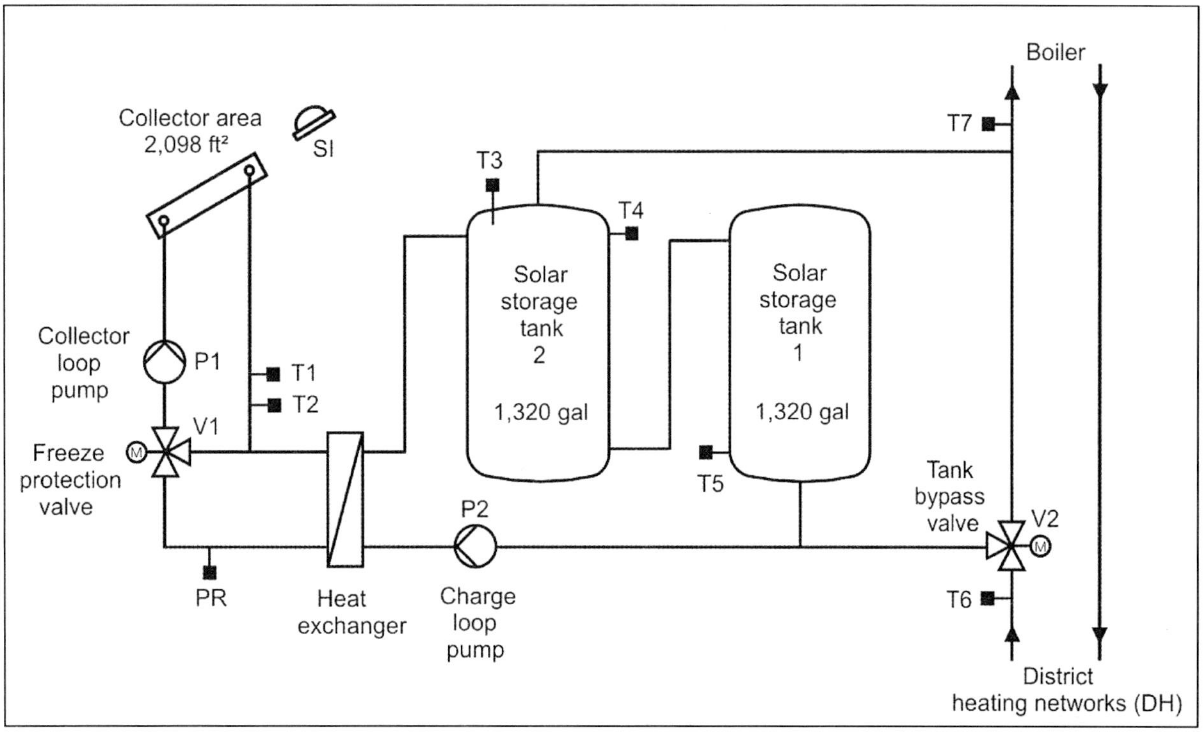

Figure A-51. Highly simplified schematics showing the positioning of the control sensors.

Description of control strategy

Collector manufacturer:	Viessmann
Type of collector:	Vitosol 100 SH1 flat plate collector
Orientation:	26 from south deviated to west
Tilt angle:	First row (seen from the street): 30°; Second row (seen from the street): 35°
Collector area:	2,098 sq ft (195 m^2) (Aperture Area)
Solar storage tank:	2 x 1320 gal (4,996 L) upright cylindrical steel tank
Freeze protection:	Collector loop filled with 40% glycol, 60% water; Freeze protection valve
Heat exchanger:	Brazed flat plate type, area 115.5 sq ft (11 m^2)

To prevent the solar storage tanks from overheating and boiling, the collector loop and the solar storage tanks are fitted with temperature sensors (T1, T3). When temperatures in the collector loop (T1) rise higher than 248 °F (120 °C), or when temperatures in the solar storage tank 2 (T3) surpass 203 °F (95 °C), the pump control functions are halted and pump P1 stops. When temperatures at sensor T1 fall below 248 °F (120 °C) and temperatures at sensor T3 fall below 203 °F (95 °C), the bolted control functions clear. Also, a pressure sensor is installed (PR) for the safety of the collector loop. Pressure measured (PR) lower than 7.3 psi (50 kPa) stops pumps P1 and P2, and pressure higher than 7.3 psi (50 kPa) clears the halted functions.

To prevent the collector loop heat exchanger from freezing, valve V1 opens the path from the collector area to the heat exchanger only when sensor T2 measures a temperature higher than 41 °F (5 °C). When the temperature falls below 41 °F (5 °C), the heat exchanger is uncoupled from the collector loop, and the volume flows through the bypass.

The running of the collector loop pump (P1) is controlled by measured solar irradiation (SI). Solar irradiation more than 63.4 Btu/(hr * sq ft) (0.2 kWh/[hr*m^2]) makes the pump running, solar irradiation lower than 57.1 Btu/(hr * sq ft) (0.18 kWh/[hr * $m^{2]}$) stops the pump.

The charge loop pump (P2) is controlled by the difference in temperature between the temperature in the collector loop (T2) and the temperature (T5) in lower area of solar storage tank 1. A difference in temperature (T2 – T5) of more than 9 °F (5 °C) makes the pump run, a difference of less than 3.6 °F (2 °C) stops the pump. Under the same conditions, valve V1 opens to guide the volume flow from the collector area to the collector loop heat exchanger or to bypass the heat exchanger.

The discharge of the storage tanks executed by valve V2 is controlled by the difference in temperature between the upper area of solar storage tank 2 (T4) and the DH return temperature (T6). Difference in temperature (T6 – T4) of more than 5.4 °F (3 °C) opens valve V2 to allow the volume to flow to the tanks, and a temperature difference of less than 1.8 °F (-1 °C) shuts V1 to bypass the tanks.

Valve V2 has a second function. It controls the temperature T7 of the DH return volume flow behind the mixing point when solar storage tanks are discharged. Valve V2 mixes discharge flow and DH return flow in such a way that the temperature T7 behind the mixing point never exceeds 167 °F (75 °C). This technique helps to avoid feeding in a volume flow from the tanks with a temperature significantly higher (e.g., 194 °F [90 °C]) than needed in the DH advance (e.g., 158 °F [70 °C]).

Table A-5. Summary of control activities and control conditions.

Control activity	Control Conditions
Collector loop/Charge loop Clearance of running pump P1 General stop running pump P1 Clearance of running pumps P1, P2 General stop running pump P1, P2 Collector loop pump P1 Charge loop pump P2	T1 < 248 °F (120 °C), T3 < 203 °F (95 °C) T1 > 248 °F (120 °C), T3 > 203 °F (95 °C) PR > 7.3 psi PR < 7.3 psi On: SI > 63.4 Btu/(hr * sq ft),Off: SI < 57.1 Btu/(hr * sq ft) On: T2 – T5 > 9.0 °F (5 °C), Off: T2 – T5 < 3.6 °F (2 °C)
Freezing protection valve Valve1	Bypass heat exchanger: T2 < 41 °F (5 °C) Open to heat exchanger: T2 > 41 °F (5 °C)
Loading solar storage tanks Charge loop pump P2 Valve 1	On: T2 – T5 > 9.0 °F (5 °C), Off: T1 – T5 < 3.6 °F (2 °C) Open to heat exchanger: T2 – T5 > 9 °F (5 °C) Bypass heat exchanger: T2 – T5 < 3.6 °F (2 °C)
Discharging solar storage tanks Valve 2 in DH return	Open to tanks: T6 – T4 > 5.4 °F (3 °C) Bypass tanks: T6 – T4 < 1.8 °F (1 °C) Max. temperature of DH volume flow behind mixing point: 167 °F (75 °C)

Economics

The costs of the solar systems include only the costs for solar collectors, piping, solar storage tanks and controls, but do not include costs for the district heating network, boiler or the heating central building (Table A-6).

Table A-6. Economics.

Costs solar system - Costs solar system, including statics - Costs planning solar system. - Steel collector support construction - Costs solar system, statics and planning - Costs solar system, statics and planning including 19% tax	$177,500 (125,000 €) $30,963 (21,805 €) $440,754 (28,700 €) $249,217 (175,505 €) **$296,568 (208,851 €)**	
Annual costs for loan living period 20 years, 6% rate, → annuity: 8.72%	$25,861 (18,212 €)	
Solar energy output	Per year	Sum total in 2 yrs
- Planned solar energy output from solar storage tanks - Measured solar energy output from storage tanks	249 *10^6 Btu (729 *10^6 kWh) 190.6 *10^6 Btu (558 *10^6 kWh) 205.6 *10^6Btu (602 *10^6kWh)	498 *10^6 Btu (1,458 *10^6 kWh) 395.8 *10^6 Btu (1,159 *10^6 kWh)
Relation measured solar energy output/ planned energy output		79.5%

Savings of gas and CO_2 calculated with measured solar energy from solar storage tanks with following assumptions: boiler efficiency: 90%; energy of natural gas: 27.4*10^3 Btu/cu yd_{Gas} emission factor: 0.129 $lbCO_2/10^3$ Btu_{Gas} - saving amount of natural gas - avoidable amount of CO_2		16,050 cu yd 28.3 (short) ton
Costs of solar energy from solar storage tanks with 8.72% annuity, including solar system, statics, planning and tax - costs and planned solar energy output - costs and measured solar energy output, without costs for maintenance and repairs	0.073 €/10^3 Btu 0.092 €/10^3 Btu	

User evaluation: No information available from Wirtschaftsbetriebe Norderney.

District heating network

Table A-7. Expected data during planning

Number of buildings supplied by DH (row house, multi-family house):	23 + 1
Number of people supplied by DH:	124
Maximum energy demand for space heating supplied from DH:	1,000*10^6 Btu/yr (2,928 *10^6 kWh/yr)
Maximum energy demand for domestic hot water supplied from DH:	340*10^6 Btu/yr (996 *10^6 kWh/yr)
Maximum energy losses in DH underground piping:	140*10^6 Btu/yr (410 *10^6 kWh/yr)
Total energy demand from heating central:	1,500*10^6 Btu/yr (4,392*10^6 kWh/yr)
Total energy supplied from solar system to DH:	249*10^6 Btu/yr (729*10^6 kWh/yr)
Share of solar thermal production of total demand DH:	17.0%
Solar system efficiency (energy output from solar storage tanks/irradiation energy):	31.6%
Max. DH advance temperature (depends on ambient temperature):	149–176 °F (65–80 °C)
Max. DH return temperature	122 °F (50 °C)

Table A-8. Measured data from 2008 to 2009.

Parameter	2008		2009	
Energy demand of DH	1,511*10^6 Btu/yr	(4,424 *10^6 kWh/yr)	1,521*10^6 Btu/yr	(4,453*10^6 kWh/yr)
Solar energy output solar storage tanks	190.4*10^6 Btu/yr	(557*10^6 kWh/yr)	205.3*10^6 Btu/yr	(601*10^6 kWh/yr)
Solar fraction of energy demand DH	12.6%		13.5%	
DH advance temperature, yearly average	156.4 °F	(69 °C)	152.2 °F	(67 °C)
DH return temperature, yearly average	134.6 °F	(57 °C)	131.7 °F	(55 °C)

Figure A-52 shows the measured demand of energy DH from 2008 to 2009, which is constant in both years. Solar energy output from the solar storage tank is slightly rising because the irradiation in 2009 is slightly higher in 2008 and the DH return temperature fell from 134.6 to 131.7 °F (57 to 55 °C) in 2009. The Solar fraction of energy demand DH rose from 12.6% in 2008 to 13.5% in 2009, but did not reach 17% as planned.

The DH advance temperature is in the field as planned, but not the DH return temperature. The DH return temperature between 131.7 and 134.6 °F (yearly average) is clearly above the planned 122 °F (50 °C). The public service tried several times to reduce the DH return temperature to the aspired 122 °F (50 °C) by adjusting the heat transfer stations in the houses, but ultimately failed. The lesson learned here is that that the integrated heat transfer stations were not optimal matched to the solar system.

The building landlords own the heat transfer stations. Consequently, no information was available regarding manufacturing uses and types of stations; nor was measured data available from inside the stations.

Figure A-52 shows a typical energy demand curve in a DH in Germany, here year 2008. The demand in the winter is about three times higher than in the summer. In the summer, the solar fraction of the energy demand of the DN can reaches up to 100%; in the winter, it is close to zero. Designing the solar system it was one goal, **not** to produce an excess of solar energy in the summer, which means a waste of expensive generated solar energy. Figure A-52 clearly shows that, with a maximum solar fraction not more than 100% in the summer, this goal was reached.

Figure A-53 shows the volume flow of the DH and the volume flow through the solar storage tank during discharge. Only a small amount of volume flow of the DH is guided through the solar storage tank for discharge. Through flow is stopped and tanks are bypassed when solar storage tanks are discharged (means temperature in storage tanks is lower than DH return temperature). Advance and return temperatures of DH are shown. In the winter, the advance temperature rises depending on ambient temperatures (explained in Chap. Control). Return temperatures rise in the summer because energy is not needed for space heating in the buildings. This causes rising return temperatures in heat transfer stations and in the DH return. The temperature limiter in the heat transfer stations influences the return temperature of the space heating loops, but not the return temperature of domestic hot water facility. A temperature limiter in the domestic hot water facility could cause water temperature to fall too low, thereby allowing the dangerous Legionella bacteria to grow.

Experiences/lessons learned

Energy use reduction

In 2 years from 2007 to 2009 energy output was measured from the solar storage tanks at 395.8*10^6 Btu or 197.9*10^6 Btu on a yearly average. Assuming a boiler efficiency of 90%, an energy content of natural gas of 27.4*10^3 Btu/cu yd_{Gas} there is a saving of natural gas of 16,050 cu yd in 2 years. Per year this is an average saving of 8025 cu yd.

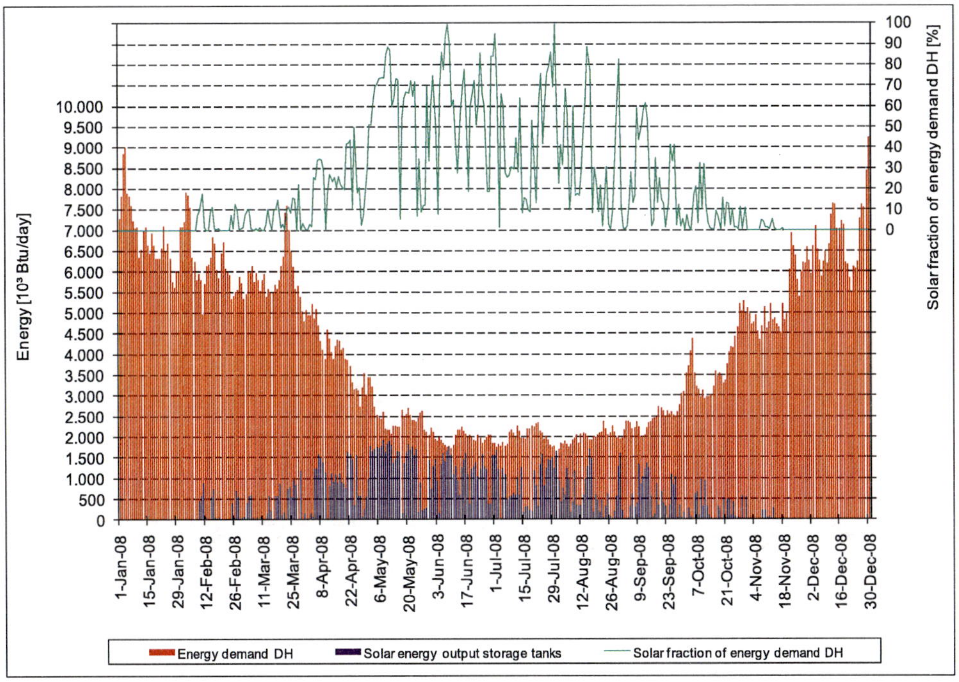

Figure A-52. Energy demand DH, energy output solar system, and solar fraction measured in 2008.

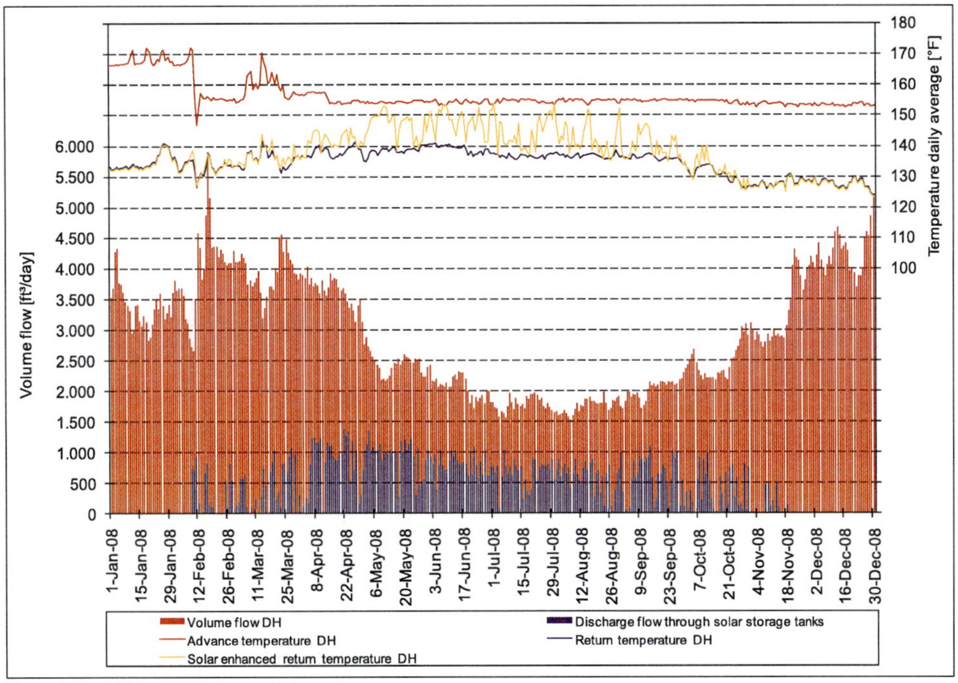

Figure A-53. Volume flow DH, volume flow through solar storage tanks, and temperatures DH.

Lessons learned

From the data we measured and the monitoring of the solar system we learned that:

- In general the solar system in combination with the district heating network operated without severe problems. The input of solar energy in the return pipe of the DH is the most promising way.
- The solar system, the connection to the DN return, and the controls were (in comparison to other solar systems) designed and arranged as simply as possible. That makes it easier for the

maintenance staff to understand and repair the system. We recommend not to attempt to reach the last percentage of efficiency by complicating the system.

- The difference between the planned DH return temperature (122 °F [50 °C]) and the measured from 131.7 to 134.6 °F (55 to 57 °C) in yearly averages from 2007 to 2008 is not desirable. The heat transfer stations in the houses are not optimal fitting to the requirements of the solar system.
- Valve V2 successfully performed its double function (1. Guide the return flow of the DH return through the solar storage tanks or bypass them. 2. Control the DH return temperature behind mixing point to not more than 167 °F [75 °C]). This technical arrangement can be revised.
- The thermal losses of the solar storage tanks are excessive. Investigations are under way to find out the reason. We have the suspicion that the thermal losses are not produced by the storage tanks itself, but by the piping connections to the storage tanks. It may be that a circulating volume flow in this piping driven by gravity has built up (known as thermosiphon flow), which can produce such thermal losses.
- Because of the strong winds and the salty air on the island of Norderney the collector array construction was planned and carried out very carefully. Corrosion protection and static firmness were enhanced here in comparison to less endangered locations.

General Data

Address of the project

Solar water heating connected to a district heating network
Residential Area "Gorch-Fock-Weg"
Norderney, Germany

Date of report

Measured data period: 1st January 2008 to 31st December 2009
Date of report: June 2010

Acknowledgement

Promoting department

Bundesministerium für Umwelt, Naturschutz und Reaktorsicherheit (BMU)
(Federal department of environment, nature conservation and nuclear reactor safety)
Alexanderstr. 3
10178 Berlin
Germany

Operating company (owner)

Wirtschaftsbetriebe Norderney
Jann-Berghaus-Str. 34
26548 Norderney
Germany

Planning company

Ingenieurgesellschaft Bannert GmbH
Flughafenallee 3
28199 Bremen
Germany

Mounting company

Haustechnik Rosenboom
Lippestr. 24
26548 Norderney
Germany

Measuring company

ZfS-Rationelle Energietechnik GmbH
Verbindungsstr. 19
40723 Hilden
Germany

References

Mies, M.; Rehrmann, U.; Zwischenbericht für das Projekt Nahwärme Gorch-Fock-Weg Norderney May 2009, http: //www.zfs-energietechnik.de/main.php?RND=33644&SESS=&LANG=de&ID=228 (More detailed information about the solar system in Norderney are outlined in this report).

Peuser, Felix A.; Remmers, Karl-Heinz; Schnauss, Martin. Solar Thermal Systems. Successful Planning and Construction. Solarpraxis AG Berlin Germany in association with James & James London UK, 2002. ISBN: 3-934595-24-3

VDI 6002, part 1, September 2004 (technical guideline). Solar heating for domestic water. General principles, system technology and use in residential buildings. Distributer: Beuth Verlag, 10722 Berlin, Germany

DVGW W551, April 2004 (technical guideline)

Trinkwassererwärmungs- und Leitungsanlagen; Technische Maßnahmen zur Verminderung des Legionellenwachstums (Hot water systems, technical arrangements to reduce developing of legionella bacteria), no English translation available. Distributer: Wirtschafts- und Verlagsgesellschaft Gas und Wasser mbH. Josef-Wimmer-Str. 1 - 3, 53123 Bonn, Germany

BINE Project Info 12/2009: Solar retrofitting of local heating, http: //www.bine.info/en/hauptnavigation/publications/projektinfos/publikation/nahwaerme-solar-nachruesten/?artikel=1451

FPC – 13. Solar Water Heating (SWH) Connected to District Heating Networks (DH)

Title: Residential area "Cohnsches Viertel," Hennigsdorf, Germany

Location: Hennigsdorf, Germany

Photo of the installation

Figure A-54. One of five collector areas partly roof-integrated.

Figure A-55. Heating central with in-housed solar storage tank (left), collector loop flat plate heat exchanger (red box) with piping (right).

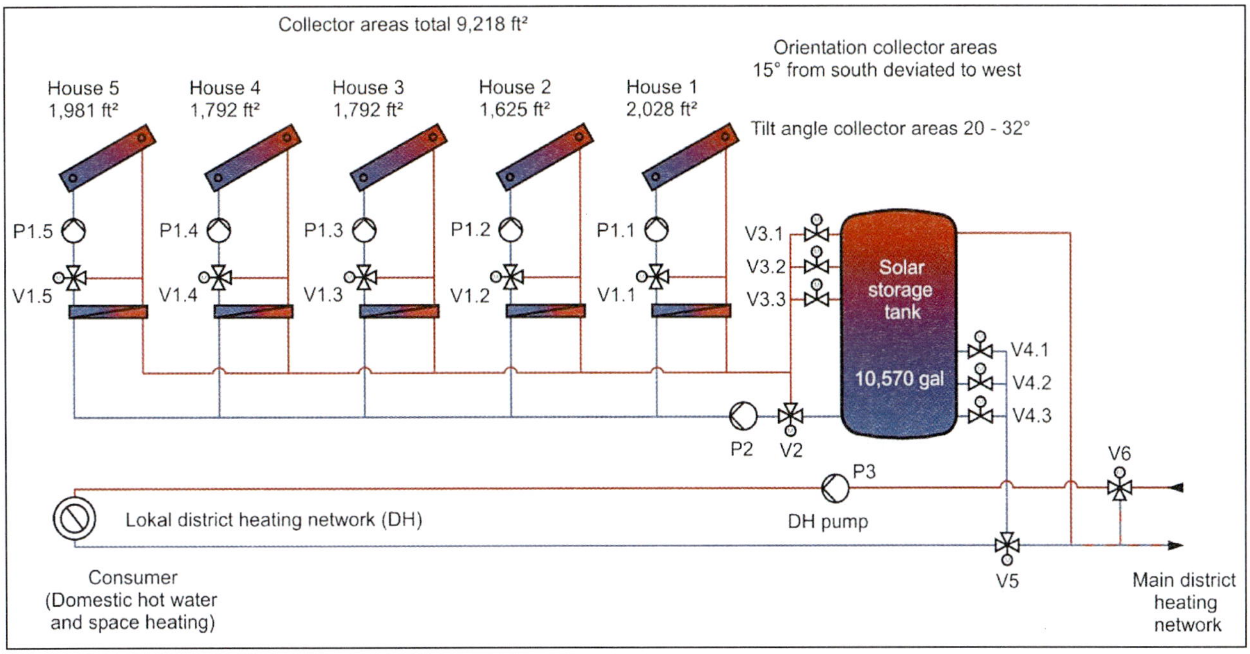

Figure A-56. Highly simplified schematics of the solar system, solar storage tank arrangement, and the integration in the district heating networks.

Project summary

The residential area "Cohnsches Viertel" in Hennigsdorf filled with multi-family houses (1300 flats) between 1940 and 1960. In 2000, a reconstruction program was begun to gradually improve the technical and convenience standard of the houses, which are owned by Hennigsdorfer Wohnungsbaugesellchaft (HWB). The space heating and the domestic hot water were formerly supplied by coal-fired ovens and are now connected to a (local) district heating network (DH).

This district heating network is owned by Stadtwerke Hennigsdorf (Public service Hennigsdorf), which runs a wide-ranging main district heating network in the town of Hennigsdorf. The network in the residential area "Cohnsches Viertel" is connected to this wide-ranging main network as a sub-network with lower advance temperatures.

A solar system, which supplies the energy to the DH return pipe, is integrated. The collector area (9218 sq ft [857.3 m^2]) is divided in five subareas with separate collector loops, which are connected by an underground charge loop to the solar storage tank (10,570 gal [40,008 L]) in the heating central. Here the solar energy is fed in the DH return.

The power supply and the DH advance temperature (in the range between 149 and 203 °F (65 and 95 °C) depending on ambient temperature) are guaranteed by the Stadtwerke Hennigsdorf. The DH return temperature should not exceed 96.8 °F (36 °C) in summer and 105.8 °F (41 °C) in winter.

The arrangement of solar system, boilers, and district heating network works well. No severe problems were found in the concept. One lesser problem (no controlled solar storage tank outlet temperature) was later solved.

Site

Location:	Residential Area "Cohnsches Viertel"
Town:	Hennigsdorf
Country:	Germany
Latitude:	52 38' North
Longitude:	13 12' East

Project description

A collector area of 9218 sq ft is divided into five separate collector loops that are connected by an underground piping to a central solar storage tank in the heating central. Each collector fraction is installed on a separate house, numbered here from 1 to 5. The distance between the heating central and the farthest collector fraction on house 5 is 1150 ft. The collector loops are driven by collector loop pumps and are equipped with a freeze protection valve. The solar energies from the collector loops are transferred to the common charge loop by flat plate heat exchangers. The charge loop to the solar storage tank in the heating central with a capacity of 10,570 gal is carried out by underground piping and driven by one central charge pump.

Depending on the temperature of DH return the volume flow can be guided by shutoff valves to discharge or bypass the solar storage tanks. To guarantee the DH advance temperature in periods with low irradiation and less output of solar energy, additional energy from the main district heating network can be fed in. The correct DH advance temperature is controlled by a control valve. An overflow of solar energy, which can happen in periods of little energy demand DH in the summer, can be conducted into the main district heating network. So overheating and standstill of the solar system can be avoided. The solar system was connected to the DH in 2001.

Expected data during planning

Number of flats supplied by DH:	460
Number of people supplied by DH:	1,150
Maximum energy demand for space heating supplied from DH:	$9{,}420 \times 10^6$ Btu/yr ($27{,}582 \times 10^6$ kWh/yr)
Maximum energy demand for domestic hot water supplied from DH:	$3{,}210 \times 10^6$ Btu/yr ($9{,}399 \times 10^6$ kWh/yr)
Maximum energy losses in DH underground piping:	$1{,}160 \times 10^6$ Btu/yr ($3{,}396 \times 10^6$ kWh/yr)
Total energy demand from heating central:	$13{,}790 \times 10^6$ Btu/yr ($40{,}377 \times 10^6$ kWh/yr)
Total energy supplied from solar system to DH:	$1{,}190 \times 10^6$ Btu/yr ($3{,}484 \times 10^6$ kWh/yr)
Solar fraction of total energy demand DH:	9.0%
Solar system efficiency (energy output from solar storage tank/irradiation energy):	37.1%
Max. DH advance temperature (depends on ambient temperature):	149–203 °F (65–95 °C)
Max. DH return temperature	winter: 96.8 °F (36 °C) summer 105.8 °F (41 °C)

Table A-9 lists the measured data from 2003 to 2009. Irradiation in the collector areas differed in a small range with exception of year 2003, which is known in Germany as the "century summer." Solar output from the collector loops or solar output from the solar storage tank (which includes thermal losses through the underground charge loop piping and the solar storage tank) Is related directly to solar irradiation. This proportion between irradiation and solar output of the collector loops can be shown by the "collector loop efficiency," which operates in the close range between 27.12 and 31.45%. The proportion between irradiation and solar output of the storage tank is called the "solar system efficiency," which operates in the close range between 21.36 and 25.50%. Important to mention is the difference between collector loop and solar system efficiency, which varies between 5.1 and 7.2% points. This unanticipated number shows the thermal loss, which the solar system sustains in the underground piping of the charge loop and in the solar storage tank. Closer investigations showed that the main losses were caused by the underground piping.

Table A-9. Measured data from 2003 to 2009.

	IP	2003	2004	2005	2006	2007	2008	2009
Irradiation in horizontal area	10^3 Btu/(sq ft*yr)	355.7	320.5	333.2	331.6	327.1	328.1	335.1
Irradiation in collector area	10^6 Btu/yr	3,774	3,261	3,508	3,391	3,397	3,322	3,450
Irradiation in collector area, specific	10^3 Btu/(sq ft*yr)	409.6	353.8	380.4	368.0	368.4	360.4	374.1
Solar energy output collector loop	10^6 Btu/yr	1,140	884.8	1,102	1,006	1,058	1,034	998.7
Solar energy output solar storage tank	10^6 Btu/yr	947.9	696.8	894.3	780.4	812.4	807.3	767.4
Energy demand of DH	10^6 Btu/yr	Not measured						
Collector loop efficiency	%	30.19	27.12	31.45	29.66	31.16	31.11	28.96
Solar system efficiency	%	25.11	21.36	25.50	23.01	23.92	24.29	22.25
Solar fraction of total energy demand DH	%	Not measured						
DH advance temperature, yearly average	°F	157.8	154.2	155.3	158.0	156.6	154.2	154.8
DH return temperature, yearly average	°F	126.3	125.8	123.3	122.0	122.9	121.6	122.2
	SI	**2003**	**2004**	**2005**	**2006**	**2007**	**2008**	**2009**
Irradiation in horizontal area	10^3 kWh/(m^2*yr)	1.12	1.01	1.05	1.05	1.03	1.03	1.06
Irradiation in collector area	10^6 kWh/yr	11,050.27	9,548.21	10,271.42	9,928.85	9,946.42	9,726.82	10,101.60
Irradiation in collector area, specific	10^3 kWh/(m^2*yr)	1.29	1.12	1.20	1.16	1.16	1.14	1.18
Solar energy output collector loop	10^6 kWh /yr	3,337.92	2,590.69	3,226.66	2,945.57	3,097.82	3,027.55	2,924.19
Solar energy output solar storage tank	10^6 kWh /yr	2,775.45	2,040.23	2,618.51	2,285.01	2,378.71	2,363.77	2,246.95
Energy demand of DH	10^6 kWh /yr	Not measured						
Collector loop efficiency	%	30.19	27.12	31.45	29.66	31.16	31.11	28.96
Solar system efficiency	%	25.11	21.36	25.50	23.01	23.92	24.29	22.25
Solar fraction of total energy demand DH	%	Not measured						
DH advance temperature, yearly average	°C	69.89	67.89	68.50	70.00	69.22	67.89	68.22
DH return temperature, yearly average	°C	52.39	52.11	50.72	50.00	50.50	49.78	50.11

Since the energy demand of the DH was not measured, the solar fraction of energy demand DH cannot be calculated.

The DH advance temperature runs in a normal and planned range between 154.2 and 158.0 °F (68 and 70 °C) on a yearly average. The DH return temperature is between 121.6 and 126.3 °F (50 and 52 °C) on a yearly average with a slightly fallen tendency over the years. We relate this effect to the efforts made to adjust the heat transfer stations in the connected houses. Despite this tendency, the DH return temperature is much too high compared with a planned return temperature (96.8 °F [36 °C] in summer and 105.8 °F [41 °C] in winter). For clarity, the measured data from 2003 to 2009 listed in Table A-9 are also shown Figures A-57 to A-59.

Figure A-60 shows the in 2008 measured daily data of irradiation, solar energy output from solar storage tank, and the calculated efficiency of the solar system. The irradiation over the year is typical shape for Germany, with explicit differences between summer and winter. The solar energy output is similar; the main harvest of energy is found from March to October, energy output in November to February is almost negligible. This effect is most noticeable in more northern regions, in more southern regions the effect dwindles. In the summer, solar system efficiency reaches nearly 35%; poor data in the winter reduces the yearly average to only 24.3%. The peaks up to 60% are atypical because they occur when a (partly) loaded solar storage tank is discharged; the next day is characterized with low irradiation.

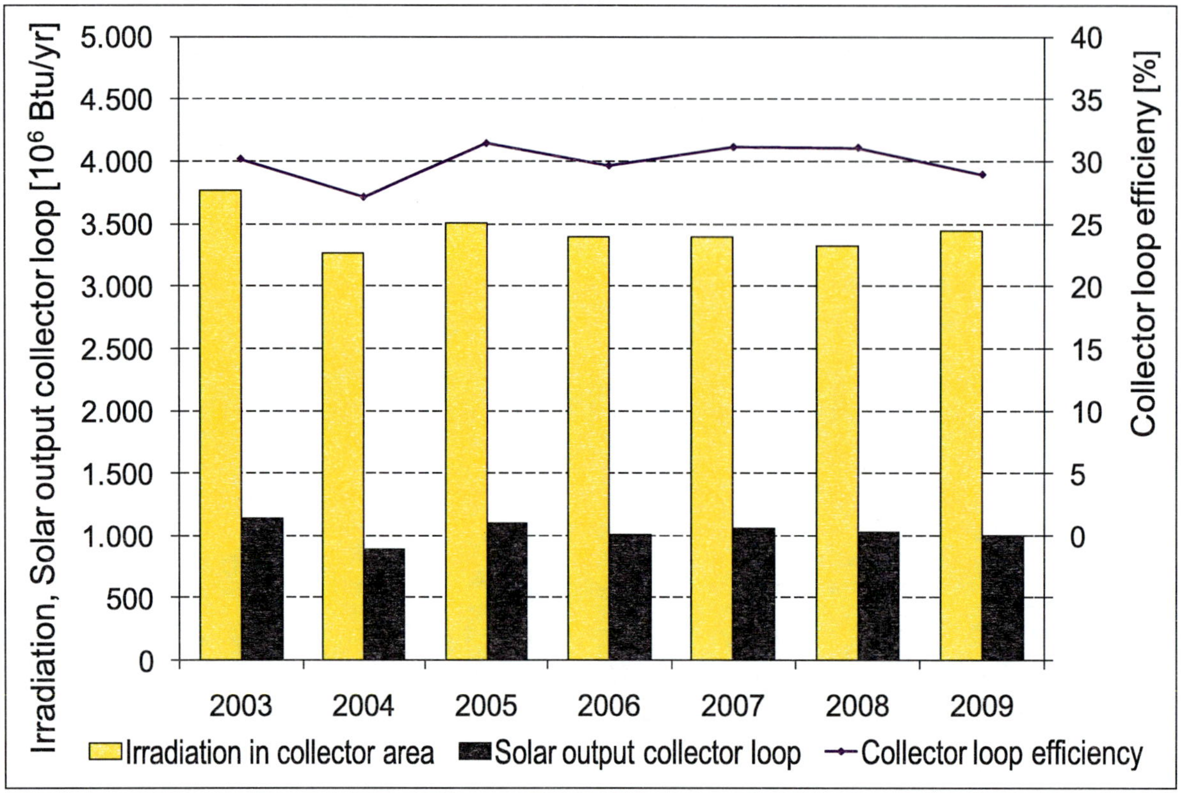

Figure A-57. Irradiation in collector area, solar energy output from collector loop, and collector loop efficiency.

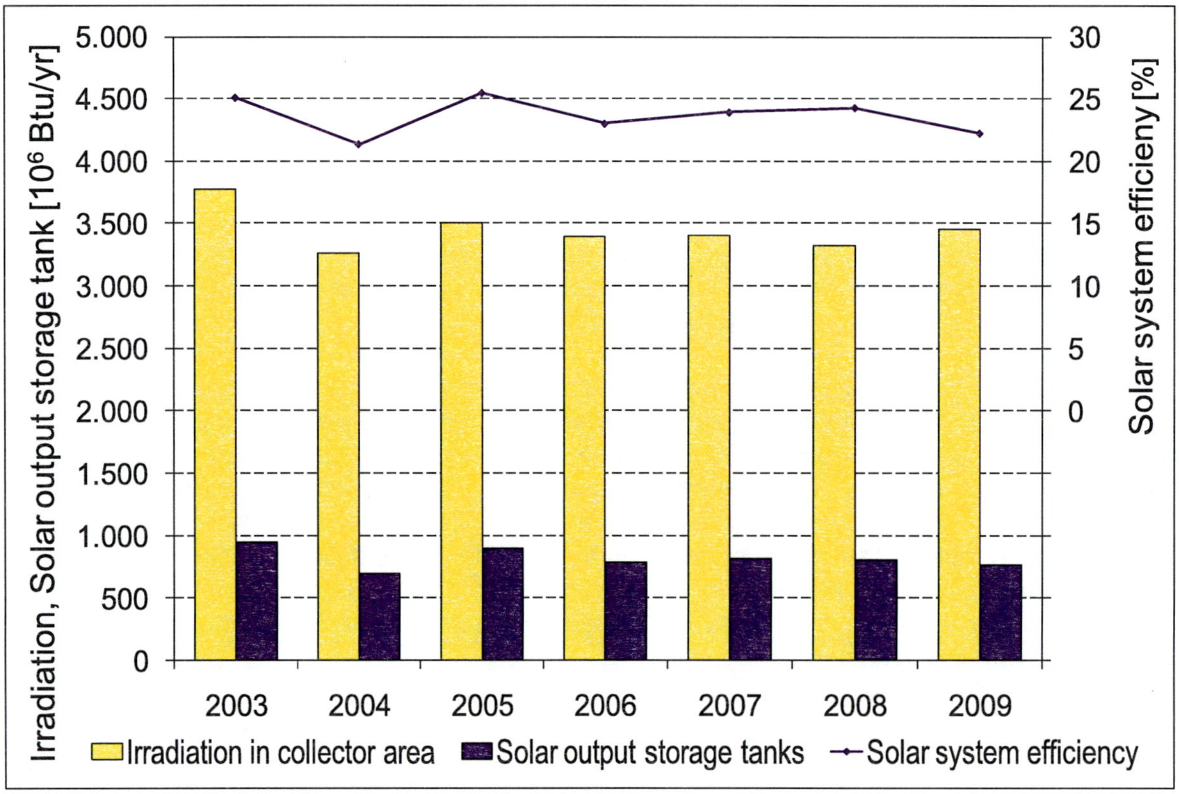

Figure A-58. Irradiation in collector area, solar energy output from the solar storage tank, and solar system efficiency.

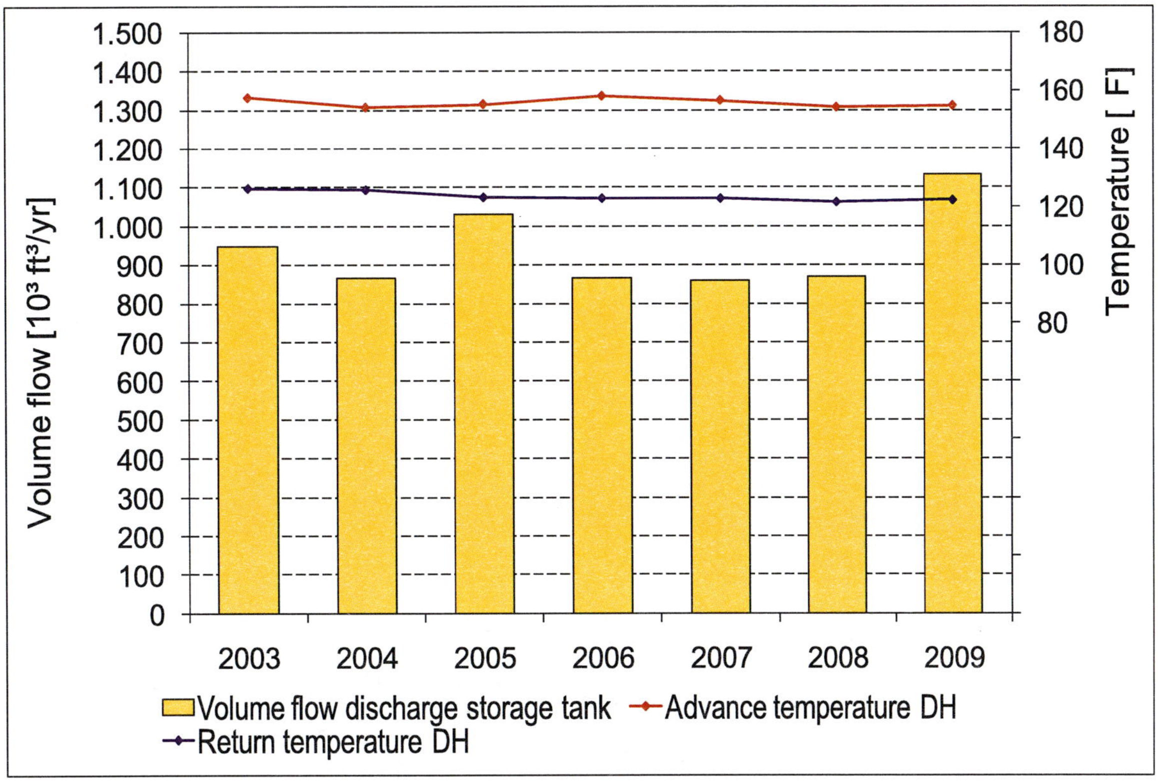

Figure A-59. Volume flow discharge solar storage tank, advance and return temperature DH.

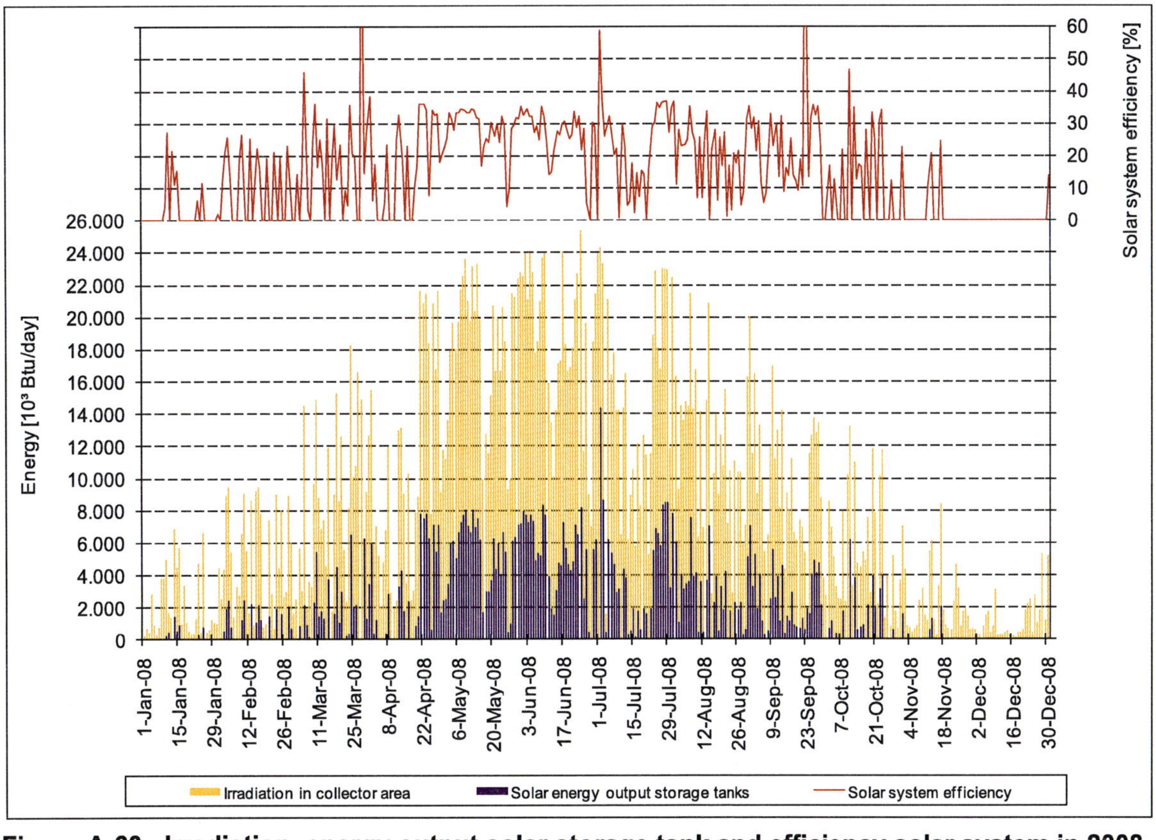

Figure A-60. Irradiation, energy output solar storage tank and efficiency solar system in 2008, daily data solution.

System details

Connection of SWH to DH

The solar system is connected to the DH in the return pipe, located in the heating central building (Figure A-61).

Solar system

The solar system contains a collector area divided into five fields situated on the sloped roof of four reconstructed houses and one newly built house. Five collector loops with a flat plate heat exchanger each are connected by an underground charge loop to the solar storage tank in the heating central. Here the solar storage tank is connected to the DH return pipe. The central heating plant houses most of the control equipment needed for the solar system. Satellite control equipment is located in the five houses where the collector loops are found.

Collector manufacturer:	UFE Solar
Type of collector:	Jumbostar, Ecostar flat plate collector
Orientation:	15 from south deviated to west
Tilt angle:	House 1: 20°, House 2 – 5: 30 – 32°
Collector areas:	2028 sq ft + 1625 sq ft + 1792 sq ft + 1792 sq ft + 1981 sq ft = 9218 sq ft ($189\ m^2 + 151\ m^2 + 167\ m^2 + 167\ m^2 + 184\ m^2 = 857\ m^{2)}$
Total collector area:	9218 sq ft ($857\ m^2$) (aperture area)
Solar storage tank:	1 x 10,570 gal (40,007 L) upright cylindrical steel tank
Freeze protection:	Collector loop filled with 40% glycol, 60% water Freeze protection valve in each collector loop
Heat exchanger:	Brazed flat plate type to each collector loop, 81.3–108.8 sq ft (8–10 m^2)

Description of control Strategy

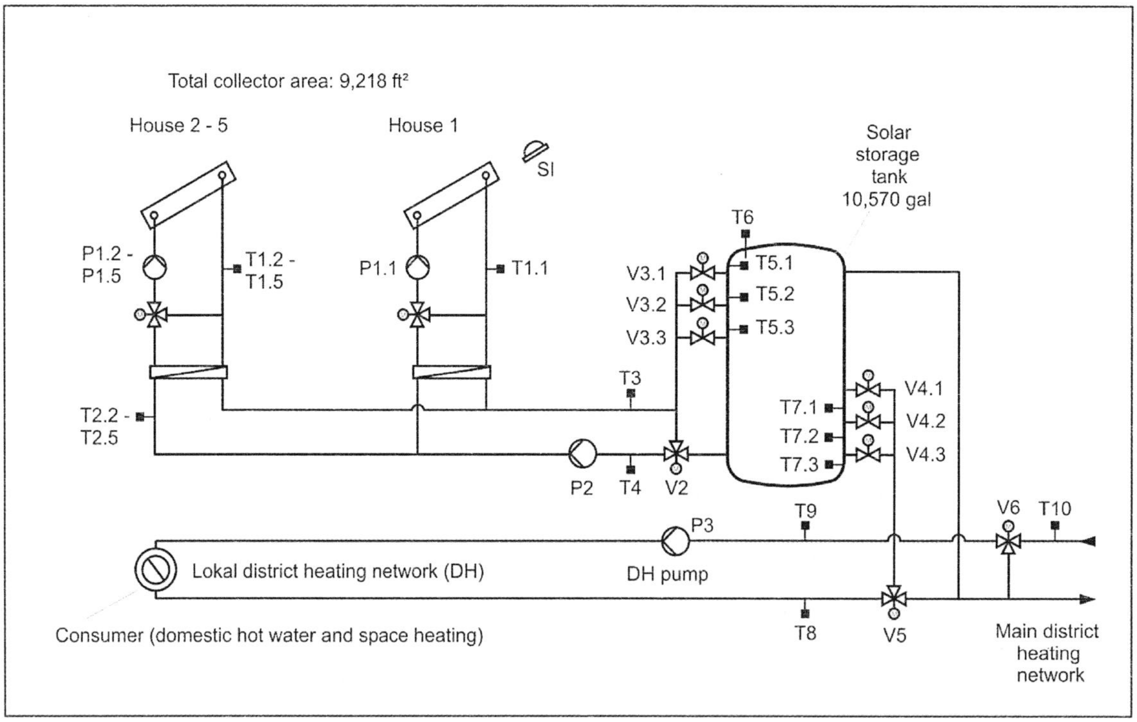

Figure A-61. Highly simplified schematics showing the positioning of the control sensors.

To prevent it from overheating and boiling, the solar storage tank is fitted with a temperature sensor (T5.1). When temperatures exceed 212 °F (100 °C) the collector loop pump control functions are halted and the pumps (P1.1 to P1.5) generally stop; temperatures below 212 °F (100 °C) clear the halted control functions. To conform to boiler regulations, the tank is additionally fitted with a temperature switch (T6). When temperatures rise above 230 °F (110 °C), the charge loop pump P2 control functions are halted, the pump generally stops, and valve V2 bypasses the solar storage tank. Temperatures lower than 230 °F (110 °C) clear the halted control functions.

The collector loop pumps (P1.1 to P1.5) are controlled by measured solar irradiation (SI). Solar irradiation more than 63.4 Btu/(hr*sq ft) [185.6 kWh/(hr*m^2)] make the pumps run, solar irradiation lower than 57.1 Btu/(hr*sq ft) [167.2 kWh/(hr*m^2)] stop the pumps.

The charge loop pump (P2) is controlled by the differences in temperatures between the temperature in the collector loop (T1.1 to T1.5) and the temperature (T7.3) in lower area of solar storage tank. Differences in temperatures (T1.1 to T1.5 – T7.3) of more than 10.8 °F (-12 °C) make pump P2 run; differences less than 3.6 °F (-16 °C) or a halt of the collector loop pumps stop pump P2 (for details see summary).

To protect the charge loop underground piping from freezing, charge loop pump (P2) runs when the ambient temperature is below 32 °F (0 °C) and stops when ambient temperature is above 32 °F (0 °C) unless other conditions make the pump run.

To prevent the collector loop heat exchangers from freezing, freeze protection valve V1.1 opens when the difference in temperature of the collector loop (T1.1) and the temperature (T7.3) in the lower area of the solar storage tank is more than 10.8 °F (-12 °C); the heat exchanger is bypassed when difference is less than 3.6 °F (-16 °C). Valve V1.2 (analog valve 1.3 to V1.5) opens the collector loop when T1.2 – T2.2 is more than 10.8 °F (-12 °C); it bypasses the heat exchanger when T1.2 – T2.2 is less than 3.6 °F (-16 °C) or pump P1.2 is off.

The solar storage tank can be charged by switching valve V2. Temperature in the charge loop (T3) higher than temperature in the lower part of the tank (T7.3) opens valve V2 to the tank. Temperature T3 lower than temperature T7.3 bypasses the tank. An additional condition is that one of the valves V3.1 or V3.2 or V3.3 is open.

To achieve a well developed layering in the solar storage tank, the three valves V3.1 to V3.3 can be controlled such that the temperature from the charge loop matches to the temperature in the tank as much as possible. If the temperature in the charge loop (T3) is above temperature T5.2, the flow is fed through the upper valve V3.1. If the temperature T3 is between temperature T5.2 and T5.3, the flow is fed through valve V3.2. If temperature T3 is below temperature T5.3, the flow is fed through the lower valve V3.3.

To discharge the solar storage tank to the DH return, the temperature in the upper tank (T5.1) and temperature in the DH return (T8) are compared. A condition where T5.1 – T8 > 10.8 °F (-12 °C) opens valve V5 to discharge the tank; a condition where T5.1 – T8 < 3.6 K bypasses the tank. When the temperature in the solar storage tank (T5.1) exceeds the temperature of the main DH advance (T10), valve V6 opens to the main district network to prevent the DH advance from overheating.

To achieve a well developed layering in the solar storage tank, the three valves V4.1 to V4.3 can be controlled such that the temperature from the DH return matches to the temperature in the tank as much as possible. If the temperature in the DH return (T8) is higher than temperature T7.1, the flow is fed through the upper valve V4.1. If the temperature T8 is between temperature T7.1 and T7.2, the flow is fed through valve V4.2. When temperature T8 is below temperature T7.2, the flow is fed through the lower valve V4.3.

Table A-10. Summary of control activities and control conditions.

Control activity	Control Conditions
Clearances P1.1 to P1.5 V2 P2 Boltings P1.1 to P1.5 Off V2 bypass solar storage tank, P2 Off	 T5.1 < 212 °F (100 °C) T6 < 230 °F (110 °C) T6 < 230 °F (110 °C) T5.1 > 212 °F (100 °C) T6 > 230 °F (110 °C) T6 > 230 °F (110 °C)
Collector loop pumps P1.1 to P1.5	 On: SI > 63.4 Btu/(hr * sq ft) Off: SI < 57.1 Btu/(hr * sq ft)
Charge loop pump P2	 On: T1.1 – T7.3 or T1.2 – T7.3 or T1.3 – T7.3 or T1.4 – T7.3 or T1.5 – T7.3 > 10.8 °F (-12 °C) Off: T1.1 – T7.3 and T1.2 – T2.2 and T1.3 – T2.3 and T1.4 – T2.4 and T1.5 – T2.5 < 3.6 °F (-16 °C) or P1.1 to P1.5 Off
Freeze protection charge loop underground piping: P2	On: Ambient temperature < 32 °F (0 °C) Off: Ambient temperature > 32 °F (0 °C) unless other conditions make the pump run
Freeze protection valves V1.1 V1.2 V1.3, V1.4, V1.5	 Open: T1.1 – T7.3 > 10.8 °F (-12 °C) and P1.1 4 min running Bypass: T1.1 – T7.3 < 3.6 °F (-16 °C) Open: T1.2 – T2.2 > 10.8 °F (-12 °C) and P1.2 4 min running Bypass: T1.2 – T2.2 < 3.6 °F (-16 °C) or P1.2 Off Analog to V1.2
Connecting charge loop to solar storage tank V2	 Open : T6 < 230 °F (110 °C) and T3 > T7.3 and V3.1 open or V3.2 open or V3.3 open Bypass: T6 > 230 °F (110 °C) or T3 < T7.3 or V3.1 shut and V3.2 shut and V3.3 shut
Layering in solar storage tank, charge V3.1 open, V3.2 shut, V3.3 shut V3.1 shut, V3.2 open, V3.3 shut V3.1 shut, V3.2 shut, V3.3 open	 T3 > T5.2 T3 < T5.2 and T3 > T5.3 T3 < T5.3
Connection solar storage tank to DH return V5 V6	 Open: T5.1 > T8 + 10.8 °F (-12 °C) Bypass: T5.1 < T8 + 3.6 °F (-16 °C) Controlling: T5.1 < T10 Open to main DH: T5.1 > T10
Layering in solar storage tank, discharge V4.1 open, V4.2 shut, V4.3 shut V4.1 shut, V4.2 open V4.3 shut V4.1 shut, V4.2 shut V4.3 open	 T8 > T7.1 T8 < T7.1 and T8 > T7.2 T8 < T7.2

Economics

The costs of the solar system include only the costs for solar collectors, piping, solar storage tanks and controls, but do not include costs for the district heating network, boiler or the heating central building (Table A-11).

Table A-11. Economics.

Costs solar system		
- costs solar system	$580,186 (408,582 €)	
- costs planning solar system.	$ 92,563 (65,185 €)	
costs solar system including planning	$672,749 (473,767 €)	
costs solar system, statics and planning including 16% tax	**780,389 $ (549,570 €)**	
Annual costs for loan living period 20 years, 6% rate, → annuity: 8.72%	$68,051 (47,923 €)	
	Per year	Sum total in 7 years
Solar energy output		
- planned solar energy output from solar storage tank	1189 Btu *10^6 Btu (3,481 *10^6 kWh)	8322*10^6 Btu
- measured solar energy output from storage tank	69–948*10^6 Btu (2,041 kWh–2,776*10^6 kWh)	5705*10^6 Btu
Relation measured solar energy output/ planned energy output		68.6%
Savings of gas and CO_2 calculated with measured solar energy from solar storage tank with following assumptions: boiler efficiency: 90% energy of natural gas: $27.4*10^3$ Btu/cu yd_{Gas} emission factor: 0.129 lbs $CO_2/10^3$ Btu_{Gas}		
- saving amount of natural gas	229,150 cu yd (175,208,084 L)	
- avoidable amount of CO_2	409 (short) ton (271,038 kg)	
Costs of solar energy from solar storage tank with 8.72% annuity, including solar system, planning and tax		
- costs and planned solar energy output	0.040 €/10^3 Btu ($0.02/$10^3$ kWh)	
- costs and measured solar energy output, without costs for maintenance and repairs	0.059 €/10^3 Btu ($0.027/$10^3$ kWh)	

User evaluation

No information from operating company (Stadtwerke Hennigsdorf) available.

District heating network

Number of flats supplied by DH:	460
Number of people supplied by DH:	1,150
Maximum energy demand for space heating supplied from DH:	9,420*10^6 Btu/yr (1,229*10^6 kWh)
Maximum energy demand for domestic hot water supplied from DH:	3,210*10^6 Btu/yr (614*10^6 kWh)
Maximum energy losses in DH underground piping:	1,160*10^6 Btu/yr (468*10^6 kWh)
Total energy demand from heating central:	13,790*10^6 Btu/yr (11,097*10^6 kWh)
Total energy supplied from solar system to DH:	1,190*10^6 Btu/yr (556*10^6 kWh)
Solar fraction or total energy demand DH:	9.0%
Solar system efficiency (energy output from solar storage tank/irradiation energy):	37.1%
Max. DH advance temperature (depends on ambient temperature):	149– 203 °F (9–95 °C)
Max. DH return temperature	winter: 96.8 °F (36 °C), summer 105.8 °F (41 °C)

Table A-12. Measured data from 2003 to 2009.

	IP	2003	2004	2005	2006	2007	2008	2009
Energy demand of DH	10^6 Btu/yr	Not measured						
Solar energy output solar storage tank	10^6 Btu/yr	947.9	696.8	894.3	780.4	812.4	807.3	767.4
Solar fraction of total energy demand DH	%	Not measured						
DH advance temperature, yearly average	°F	157.8	154.2	155.3	158.0	156.6	154.2	154.8
DH return temperature, yearly average	°F	126.3	125.8	123.3	122.9	122.9	121.6	122.2
	SI	**2003**	**2004**	**2005**	**2006**	**2007**	**2008**	**2009**
Energy demand of DH	10^6 Btu/yr	Not measured						
Solar energy output solar storage tank	10^6 kWh/yr	2,775	2,040	2,619	2,285	2,379	2,364	2,247
Solar fraction of total energy demand DH	%	Not measured 780.4						
DH advance temperature, yearly average	°C	70	68	69	70	69	68	68
DH return temperature, yearly average	°C	52	52	51	51	51	50	50

The energy demand of the DH is not measured, so no solar fraction of total energy demand DH can be determined. Solar energy output from the solar storage tank to DH reached from 696.8*10^6 to 947.9*10^6 Btu/yr (2*10^6 MWh to 2.8*10^6 MWh/yr) in the reported years, but failed 1189*10^6 Btu/yr (3.5*10^6 MWh) as planned. The harvest in 2003 is abnormal high compared to the other years. The reason is found in the exceptional irradiation in 2003, which is known in Germany as the "Century summer."

The DH advance temperature runs in a normal and planned range between 154.2 and 158.0 °F (68 and 70 °C) on a yearly average. The DH return temperature is between 121.6 and 126.3 °F (50 and 52 °C) on a yearly average with a slightly fallen tendency over the years. We relate this effect to the efforts made to adjust the heat transfer stations in the connected houses. Despite this tendency the return temperature is much too high compared with the planned DH return temperature (96.8 °F [36 °C] in summer and 105.8 °F [41 °C] in winter).

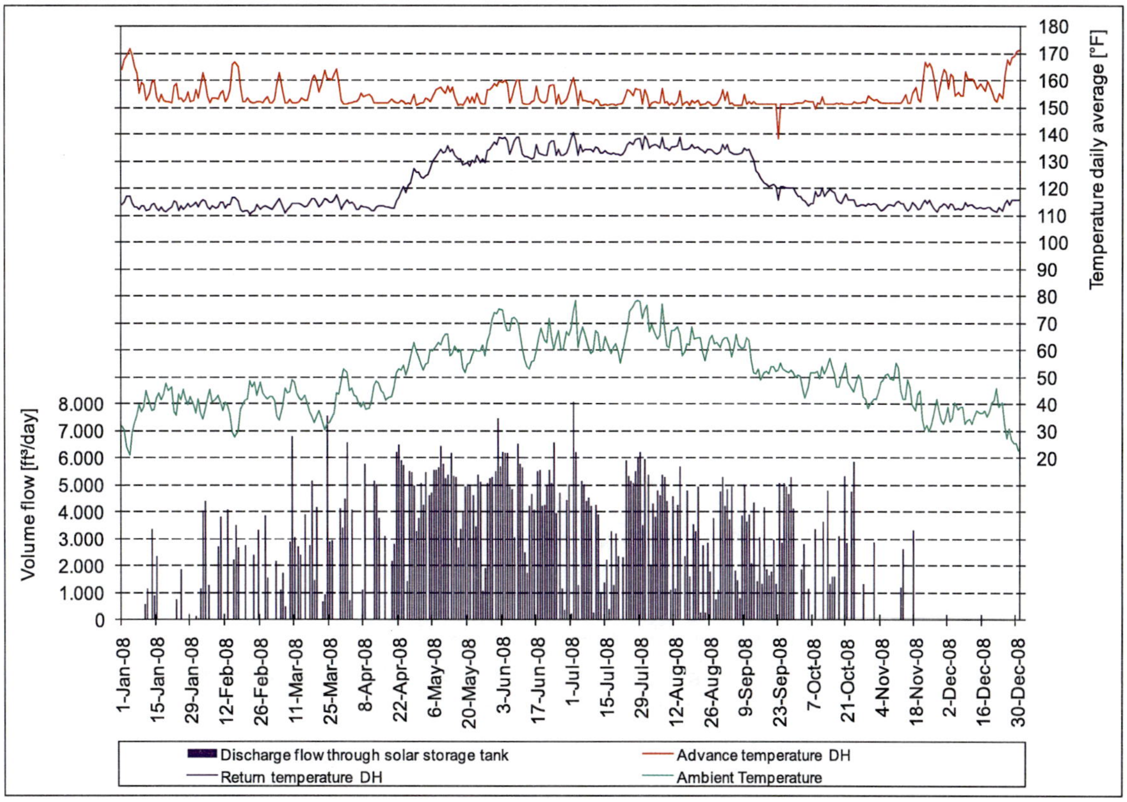

Figure A-62. Discharge flow through solar storage tank and temperatures DH in 2008.

Figure A-62 shows the discharge flow through the solar storage tank in 2008. Only a small amount of volume flow of the DH is guided through the solar storage tank for discharge. The main stream in the DH return is not measured so the percentage of volume that finds its way through the tank cannot be determined. Through flow is stopped and tank is bypassed when it is not discharged (i.e., when the temperature in the storage tank is lower than DH return temperature). DH advance and return temperatures are also shown. In the winter, the DH advance temperature rises depending on ambient temperature. DH return temperature in the summer rises because of the absence of energy requirement for building space heating. This causes rising return temperatures in heat transfer stations, and consequently, rising temperature in the DH return.

Experiences

Energy use reduction

In the 7 years from 2003 to 2009, an energy output was measured from the solar storage tank ($5705*10^6$ Btu or $815*10^6$ Btu [16.7*106 kWh or 2.4*106 kWh] on average per year). Assuming a boiler efficiency of 90%, and an energy content of natural gas of $27.4*10^3$ Btu/cu yd_{Gas} (1040 kWh/L_{gas}) there is a saving of natural gas of 229,150 cu yd (175,208,084 L) in 7 years. Per year there is an average saving of 32,735 cu yd (25,029,180 L).

Lessons learned

From the data we measured and the monitoring of the solar system we learned that:

- In general, the solar system in combination with the district heating network operated without severe problems. The input of solar energy in the return pipe of the DH is the most promising way.

- The thermal loss through the underground charge loop piping was much larger than calculated. This explains why the measured solar energy output of the solar storage tank did not reach the planned output. We recommend to calculate the losses through piping very carefully by a proven calculation program and to select a high-grade isolation of the piping.
- The solar system arrangement with five collector loop pumps and only one charge loop pump worked without hydraulic problems. The alternative way would have been the installation of five charge loop pumps. The decision of installing only one common loop pump to save costs was correct.
- In principal, the control strategy worked correctly. Because of the manifold control conditions a lot of control hardware was necessary, which failed from time to time. We recommend to select high-grade control units and to assure that the cables between sensors and control units are laid very carefully without damages. The in some aspects different handling of the control functions of collector loop 1 in comparison to loop 2 – 5 makes the control system unnecessarily complicated, simplification is recommended.
- The difference between the planned DH return temperature (96.8 °F [36 °C] in summer, 105.8 °F [41 °C] in winter) and the measured from 121.6 to 126.3 °F (50 to 52 °C) on yearly average in the years from 2003 to 2009 is not acceptable. All attempts to lower the DH return temperature significantly were not effective.
- A design to control the outlet temperature of the solar storage tank was not envisioned. We recommend incorporating such an arrangement to limit the outlet temperature of the tank to the maximum temperature of the DH advance temperature. That can be achieved by fitting valve V5 with two functions: 1st Discharge or bypass the tank, 2nd Control the temperature behind the mixing point.
- The collector loop heat exchangers were contaminated at the charge loop side. Be sure that the water quality in the charge loop is appropriate to the demand of a flat plate heat exchanger.

General date

Address of the project

Solar water heating connected to a local district heating network
Residential Area "Cohnsches Viertel"
Hennigsdorf, Germany

Date of report

Measured data period: 1st January 2003 to 31st December 2009
Date of report: June 2010

Acknowledgement

Promoting department

Bundesministerium für Umwelt, Naturschutz und Reaktorsicherheit (BMU)
(Federal department of environment, nature conservation and nuclear reactor safety)
Alexanderstr. 3
10178 Berlin, Germany

Housing society

Hennigsdorfer Wohnungsbaugesellschaft GmbH
Edisonstraße 1
16761 Hennigsdorf

Operating company

Stadtwerke Hennigsdorf GmbH
Rathenaustraße 4
16761 Hennigsdorf, Germany

Planning company

Tetra-Ingenieure GmbH
Rosa-Luxemburg-Straße 30
16816 Neuruppin, Germany
Planungsbüro Roth & Grube
Hegermühlenstraße 19
15344 Strausberg

Mounting company

Heizungsbau Wolfgang Schiemann
Bahnhofstr. 1
16816 Neuruppin, Germany

Measuring company

ZfS-Rationelle Energietechnik GmbH
Verbindungsstr. 19
40723 Hilden, Germany

References

Mies, M.; Rehrmann, U. Abschlussbericht für das Projekt Cohnsches Viertel Hennigsdorf August 2007, http: //www.zfs-energietechnik.de/main.php?RND=33644&MANDANT_ID=1&ID=167 (more and detailed information about the solar system in Hennigsdorf are outlined in this report).

Peuser, Felix A.; Remmers, Karl-Heinz; Schnauss, Martin. Solar Thermal Systems. Successful Planning and Construction. Solarpraxis AG Berlin Germany in association with James & James London UK, 2002. ISBN: 3-934595-24-3

VDI 6002, part 1, September 2004 (technical guideline). Solar heating for domestic water. General principles, system technology and use in residential buildings. Distributer: Beuth Verlag, 10722 Berlin, Germany

DVGW W551, April 2004 (technical guideline). Trinkwassererwärmungs- und Leitungsanlagen; Technische Maßnahmen zur Verminderung des Legionellenwachstums (Hot water systems, technical arrangements to reduce developing of legionella bacteria), no English translation available. Distributer: Wirtschafts- und Verlagsgesellschaft Gas und Wasser mbH. Josef-Wimmer-Str. 1 - 3, 53123 Bonn, Germany

FPC – 14. Solar Water Heating (SWH) Connected to District Heating Networks (DH)

Title: Residential Area "Badener Hof," Heilbronn, Germany

Location: Heilbronn, Germany

Photo of installation

Figure A-63. Front view of the heating central with two roof-integrated collector areas.

Figure A-64. Solar storage tanks.

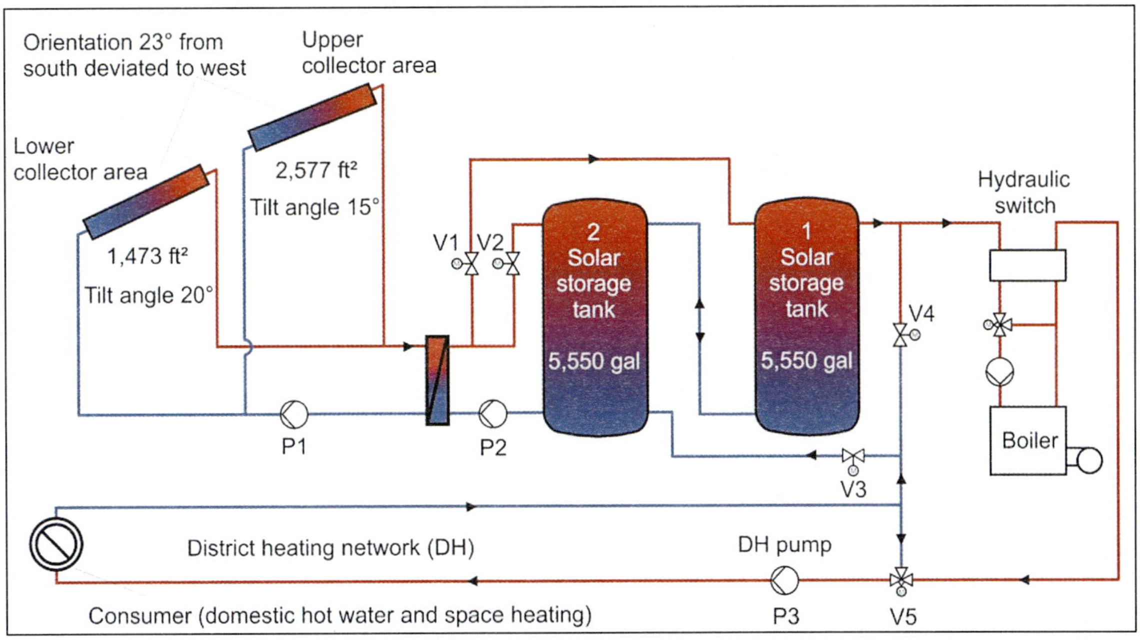

Figure A-65. Highly simplified schematics of the solar system, solar storage tanks arrangement and the integration in the district heating network (DH).

Project summary

About 540 homes were planned to be built on the site of former US barracks on the eastern outskirts of the city of Heilbronn (Germany), beginning in 2000. The plots were sold by the city of Heilbronn to individual clients, so there was no award of the entire region to a carrier or a construction contractor. The design, building, and implementation of the houses were left to the builders. According to the developing plan of the site, the following house types should be realized:

- one family houses
- twin houses
- row houses
- multi-family houses.

The public services of the city of Heilbronn (Stadtwerke Heilbronn) operate an underground district heating network (DH) to supply the area with heat for domestic hot water and space heating. For this purpose, a heating central with two natural gas-fired boilers, one oil-fired boiler, a solar energy system with a collector area of 4049 sq ft (376.56 m^2) and a solar storage tank capacity of 2 x 5550 gal (21,007 L) are provided. In the planning phase additionally a wood chip boiler was intended, but the installation has not been realized. In the sales contracts for the land plots the connection and the coercion to use the DH were defined. In addition, strong guidelines to the owner of the houses affecting the thermal insulation of buildings were made.

The district heating network, the solar system and the heat transfer stations (HTS) in the houses have been dimensioned by the Steinbeis Transfer Center in Stuttgart. Carried out were indirect working HTS, which separates the DH from the heating loops inside the houses by heat exchanger. The manufacturer of the HTS was selected by Stadtwerke Heilbronn; the purchase and maintenance lies within the responsibility of the house owners.

The power supply and the DH advance temperature (in the range between 149 and 176 °F (65 and 80 °C) depending on ambient temperature) are guaranteed by the Stadtwerke Heilbronn. The installed return temperature limiters restrict the return temperature of space heating loops in the houses to no more than 113 °F (45 °C) to protect against Legionella bacteria, for which domestic hot water must be at least 140 °F (60 °C). Therefore, it is not advised to integrate return temperature limiters in the domestic hot water system.

The arrangement of solar system, boilers, and district heating network works well. No severe problems were found in the concept. Lesser problems (no solar storage bypass valve, suboptimal located control sensors in solar storage tanks) were later solved.

Site

Location:	Residential Area "Badener Hof"
Town:	Heilbronn
Country:	Germany
Latitude:	49 08' North
Longitude:	9° 14' East

Project description

A collector area of 4049 sq ft is installed as a roof-integrated construction on top of the heating central, which houses the boilers, the solar storage tanks, and the control facilities. The energy from the collector loop is transferred by a flat plate heat exchanger to the charge loop of the solar storage tanks. Depending on the temperature of the charge loop, tanks 1 and 2 are loaded in sequence, or only tank 2 is loaded. Both tanks have a capacity of 5550 gal (21,007 L). In proportion to the collector area and to climate conditions in this location, the installed volume of the storage tanks is too large. This volume was realized in hopes to enlarge the collector area in future, which has not yet happened. Under normal conditions one storage tank with a capacity of 5550 gal (21,007 L) would be suitable.

Depending on the DH return temperature, the volume flow can be guided by shutoff valves to discharge the solar storage tanks or bypass them. Natural gas and oil boilers can feed in energy to guarantee the DH advance temperature in periods with low irradiation and less output of solar energy. The correct DH advance temperature is adjusted by a control valve, which compensates for the fluctuations waves in temperature generated by the stop-and-go operation of the boiler. The solar system was connected to the DH in 2000.

Table A-13. Expected data during planning

Number of buildings supplied by DH:	129	
Number of flats supplied by DH:	538	
Number of people supplied by DH:	1,000	
Maximum energy demand for space heating supplied from DH:	$9{,}900*10^6$ Btu/yr	($28{,}987*10^6$ kWh/yr)
Maximum energy demand for domestic hot water supplied from DH:	$2{,}760*10^6$ Btu/yr	($8{,}081*10^6$ kWh/yr)
Maximum energy losses in DH underground piping:	$1{,}880*10^6$ Btu/yr	($5{,}505*10^6$ kWh/yr)
Total energy demand for DH (from heating central):	$14{,}540*10^6$ Btu/yr	($42{,}573*10^6$ kWh/yr)
Total energy supplied from solar system to DH:	$570*10^6$ Btu/yr	($1{,}669*10^6$ kWh/yr)
Solar fraction of total energy demand DH:	4.0%	
Solar system efficiency (energy output from solar storage tanks/irradiation energy):	35.6%	
Max. DH advance temperature (depends on ambient temperature):	149 – 176 °F	(65 – 80 °C)
Max. DH return temperature	113 °F	(45 °C)

Data in Table A-13 were calculated using the following equations:

Solar system efficiency = Solar energy output from solar storage tanks/Irradiation in collector area

Solar fraction of total energy demand DH = Solar energy output from solar storage tanks/Total energy demand for DH

Table A-14 lists the measured data in the years between 2002 and 2009. Irradiation in the collector areas differ in a small range with exception of year 2003, which is known in Germany as the "century summer." So solar output from the collector loop or solar output from the solar storage tanks (which includes the thermal losses of the tanks) developed similar to the solar irradiation. This proportion between irradiation and solar output, termed "solar system efficiency," operates in the range between 22.1 and 28.1%. The decline of the solar efficiency from 2007 to 2009 in comparison to the years before may be attributed to shadowing caused by the trees in front of the central heating plant (Figure A-63).

Table A-14. Measured data from 2002 to 2009.

	IP	2002	2003	2004	2005	2006	2007	2008	2009
Irradiation in horizontal area	10^3 Btu/(sq ft*yr)	336.6	389.9	352.2	356.9	356.9	399.4	387.7	355.4
Irradiation in collector area	10^6 Btu/yr	1,522	1,771	1,586	1,611	1,651	1,631	1,578	1,627
Irradiation in collector area, specific	10^3 Btu/(sq ft*yr)	376.0	437.5	391.5	397.8	407.7	402.9	389.6	401.6
Solar energy output collector loop	10^6 Btu/yr	471.6	553.4	427.9	475.0	462.0	430.6	408.4	415.3
Solar energy output solar storage tanks	10^6 Btu/yr	423.1	498.2	385.6	430.3	419.0	381.8	353.8	359.6
Total energy demand of DH	10^6 Btu/yr	5,678	5,951	7,950	8,762	9,643	9,383	10,206	10,373
Solar system efficiency	%	27.8	28.1	24.3	26.7	25.4	23.4	22.4	22.1
Solar fraction of energy demand DH	%	7.5	8.4	4.9	4.9	4.3	4.1	3.5	3.5
DH advance temperature, yearly average	°F	158.5	159.3	154.6	159.1	160.7	161.4	161.4	160.3
Dh return temperature, yearly average	°F	113.2	117.9	123.3	116.2	114.8	115.3	115.9	117.5
	SI	**2002**	**2003**	**2004**	**2005**	**2006**	**2007**	**2008**	**2009**
Irradiation in horizontal area	10^3 kWh/(m^2*yr)	1.06	1.23	1.11	1.12	1.12	1.26	1.22	1.12
Irradiation in collector area	10^6 kWh/yr	4,456	5,185	4,644	4,717	4,834	4,776	4,620	4,764
Irradiation in collector area, specific	10^3 kWh/(m^2*yr)	1.19	1.38	1.23	1.25	1.28	1.27	1.23	1.27
Solar energy output collector loop	10^6 kWh/yr	1,381	1,620	1,253	1,391	1,353	1,261	1,196	1,216
Solar energy output solar storage tanks	10^6 kWh/yr	1,239	1,459	1,129	1,260	1,227	1,118	1,036	1,053
Total energy demand of DH	10^6 kWh/yr	16,625	17,425	23,278	25,655	28,235	27,473	29,883	30,372
Solar system efficiency	%	27.8	28.1	24.3	26.7	25.4	23.4	22.4	22.1
Solar fraction of energy demand DH	%	7.5	8.4	4.9	4.9	4.3	4.1	3.5	3.5
DH advance temperature, yearly average	°C	70	71	68	71	72	72	72	71
Dh return temperature, yearly average	°C	45	48	51	47	46	46	47	48

The solar fraction of the total energy demand DH declines from 7.5% in 2002 to 3.5% in 2009. This is understandable because more plots were sold in these years and the energy demand of the DH rose from 5678*10^6 Btu/yr (16.6*10^6 MWh/yr) in 2002 up to 10,373*10^6 Btu/yr (30.4*10^6 MWh/yr) in 2009. There was also a decline in solar energy output.

The DH advance temperature runs in a normal and planned range between 154.6 and 161.4 °F (68 and 72 °C) on a yearly average. The return temperature between 113.2 and 123.3 °F (45 and 51 °C) on yearly average lies a little bit above the planned temperature of 113 °F (45 °C). Compared with other DH this data however can be evaluated as "just good." For clarity, the measured data from 2002 to 2009 listed in Table A-14 are also shown in Figures A-66 to A-68.

Figure A-69 shows the 2008 measured daily data of irradiation, solar energy output from solar storage tanks, and the calculated solar system efficiency. The irradiation over the year is typical for Germany, with explicit differences between summer and winter. The solar energy outputs from the solar storage tanks follows a similar pattern; the main harvest of energy is found from March to October, and the harvest from November to February is almost negligible. In more northern regions this effect is most striking, and in more southern regions, the effect dwindles. In the summer, the solar system efficiency reaches nearly 37%, but the poor winter data reduces the yearly average to only 22.4%.

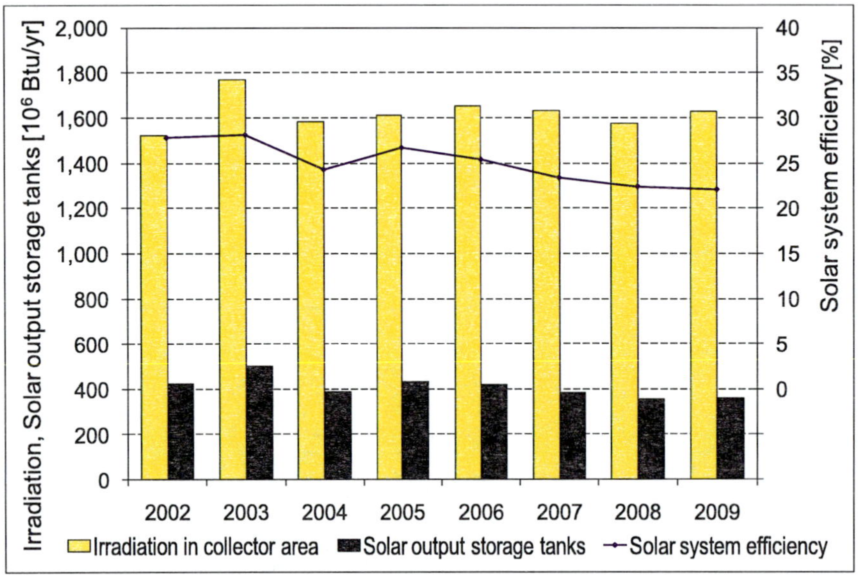

Figure A-66. Irradiation in collector area, solar energy output from solar storage tanks and solar system efficiency.

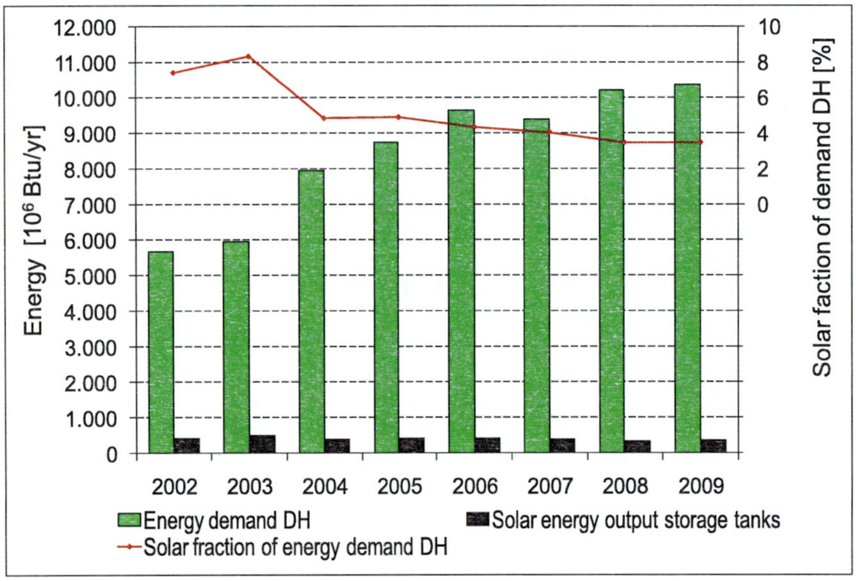

Figure A-67. Energy demand DH, solar energy output from solar storage tanks and solar fraction of total energy demand DH.

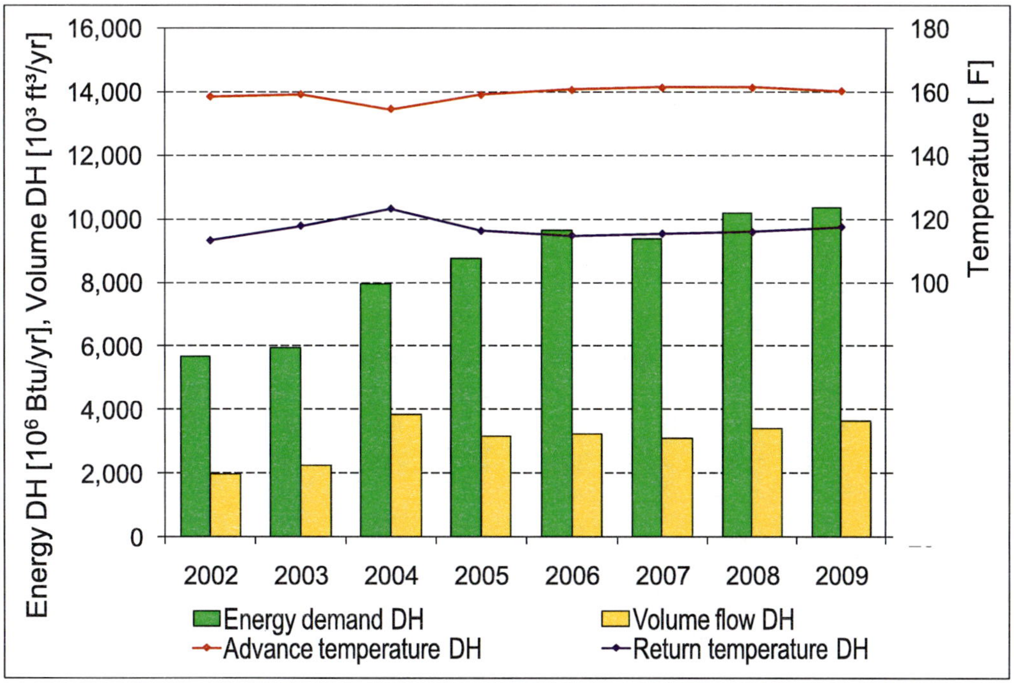

Figure A-68. Energy demand DH, volume flow DH, advance and return temperature DH.

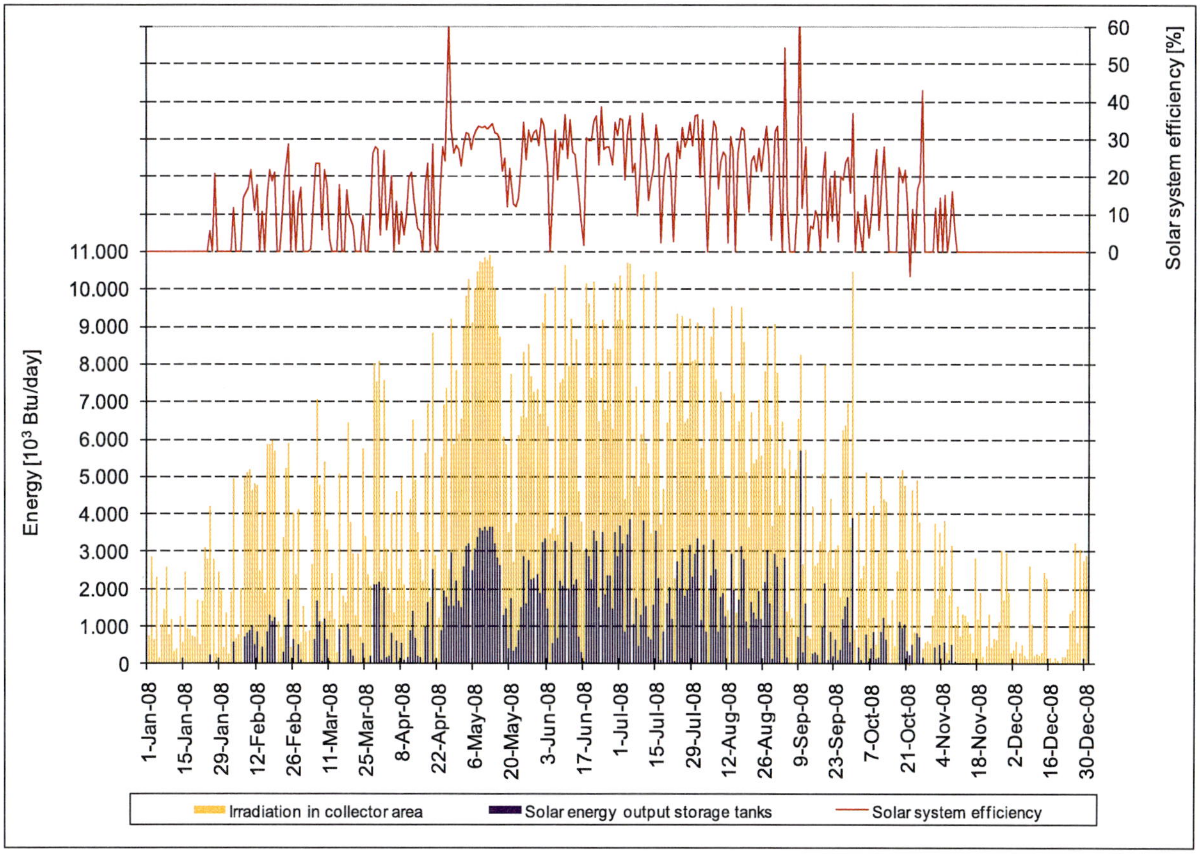

Figure A-69. Irradiation in collector area, solar energy output from solar storage tanks, and solar system efficiency in 2008, daily data solution.

System details

Connection of SWH to DH

The solar system is connected to the DH in the return pipe, located in the central heating building (Figure A-70).

Solar system

The solar system contains a collector area divided in two fields, situated on the sloped roof of the heating central building, piping, a flat plate heat exchanger, two solar storage tanks, the connection to the DH and the control technique.

Collector manufacturer:	Sonnenkraft (Austria)
Type of collector:	IMK flat plate collector, function completely as a roof
Orientation:	23° from south deviated to west
Tilt angle:	Upper collector area: 15°; Lower collector area: 20°
Collector area:	4,049 sq ft (376.56 m^2) (aperture area)
Solar storage tank:	2 x 5550 gal (21,006.75 L) upright cylindrical steel tank
Freeze protection:	Collector loop filled with 40% glycol, 60% water
Heat exchanger:	Screwed flat plate type, area 266 sq ft

Description of control strategy

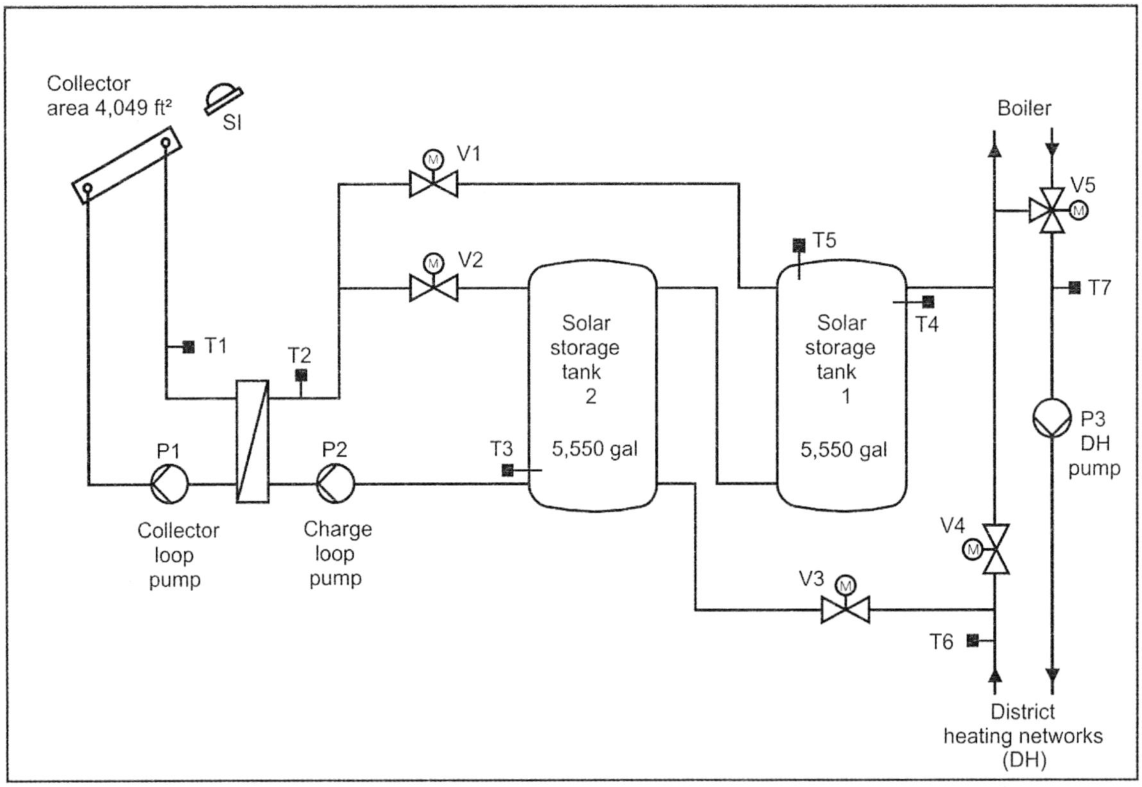

Figure A-70. Highly simplified schematic showing the positioning of the control sensors

To prevent them from overheating and boiling, the solar storage tanks are fitted with a safety temperature switch (T5). When temperatures in tank 1 rise higher than 208.4 °F (98 °C), the pump control functions are halted and the pumps generally stop; when temperatures fall lower than 208.4 °F (98 °C), the bolted control functions clear.

The collector loop pump (P1) is controlled by measured solar irradiation (SI). Solar irradiation greater than 64 Btu/(hr*sq ft) (187kWh/[hr*m^2]) makes the pump run; solar irradiation lower than 47.55 Btu/(hr*sq ft) (139.23 kWh/[hr*m^2]) stops the pump.

The charge loop pump (P2) is controlled by the difference in temperature between the temperature in the collector loop (T1) and the temperature (T3) in lower area of solar storage tank 2. A difference in temperature (T1 – T3) of more than 9.0 °F (5 °C)makes the pump run; a difference less than 3.6 °F (2°C) stops the pump.

The difference in temperature between the temperature in the charge loop (T2) and the temperature (T4) in upper area of solar storage tank 2 is used to control the valves V1 and V2 to load the solar storage tanks. A difference in temperature (T2 – T4) of more than 3.6 °F (2°C) opens V1 and shuts V2 (tank 1 + 2 are charged); a difference of less than – 3.6 °F (–2°C)shuts V1 and opens V2 (tank 2 is charged).

The discharge of the storage tanks executed by valve V3 and V4 is controlled by the difference in temperature between the upper area of solar storage tank 1 (T4) and the DH return temperature (T6). A difference in temperature (T6 – T4) of more than 9.0 °F (5 °C) opens V3 and shuts V4; a difference less than 3.6 °F (2 °C) shuts V3 and opens V4.

The DH advance temperature (T7) is controlled by valve V5 depending to the ambient temperature. The advance temperature is 176 °F (80 °C) when ambient temperature is lower than 10.4 °F (-12 °C), and the advance temperature is 149 °F (65 °C) when ambient temperature is higher than 42.8 °F (6 °C). Between 10.4 and 42.8 °F (-12 and 6 °C), advance temperature will be interpolated.

To prevent freezing of the solar loop flat plate heat exchanger, the charge loop pump (P2) is runs when the temperature in the solar loop (T1) is lower than 35.6 °F (2 °C).

Table A-15. Summary of control activities and control conditions.

Control activity	Control Conditions
Collector loop/Charge loop Clearance of running pumps General stop of running pumps Collector loop pump P1 Charge loop pump P2	T5 < 208.4 °F (98 °C) T5 > 208.4 °F (98 °C) On: SI > 63.4 Btu /(hr*sq ft) (185.6 kWh/[hr*m^2]), Off: SI < 47.55 Btu/(hr*sq ft) (139.2 kWh/[hr*m^2]) On: T1 – T3 > 9.0 °F (-5 °C), Off: T1 – T3 < 3.6 °F (-2 °C)
Freezing protection Charge loop pump P2	On: T1 < 35.6 °F (2 °C) independent from other control function
Loading solar storage tanks Loading tank 1 and 2 Loading tank 2	V1 open, V2 shut: T2 – T4 > 3.6 °F (2 °C) V1 shut, V2 open: T2 – T4 < -3.6 °F (-2 °C)
Discharging solar storage tanks Discharging Bypass DH return	V3 open, V4 shut: T6 – T4 > 9.0 °F (5 °C) V3 shut, V4 open: T6 – T4 < 3.6 °F (2 °C)
District Heating Network Advance Temperature	T7 = 176 °F (80 °C) at 10.4 °F (-12 °C) ambient temperature and lower T7 = 149 °F (65 °C) at 42.8 °F (6 °C) ambient temperature and higher When ambient temperature is between = 10.4 and 42.8 °F (-12 and 6 °C) T7 will be interpolated

Economics

The costs of the solar systems include only the costs for solar collectors, piping, solar storage tanks and controls, but does not include costs for the district heating network, boiler, or the heating central building (Table A-16).

Table A-16. Economics

Costs solar system Costs solar system, including statics Costs planning solar system. Costs solar system, statics and planning Costs solar system, statics and planning including 16% tax	 $249,659 (175,816 €) $37,477 (26,392 €) $287,134 (202,207 €) **$333,077 (234,561 €)**	
Annual costs for loan Living period 20 years, 6% rate, → annuity: 8.72%	$29,045 (20,454 €)	
	Per year	Sum total in 8 years 2002 – 2009
Solar energy output Planned solar energy output from solar storage tanks Measured solar energy output from storage tanks	 $512*10^{6}$ Btu ($1{,}499*10^{6}$ kWh) 354–$498*10^{6}$ Btu (1,037–$1{,}458*10^{6}$ kWh)	$4095*10^{6}$ Btu ($11{,}990*10^{6}$ kWh) $3250*10^{6}$ Btu ($9{,}516*10^{6}$ kWh)
Relation measured solar energy output/ planned energy output		79.4%
Savings of gas und CO_2 calculated with measured solar energy from solar storage tanks with following assumptions: Boiler efficiency: 90%; Energy of natural gas: $27.4*10^{3}$ Btu/cu yd_{Gas} ($0.1041*10^{3}$ kWh/L_{gas}) Emission factor: 0.129 lbs $CO_2/10^{3}$ Btu_{Gas} (0.0585 kg CO_2/2,928kWh_{Gas}) Saving amount of natural gas Avoidable amount of CO_2		 131,800 cu yd (100,774,277 L) 233 (short) ton (211,378 kg)
Costs of solar energy from solar storage tanks with 8.72% annuity, including solar system, statics, planning and tax Costs and planned solar energy output costs and measured solar energy output, without Costs for maintenance and repairs	 $0.0194/kWh (0.040 €/Btu) $0.0242/kWh (0.050 €/Btu)	

User evaluation

No information from operating company (Stadtwerke Heilbronn) is available.

District heating network

Expected data during planning

Maximum energy demand for space heating supplied from DH:	$9900*10^{6}$ Btu/yr	($28{,}987*10^{6}$ kWh/yr)
Maximum energy demand for domestic hot water supplied from DH:	$2760*10^{6}$ Btu/yr	($8{,}081*10^{6}$ kWh/yr)
Maximum energy losses in DH underground piping:	$1880*10^{6}$ Btu/yr	($5{,}505*10^{6}$ kWh/yr)
Total energy demand for DH (from heating central):	$14{,}540*10^{6}$ Btu/yr	($42{,}573*10^{6}$ kWh/yr)
Total energy supplied from solar system to DH:	$570*10^{6}$ Btu/yr	($1{,}669*10^{6}$ kWh/yr)
Solar fraction of total energy demand DH:	4.0%	
Solar system efficiency:	35.6%	
Max. DH advance temperature (depends on ambient temperature):	149 – 176 °F	(65–80 °C)
Max. DH return temperature	113 °F	(45°C)

Table A-17 lists the measured total demand of energy DH from 2002 to 2009. The data clearly rises from 2002 to 2006; then from 2008, one can see stagnation in energy demand. The explanation is that the buildings were completed, and fewer house were built than were planned. Differences between the years in ambient temperature and irradiation can cause differences as well, but the main reason seems to be the number of supplied buildings. Solar energy output from the solar storage tank seems considerably constant from 2002 to 2006, in 2007 and 2009 there was a decline, apparently because trees growing in front of the heating central increasingly shadow the lower collector area.

Table A-17. Measured data from 2002 to 2009.

	IP	2002	2003	2004	2005	2006	2007	2008	2009
Total energy demand of DH	10^6 Btu/yr	5,678	5,951	7,950	8,762	9,643	9,383	10,206	10,373
Solar energy output solar storage tank	10^6 Btu/yr	423.1	498.2	385.6	430.3	419.0	381.8	353.8	359.6
Solar fraction of energy demand DH	%	7.5	8.4	4.9	4.9	4.3	4.1	3.5	3.5
DH advance temperature, yearly average	°F	158.5	159.3	154.6	159.1	160.7	161.4	161.4	160.3
DH return temperature, yearly average	°F	113.2	117.9	123.3	116.2	114.8	115.3	115.9	117.5
	SI	**2002**	**2003**	**2004**	**2005**	**2006**	**2007**	**2008**	**2009**
Total energy demand of DH	10^6 kWh/yr	16,625	17,425	23,278	25,655	28,235	27,473	29,883	30,372
Solar energy output solar storage tank	10^6 kWh/yr	1,239	1,459	1,129	1,260	1,227	1,118	1,036	1,053
Solar fraction of energy demand DH	%	7.5	8.4	4.9	4.9	4.3	4.1	3.5	3.5
DH advance temperature, yearly average	°C	70	71	68	71	72	72	72	71
DH return temperature, yearly average	°C	45	48	51	47	46	46	47	48

Solar fraction of energy demand DH fell from 2002 to 2009 because the amount of energy needed in the DH was rising while the output from the solar system remained more or less constant. DH advance temperature is in the field as planned, DH return temperature as well. This is important because in other networks, the return temperature is much higher than planned. Here in Heilbronn, a good adjustment of the heat transfer stations in the buildings was done, using temperature limiter for the return temperature of the space heating loops.

The building landlords own the heat transfer stations. Consequently, so we have no information about these stations, and no measured data from inside the stations.

Figure A-71 shows a typical energy demand curve in a DH in Germany, here year 2008. The demand in the winter is about three times higher than in the summer. The solar fraction of energy demand DH reaches in the summer up to 30%, in the winter it is close to zero.

Figure A-72 shows the volume flow of the DH and the volume flow through the solar storage tank during discharge. Only a small amount of volume flow of the DH is guided through the solar storage tank for discharge. Through-flow is stopped and tanks are bypassed when solar storage tanks are discharged (means temperature in storage tanks is lower than DH return temperature). Figure A-72 also shows DH advance and return temperature. In the winter, the DH advance temperature rises depending on ambient temperature (explained in Chap. Control). DH return temperature in the summer rises because of the disappearance of energy needed for space heating in the buildings. This causes rising DH return temperatures in heat transfer stations and in the DH return. The temperature limiter in the heat transfer stations only influences the return temperature of the space heating loops, not the return temperature of domestic hot water facility. A temperature limiter in the domestic hot water facility could cause water temperature to fall too low, allowing Legionella bacteria to grow.

Experiences/Lessons learned

Energy use reduction

In the8 years from 2002 to 2009, we measured an energy output from the solar storage tanks of $3250*10^6$ Btu or $406*10^6$ Btu ($9.5*10^6$ MWh or $1.2*10^6$ MWh)on average per year. Assuming a boiler efficiency of 90%, and an energy content of natural gas of $27.4*10^3$ Btu/cu yd ($0.10*10^3$ kWh/L), we found a saving of natural gas of 131,900 cu yd (100,850 kL) in 8 years. Per year this is an average saving of 16,490 cu yd (12,608 kL) natural gas.

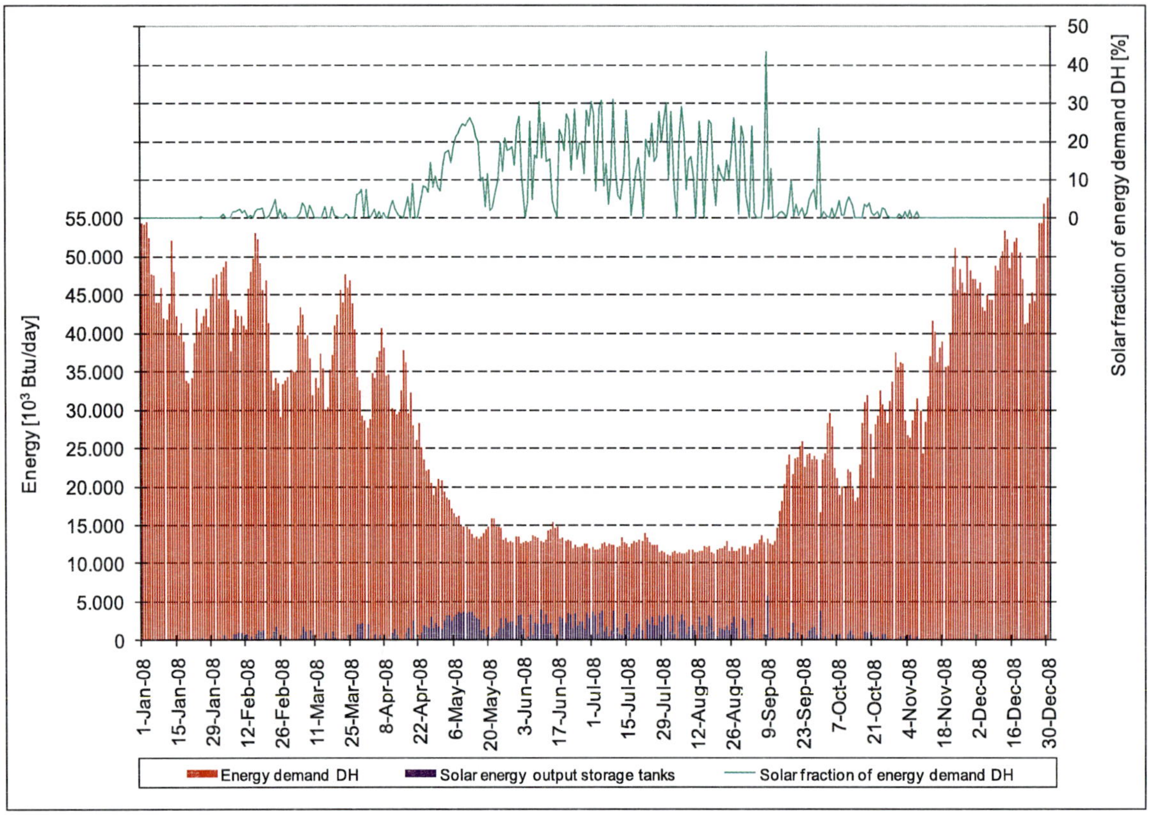

Figure A-71. Energy demand DH, solar energy output solar storage tanks and solar fraction of energy demand DH measured in 2008.

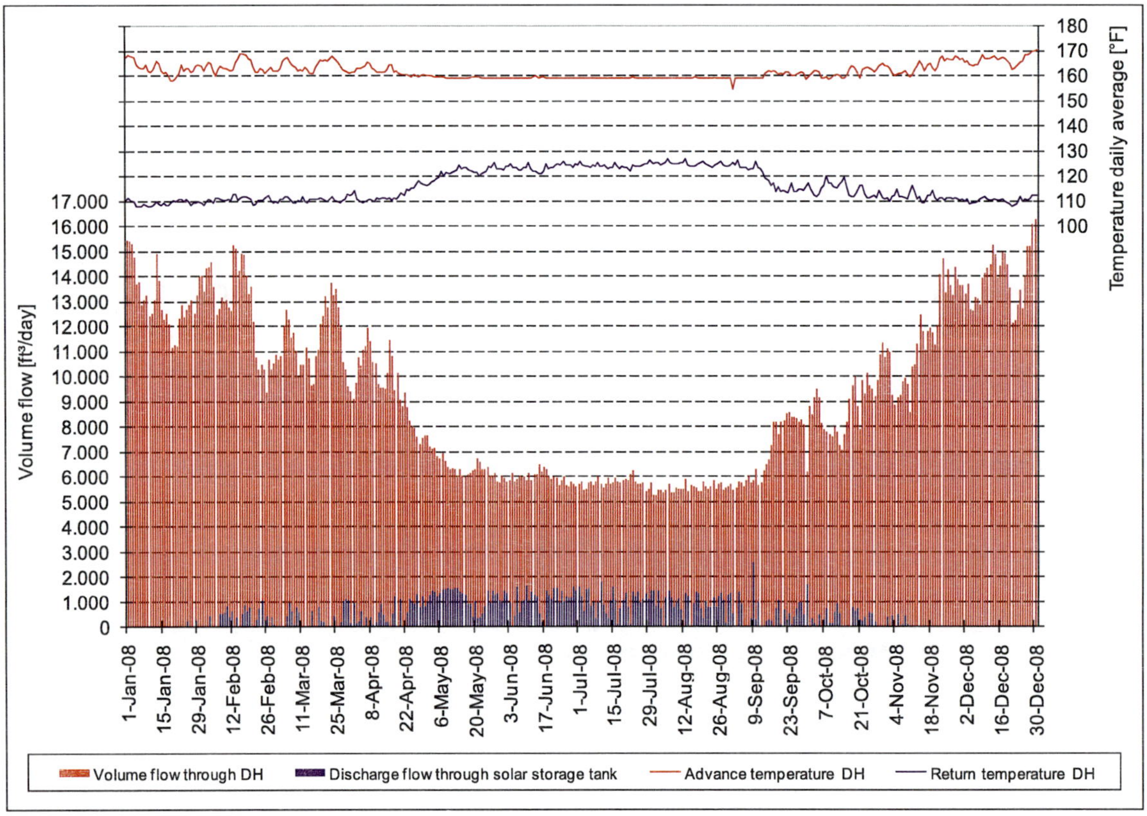

Figure A-72. Volume flow DH, discharge flow through solar storage tanks and temperatures DH.

Lessons learned

From the data we measured and the monitoring of the solar system we learned:

- Generally the solar system in combination with the district heating network operated without severe problems. The input of solar energy in the return pipe of the DH is the most promising way.
- The difference between the planned DH return temperature (113 °F [45 °C]) and the measured from 113 to 123 °F (45 to 51 °C) on a yearly average from 2002 to 2009 is clearly visible, but just about acceptable. In other DHs, much more deviation from the planned to measured temperatures is found. This relatively small deviation can be explained by the combined planning of solar system, central heating plant, and heat transfer stations connected to the DH with integrated return temperature limiter.
- There were leaks in the collector built roof. A lesson learned from this is that it is important to ensure that the collectors used for a roof-integrated collector arrangement are adequate for this purpose. We recommend to obtain a special guarantee in this case from the collector manufacturer. A very accurate installation is assumed in any case.
- The solar storage tanks should have a valve driven bypass to the DH return pipe. This reduces the energy losses from the solar storage tanks in times of poor solar irradiation because tanks were not loaded from the return flow stream of DH. The here used bypass valves (V3, V4) were refitted in 2004. Commendable (but here in the Heilbronn system not realized) is an additional temperature limiting function to the bypass valves to obtain a discharge temperature from the solar storage tanks not higher than the needed advance temperature of the DH.
- Bear in mind the political dimension of the question whether a tree can be trimmed or cut down when it shadows the collector area; This may be a special problem in Germany.

General date

Address of the project

Solar water heating connected to a district heating network
Residential Area “Badener Hof”
Heilbronn, Germany

Date of report

Measured data period: 1st January 2002 to 31st December 2008
Date of report: June 2010

Acknowledgement

Promoting department

Bundesministerium für Umwelt, Naturschutz und Reaktorsicherheit (BMU)
(Federal department of environment, nature conservation and nuclear reactor safety)
Alexanderstr. 3
10178 Berlin, Germany

Operating company (owner)

Stadtwerke Heilbronn
Weipertstr. 49

74076 Heilbronn, Germany

Planning company

EGS-plan Ingenieurgesellschaft für Energie-, Gebäude- und Solartechnik
(former Steinbeis-Transferzentrum)
Hessbrühlstr. 15
70565 Stuttgart
Germany

Mounting company

Georg Linder GmbH
Austr. 3
97996 Niederstetten
Germany

Nikolaus Gebäude- und Anlagentechnik GmbH

Rudolf-Schmidt-Str. 9
91550 Dinkelsbühl
Germany

Measuring company

ZfS-Rationelle Energietechnik GmbH
Verbindungsstr. 19
40723 Hilden
Germany

References

Mies, M.; Rehrmann, U.; Szablinski, D.; Abschlussbericht für das Projekt Neubaugebiet "Badener Hof" Heilbronn August 2006, http: //www.zfs-energietechnik.de/main.php?RND=33644&SESS=&LANG=de&ID=168, (more and detailed information about the solar system in Heilbronn are outlined in this report).

Peuser, Felix A.; Remmers, Karl-Heinz; Schnauss, Martin. Solar Thermal Systems Successful Planning and Construction. Solarpraxis AG Berlin Germany in association with James & James London UK, 2002. ISBN: 3-934595-24-3.

VDI 6002, part 1, September 2004 (technical guideline). Solar heating for domestic water General principles, system technology and use in residential buildings. Distributer: Beuth Verlag, 10722 Berlin, Germany.

DVGW W551, April 2004 (technical guideline). Trinkwassererwärmungs- und Leitungsanlagen; Technische Maßnahmen zur Verminderung des Legionellenwachstums (Hot water systems, technical arrangements to reduce developing of Legionella bacteria), no English translation available. Distributer: Wirtschafts- und Verlagsgesellschaft Gas und Wasser mbH Josef-Wimmer-Str. 1 - 3, 53123 Bonn, Germany

FPC – 15. Solar Water Heating (SWH) Connected to District Heating Networks (DH)

Title: District Heating Network, Apartment Buildings Magdeburger Straße, Hannover, Germany

Location: Hannover, Germany

Photo of installation, schematics

Figure A-73. Front view of the apartment building with nine collector rows on a flat roof.

Figure A-74. Collector row arrangement.

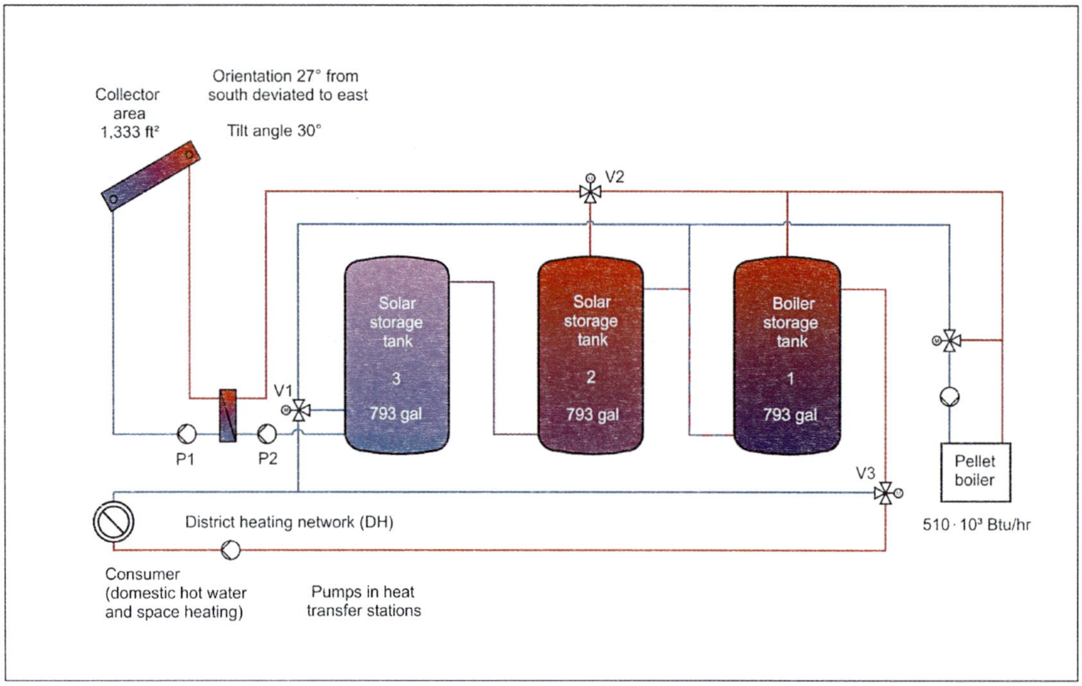

Figure A-75. Highly simplified schematics of the solar system, solar storage tanks arrangement and the integration in the district heating network (DH).

Project summary

Two nearby located apartment buildings (erected in 1960) were selected for a complete reconstruction in the city of Hannover in the north of Germany. The space heating, the domestic water supply, the boilers and the isolation of all walls should were replaced or improved. Additionally a solar system was added to the new installed pellet boiler. The two apartment buildings (reconstructed in 2006) are supplied from a heating central in one of the buildings, which works as a small district heating network. For this reason the underground piping is very short.

The two apartment buildings have in total 36 flats in different sizes to accommodate about 140 people. The owner of the apartment buildings (GBH Mieterservice Vahrenheide GmbH) operates the heating central and the solar system to supply the buildings with heat for domestic hot water and space heating. For this purpose, the central is equipped with a pellet-fired boiler ($512*10^3$ Btu/hr), a pellet store in the basement, a solar energy system with a collector area of 1333 sq ft, a solar storage tank capacity of 2 x 793 gal (3,002 L), and a storage tank of 793 gal (3,002 L) attached to the pellet boiler. The heat transfer stations (HTS) in the buildings are in the ownership of GBH too.

The arrangement of solar system, boilers, and district heating network works well. No severe problems were found in the concept. The problem of a not satisfying control valve to adjust the DH advance temperature has not yet been solved.

Site

Location:	Apartment buildings Magdeburger Straße 2 and 4
Town:	Hannover
Country:	Germany
Latitude:	52° 24’ North
Longitude:	9° 45’ East

Project description

A collector area of 1333 sq ft (123.97 m^2) is installed on a flat roof on top of one of the apartment buildings, which houses the boiler, the solar storage tanks and the control facilities. The energy from the collector loop is transferred by a flat plate heat exchanger to the charge loop of the solar storage tanks. Depending on the temperature of the charge loop, tanks 1 and 2 in sequence are loaded. Both tanks have a capacity of 793 gal (3,001 L).

Depending on the DH return temperature, the volume flow can be guided by shutoff valves to discharge the solar storage tanks or bypass them. A pellet boiler can feed in energy to guarantee the DH advance temperature in periods with low irradiation and less output of solar energy. The correct DH advance temperature is adjusted by a control valve, which compensates for the fluctuations in temperature generate by the stop-and-go operation of the boiler. To damp the stop-and-go cycles the pellet boiler is connected to a boiler storage tank with a capacity of 793 gal (3,002 L). In the summer, when the energy supply can be taken over only by the solar system, the boiler storage tank can be used from the solar system as an additionally storage capacity. The solar storage tank capacity enlarges by switching valve V2 to 3 x 793 gal (3,002 L).

Table A-18. Expected data during planning.

Number of apartment buildings supplied by DH:	2
Number of people supplied by DH:	140
Maximum energy demand for space heating supplied from DH:	775*10^6 Btu/yr (2,269 kWh/yr)
Maximum energy demand for domestic hot water supplied from DH:	287*10^6 Btu/yr (840*10^6 kWh/yr)
Maximum energy losses in DH underground piping:	20*10^6 Btu/yr (59*10^6 kWh/yr)
Total energy demand from heating central:	1,082*10^6 Btu/yr (3,168*10^6 kWh/yr)
Total energy supplied from solar system to DH:	153*10^6 Btu/yr (448*10^6 kWh/yr)
Solar fraction of total energy demand DH:	15.0%
Solar system efficiency (energy output from solar storage tanks/irradiation energy):	34.5%
Max. DH advance temperature (depends on ambient temperature):	149 – 158 °F (65–70 °C)
Max. DH return temperature	104 °F (40 °C)

Data in Table A-18 were calculated using the following equations:

Solar system efficiency = Solar energy output from solar storage tanks/Irradiation in collector area

Solar fraction of total energy demand DH = Solar energy output from solar storage tanks/Total energy demand DH

Table A-18 lists the measured data in the years between beginning of 2007 and end of 2009. Irradiation in the collector area differs within a small range. Solar output from the collector loop or solar output from the solar storage tanks (which includes the thermal losses of the tanks) develop not similar to the solar irradiation. The proportion between irradiation and solar output can be shown by the "solar system efficiency," which decreased from 29.0% in 2007 to 26.1% in 2009. We explain this descent of the solar efficiency despite rising irradiation by the rising of the DH return temperature from 116.6 to 134.6 °F (47 to 57 °C). Generally the higher DH return temperature is, the lower is the efficiency of a solar system.

Table A-19. Measured data from 2007 and 2009.

		2007		2008		2009	
Irradiation in horizontal area	10^3 Btu/(sq ft*yr) [10^3 kWh/(m^2t*yr)]	320.8	[1.01]	322.4	[1.02]	337.3	[1.06]
Irradiation in collector area	10^3 Btu/(sq ft*yr)	484.5	(1,418.62)	482.5	(1,412.76)	507.4	(1,485.67)
Irradiation in collector area, specific	10^3 Btu/(sq ft*yr) [10^3 kWh/(m^2t*yr)]	363.6	[1.15]	362.0	[1.14]	380.7	[1.20]
Solar energy output collector loop	10^6 Btu/yr (10^6 kWh/yr)	147.7	(432.47)	136.5	(399.67)	137.2	(401.72)
Solar energy output solar storage tanks (estimated)	10^6 Btu/yr (10^6 kWh/yr)	140.2	(410.51)	126.3	(369.81)	132.7	(388.55)
Total energy demand of DH	10^6 Btu/yr (10^6 kWh/yr)	810.4	(2,372.85)	873.8	(2,558.49)	924.7	(2,707.52)
Solar system efficiency	%	29.0		26.2		26.1	
Solar fraction of total energy demand DH	%	17.3		14.5		14.3	
DH advance temperature, yearly average	°F (°C)	162.0	(72)	172.8	(78)	174.6	(79)
DH return temperature, yearly averaged	°F (°C)	116.6	(47)	125.2	(51)	134.6	(57)

The solar fraction of total energy demand DH fell from 17.3 in 2007 to 14.3% in 2009. This is understandable because the energy output of the solar storage tanks declined from 140.2*10^6 Btu/yr to 132.7*10^6 Btu/yr in 2009 and the energy demand of the DH rose from 810.4*10^6 to 924.7*10^6 Btu/yr. The increase in energy demand of DH can be explained by a better allocation of the apartment buildings. Because of the occasional use of the boiler storage tank 1 as an additionally solar storage tank by switching valve V2 the energy output from the solar system can only be estimated.

The DH advance temperature on yearly average rose from 162.2 °F (72 °C) in 2007 to 174.6 °F (79 °C) in 2009. The temperature in 2007 is acceptable, but the temperature in 2009 with 174.6 °F (79 °C) is much too high. That caused a rise of the DH return temperature from 116.6 °F (47 °C) in 2007 to 134.6 °F (57 °C) in 2009, what is much too high as well. This has a negative retroactive effect to the solar system efficiency. The reason for the not matched DH advance temperature is the poor working control valve, which should adjust the correct temperature.

Figure A-76 shows the 2009 measured daily data of irradiation, solar energy output from collector loop and the calculated efficiency of the collector loop. The irradiation over the year shows the typical shape known in Germany with explicit difference between summer and winter. The solar energy output follows a similar pattern; the main harvest of energy occurs from February to October, while the harvest from November to January is almost negligible. In more northern regions this effect is more striking than in more southern regions, where the effect dwindles. In the summer, the efficiency of the collector loop reaches nearly 40%, but poor winter data reduces winter the yearly average is only 27.0%.

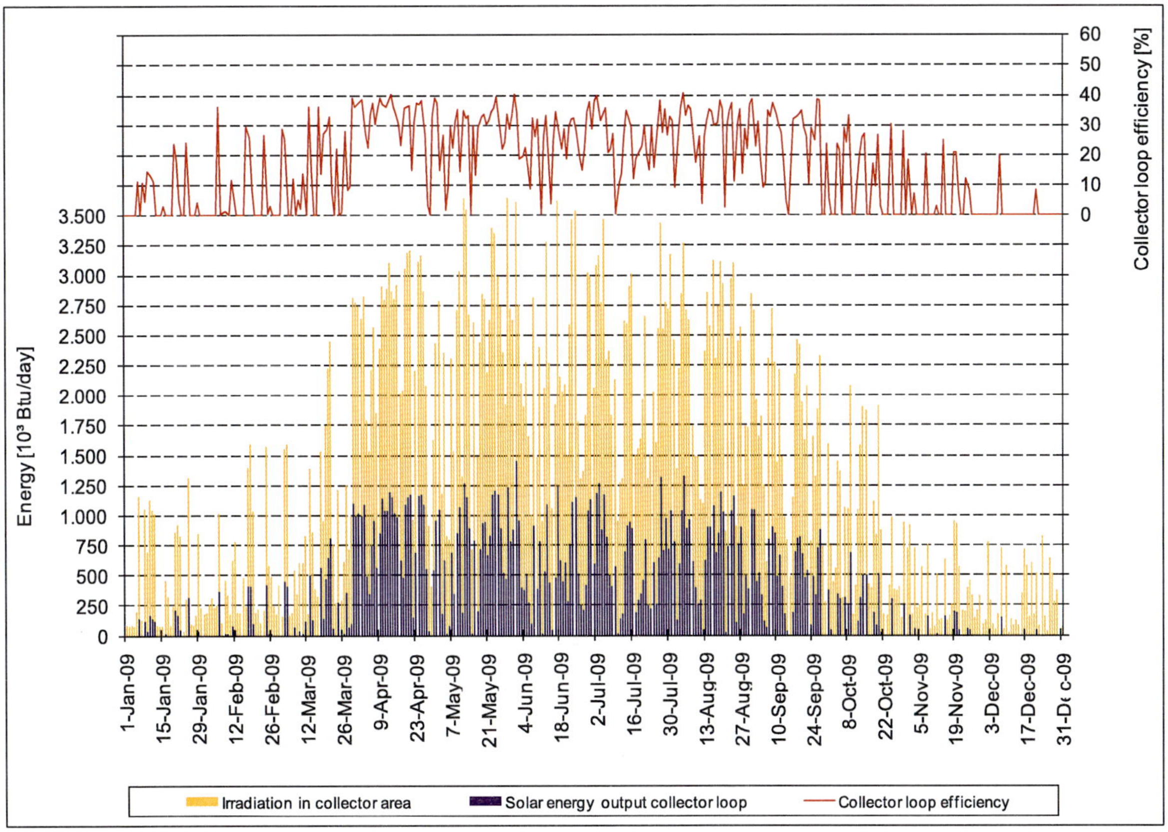

Figure A-76. Irradiation, energy output, and efficiency collector loop in 2009, daily data solution.

System details

Connection of SWH to DH

The solar system is connected to the DH in the return pipe, located in the heating central (Figure A-77).

Solar system

The solar system contains a collector area divided into nine rows (situated on the flat roof of one of the apartment buildings), piping, a flat plate heat exchanger, two solar storage tanks, the connection to the DH, and the control technique. The volume of the boiler storage tank can be added to the solar storage capacity by switching a valve.

Collector manufacturer:	Solvis
Type of collector:	Fera F552-S flat plate collector
Orientation:	27° from south deviated to east
Tilt angle:	30°
Collector area:	1,333 sq ft (123.97 m^2) (Aperture Area)
Solar storage tanks:	2 x 793 gal (3000 L)+ 1 x 793 gal (enlarged) upright cylindrical steel tank
Freeze protection:	Collector loop filled with 40% glycol, 60% water
Heat exchanger:	Brazed flat plat type, area 138 sq ft (12.83 m^2)

Description of control strategy

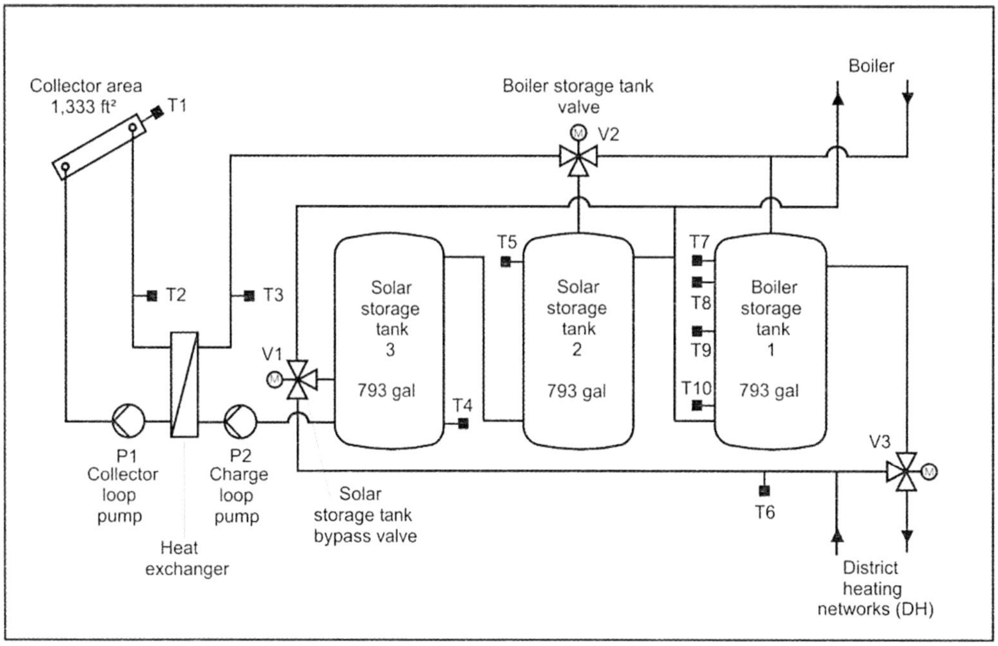

Figure A-77. Highly simplified schematics showing the positioning of the control sensors.

The boiler storage tank 1 is fitted with a temperature sensor (T7) to prevent the solar storage tanks from overheating and boiling. When temperatures (T7) are higher than 221 °F (105 °C), the pump control functions are halted and pump P1 generally stops. When temperatures at sensor T7 fall below 221 °F (105 °C), and a time lapse of 6 hours, clears the halted control functions.

Collector loop pump (P1) is controlled by the difference in temperature between the temperature in collector area (T1) and the temperature in the lower area of solar storage tank 3 (T4). A difference in temperature (T1 – T4) of more than 18.0 °F makes the pump run, a difference of less than 9.0 °F stops the pump. The charge loop pump (P2) is controlled in the same manner by the difference in temperature between the temperature in the collector loop (T2) and the temperature in the lower area of solar storage tank 3 (T4). A difference in temperature (T2 – T4) of more than 12.6 °F makes the pump run, a difference of less than 9.0 °F stops the pump.

Loading the boiler storage tank 1 additionally with solar energy is possible, when the temperature at sensor T3 exceeds 165 °F (74 °C) and a time lapse of 3 min has passed. Valve V2 than gives way to the boiler storage tank. This arrangement makes it possible to avoid running the boiler in times with good irradiation, or at least to reduce the number of stop-and-go cycles of operation of the boiler firing during the day. When the temperature at sensor T3 is lower than 158 °F (70 °C), valve V2 is switched to bypass the boiler storage tank.

The discharge of the solar storage tanks 2 and 3 executed by valve V1 is controlled by the difference in temperature between the upper area of solar storage tank 2 (T5) and the DH return temperature (T6). A difference in temperature (T5 – T6) of more than 5.4 °F opens valve V1 for the volume flow to the tanks, a difference less than 2.9 °F shuts V1 to bypass the solar storage tanks 2 and 3. Whether solar storage tanks 2 and 3 are discharged or not, the DH volume return flow passes through boiler storage tank 1.

The pellet boiler firing starts when temperature T8 < 149 °F (65 °C) and stops when T8 > 158 °F (70 °C) (70 °C). Power of firing is reduced when (T8+T9+T10)/3 > 140 °F (60 °C), power is boosted when (T8+T9+T10)/3 < 140 °F (60 °C).

Table A-20. Summary of control activities and control conditions.

Control activity	Control Conditions
Collector loop/ Charge loop Clearance of running pump P1 General stop of running pump P1 Collector loop pump P1 Charge loop pump P2	 T7 < 221 °F (105 °C) and 6 hours time lapse T7 > 221 °F (105 °C) On: T1 – T4 > 18.0 °F, Off: T1 – T4 < 9.0 °F On: T2 – T4 > 12.6 °F, Off: T2 – T4 < 9.0 °F
Loading boiler storage tank with solar energy Valve V2	 Open to boiler tank: T3 > 165 °F (74 °C) and 3 min time lapse Bypass boiler tank: T3 < 158 °F (70 °C) and 3 min time lapse
Discharging solar storage tanks Valve V1 in DH return	 Open to solar storage tanks: T5 – T6 > 5.4 °F Bypass solar storage tanks: T5 – T6 < 2.9 °F
Controlling pellet boiler	Start firing: T8 < 149 °F (65 °C) Stop firing: T8 > 158 °F (70 °C) Power of firing reduced: (T8+T9+T10)/3 > 140 °F (60 °C) Power of firing boosted: (T8+T9+T10)/3 < 140 °F (60 °C)

Economics

The costs of the solar systems include only the costs for solar collectors, piping, solar storage tanks and controls, but do not include costs for the district heating network, boiler or the heating central building (Table A-21).

Table A-21. Economics.

Costs solar system Costs solar system, including statics Costs planning solar system Steel collector support construction Costs solar system, statics and planning Costs solar system, statics and planning including 19% tax	$92,371 (65,050 €) $10,863 (7,650 €) $15,393 (10,840 €) $118,627 (83,540 €) **$141,162 (99,410 €)**	
Annual costs for loan Living period 20 years, 6% rate, → annuity: 8.72%	$12,309.98 (8,669 €)	
	Per year	Sum total in 3 years
Solar energy output Planned solar energy output from solar storage tanks Measured solar energy output from storage tanks	$149.7*10^6$ Btu ($438*10^6$ kWh) 133.1– $139.9*10^6$ Btu (390 kWh–410 $*10^6$ kWh)	$449.0*10^6$ Btu $399.2*10^6$ Btu
Relation measured solar energy output/ planned energy output		88.9%
Savings of gas und CO_2 calculated with measured solar energy from solar storage tanks with following assumptions: Boiler efficiency: 90%; Energy of natural gas: $27.4*10^3$ Btu/cu yd_{Gas} Emission factor: 0.129 lbs $CO_2/10^3$ Btu_{Gas} Saving amount of natural gas avoidable amount of CO_2		16,190 cu yd (12,378,873 L) 29 (short) ton (26,306 kg)
Costs of solar energy from solar storage tanks with 8.72% annuity, including solar system, statics, planning and tax Costs and planned solar energy output Costs and measured solar energy output, without Costs for maintenance and repairs	$0.03/10^3 kWh (0.058 €/10^3 Btu) $0.03/10^3 kWh (0.065 €/10^3 Btu)	

District heating network

Table A-22. Expected data during planning.

Number of apartment buildings supplied by DH:	2
Number of people supplied by DH:	140
Maximum energy demand for space heating supplied from DH:	$775*10^6$ Btu/yr ($2{,}269*10^6$ kWh/yr)
Maximum energy demand for domestic hot water supplied from DH:	$287*10^6$ Btu/yr ($840*10^6$ kWh/yr)
Maximum energy losses in DH underground piping:	$20*10^6$ Btu/yr ($59*10^6$ Btu kWh/yr)
Total energy demand from heating central:	$1{,}082*10^6$ Btu/yr ($3{,}168*10^6$ kWh)/yr
Total energy supplied from solar system to DH:	$153*10^6$ Btu/yr ($448*10^6$ kWh/yr)
Solar fraction of total energy demand DH:	15.0%
Solar system efficiency (energy output from solar storage tanks/irradiation energy):	34.5%
Max. DH advance temperature (depends on ambient temperature):	149 – 158 °F (65 – 70 °C)
Max. DH return temperature	104 °F (40 °C)

Table A-23. Measured data from 2007 to 2009.

		2007		2008		2009	
Total energy demand of DH	10^6 Btu/yr (10^6 kWh/yr)	810.4	(2,372.85)	873.8	(2,558.49)	924.7	(2,707.52)
Solar energy output solar storage tanks	10^6 Btu/yr (10^6 kWh/yr)	140.2	(410.51)	126.3	(369.81)	132.7	(388.55)
Solar fraction of total energy demand DH	%	17.3	17.3	14.5	14.5	14.3	14.3
DH advance temperature, yearly average	°F (°C)	162.0	(72)	172.8	(78)	174.6	(79)
DH advance temperature, yearly average	°F (°C)	116.6	(47)	125.2	(52)	134.6	(57)

Table A-23 lists the measured demand of energy DH from 2007 to 2009, which is rising from $810.4*10^6$ to $924.7*10^6$ Btu/yr ($2.4*10^6$ MWh to $2.7*10^6$ MWh/yr). Solar energy output from the solar storage tank fell from $140.2*10^6$ to $132.7*10^6$ Btu/yr ($410*10^6$ kWh to $389*10^6$ kWh/yr). Solar fraction of total energy demand DH declined from 17.3% in 2007 to 14.3% in 2009. We can explain this effect with the reduced solar efficiency related to the rising DH return temperature and the rising energy demand of the DH.

Advance temperature is much higher than planned. The planned temperature was 149 °to 158 °F (65 to 70 °C) depending on the ambient temperature; the measured temperature was up to 174.6 °F (79 °C) (on yearly average). The reason was a developing malfunction of the DH control valve V3. The DH return temperature did not match the planned temperature of 104 °F (40 °C) ; We measured from 2007 to 2009 a rising temperature from 116.6 to 134.6 °F (47 to 57 °C).

We have no information about the used manufactures and types of the heat transfer stations. As well we have no measured data from inside the stations, no further information can be given here.

Figure A-78 shows a typical energy demand curve of a DH in Germany in 2009. The demand in the winter is about 3 times higher than in the summer. The solar fraction of the energy demand DH reaches in the summer up to 80%, in the winter it is close to zero. Designing the solar system it was one goal, **not** to produce an excess of solar energy in the summer, which means a waste of expensive generated solar energy. Figure A-78 shows clearly, that with a maximum solar fraction not more than 80% in the summer this goal was reached.

Figure A-79 shows the volume flow of the DH and the volume flow through the solar storage tanks during discharge. Only a small amount of volume flow of the DH is guided through the solar storage tanks for discharge. Through flow is stopped and tanks are bypassed when solar storage tanks are not discharged (means temperature in storage tanks is lower than DH return temperature). The DH advance temperature is about constant over the year. DH return temperature in the summer rises because of the disappearance of energy demand for space heating in the buildings.

Experiences/Lessons learned

Energy use reduction

In 3 years from 2007 to 2009, we measured an energy output from the solar storage tanks of 399.2×10^6 Btu or 133.0*106 Btu (1.2×10^6 MWh or 0.39×10^6 MWh) on a yearly average. Assuming a boiler efficiency of 90%, an energy content of natural gas of 27.4×10^3 Btu/cu yd_{Gas} (0.10 kWh/L_{Gas}) there is a saving of natural gas of 16,190 cu yd (12,379 kL) in 3 years. Per year there is a saving of 5400 cu yd (4,129 kL).

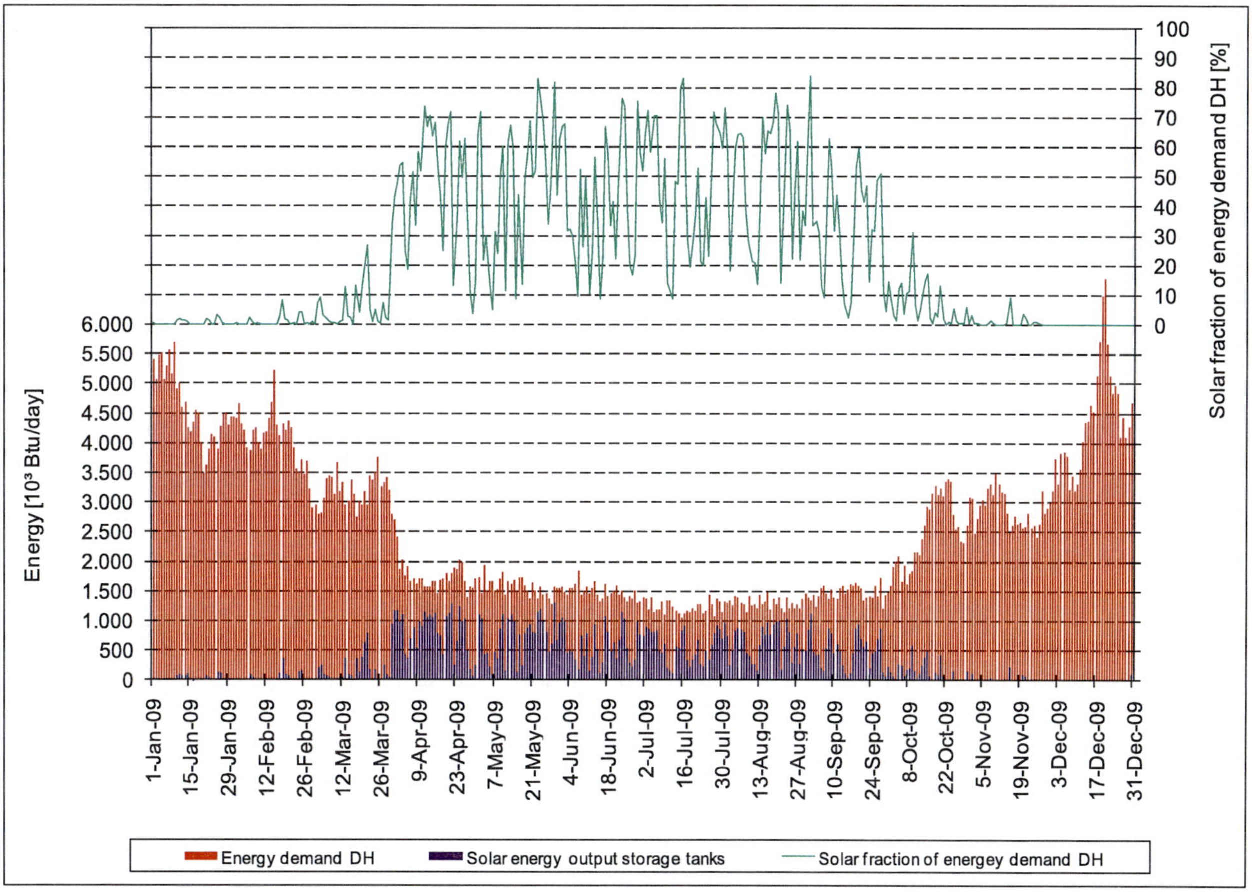

Figure A-78. Energy demand DH, energy output solar system and solar fraction of total energy demand DH measured in 2009.

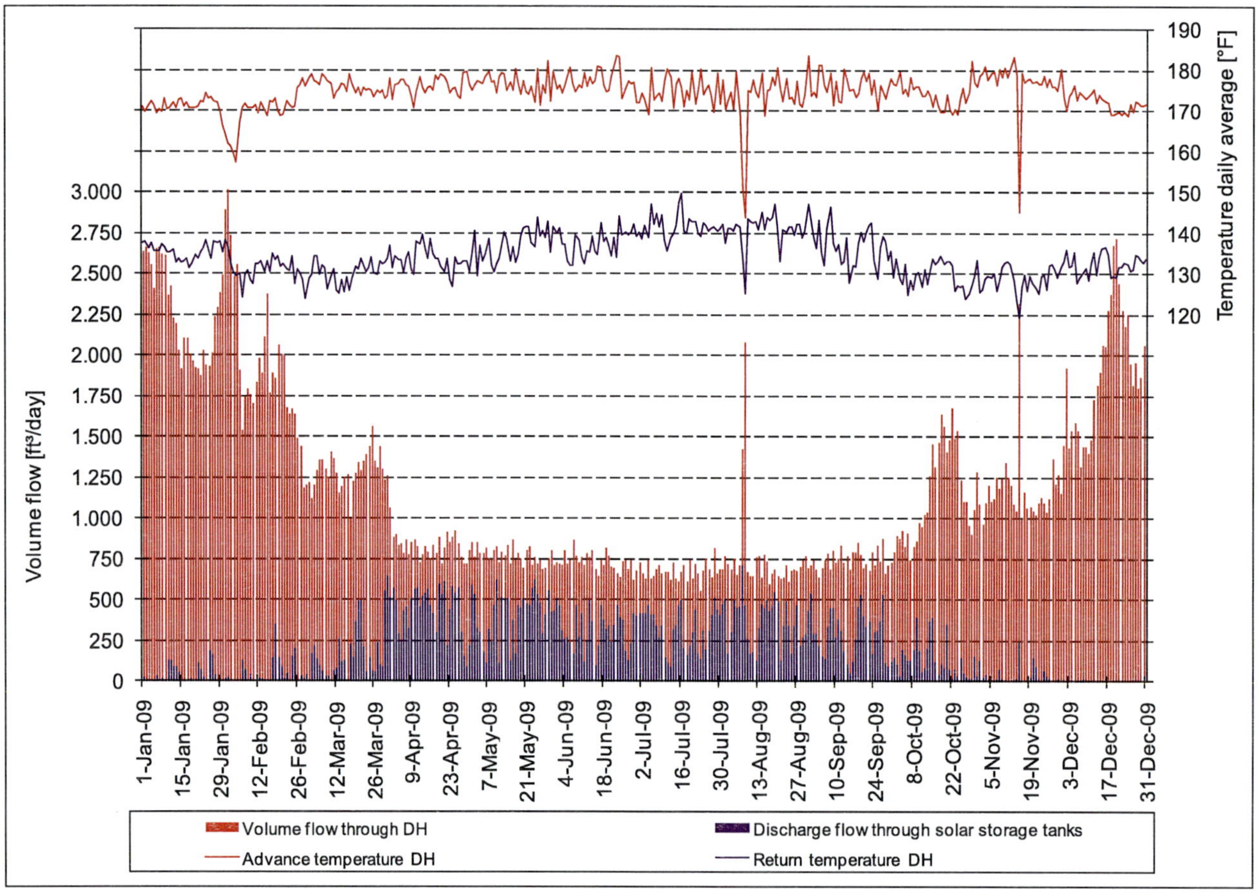

Figure A-79. Volume flow DH, volume flow through solar storage tanks and temperatures DH measured in 2009.

Lessons learned

From the data we measured and the monitoring of the solar system we learned that:

- In general, the solar system in combination with the district heating network operated without severe problems. The input of solar energy in the return pipe of the DH is the most promising way.
- The solar system and the connection to the DH return is designed more complex than in other systems. Here the designer tried to integrate the boiler storage tank as an additional solar storage tank to enlarge the storage capacity in the summer and reduce the stop-and-go operation of the pellet boiler. This goal was reached. Contrary to primarily existing concerns, this arrangement functions without problems.
- The difference between the planned DH return temperature (104 °F [40 °C]) and the measured up to 134.6 °F (57 °C) is extreme. The planned temperature of 104 °F (40 °C) was certainly too optimistic, but measured 134.6 °F (57 °C) is very bad. The reason is found in the failure of the DH advance temperature control valve, which produces a DH advance temperature to high. It is urgent to replace this valve with a better adapted one.

General date

Address of the project

Solar water heating connected to a district heating network
Apartment buildings Magdeburger Straße 2 und 4
Hannover
Germany

Date of report

Measured data period: 1st January 2007 to 31st December 2009
Date of report: June 2010

Acknowledgement

Promoting department

Bundesministerium für Umwelt, Naturschutz und Reaktorsicherheit (BMU)
(Federal department of environment, nature conservation and nuclear reactor safety)
Alexanderstr. 3
10178 Berlin
Germany

Operating company (owner)

GBH Mieterservice Vahrenheide GmbH
In den Sieben Stücken 7A
30655 Hannover
Germany

Planning company

EGS-plan Ingenieurbüro für Energie-, Gebäude- und Solartechnik
Heßbrühlstr. 15
70565 Stuttgart
Germany

Mounting company

Altmärkische Haustechnik GmbH
Düsedauer Str.
39606 Osterburg
Germany

Measuring company

ZfS-Rationelle Energietechnik GmbH
Verbindungsstr. 19
40723 Hilden
Germany

References

Mies, M.; Rehrmann, U.; 2. Zwischenbericht für das Projekt Wohngebäude Magdeburger Str. Hannover. June 2009, http: //www.zfs-energietechnik.de/main.php?RND=27410&SESS=&LANG=de&ID=226 (taking into account a boiler efficiency).

Peuser, Felix A.; Remmers, Karl-Heinz; Schnauss, Martin. Solar Thermal Systems Successful Planning and Construction. Solarpraxis AG Berlin Germany in association with James & James London UK, 2002. ISBN: 3-934595-24-3

VDI 6002, part 1, September 2004 (technical guideline). Solar heating for domestic water. General principles, system technology and use in residential buildings. Distributer: Beuth Verlag, 10722 Berlin, Germany

DVGW W551, April 2004 (technical guideline). Trinkwassererwärmungs- und Leitungsanlagen; Technische Maßnahmen zur Verminderung des Legionellenwachstums (Hot water systems, technical arrangements to reduce developing of Legionella bacteria), no English translation available. Distributer: Wirtschafts- und Verlagsgesellschaft Gas und Wasser mbH. Josef-Wimmer-Str. 1 - 3, 53123 Bonn, Germany

FPC – 16. Solar Water Heating (SWH) Connected to District Heating Networks (DH)

Title: District Heating Network, Residential Area "Burgholzhof," Stuttgart, Germany

Location: Stuttgart, Germany

Photo of the installation

Figure A-80. View from the heating central to some of the buildings carrying a roof-integrated collector area.

Figure A-81. Solar storage tank encased in a concrete structure and the stacks from the heating central.

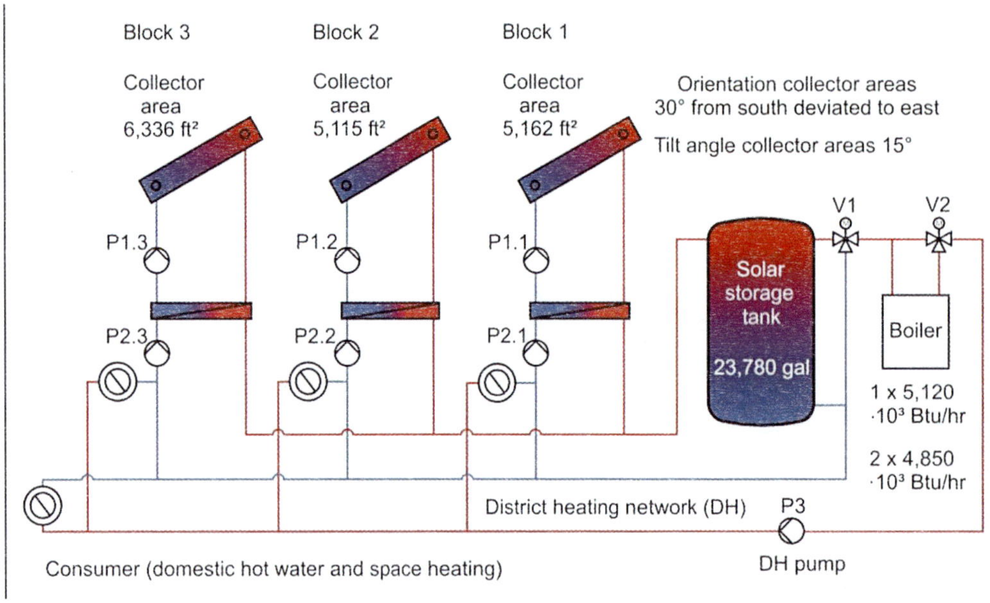

Figure A-82. Highly simplified schematics of the solar system, solar storage tank arrangement and the integration in the district heating network (DH).

Project summary

About 1360 homes for 2800 people were built on the site of former US barracks on the outskirt of the city of Stuttgart (Germany) beginning in 1996. The plots were sold to several contractors, so there was no award of the entire region to one carrier or construction contractor. According to the development plan of the site mainly multi-family houses and row houses were realized. Additionally a kindergarten, a school and shopping facilities were integrated. The former public services Neckarwerke, now EnBW, operates an underground district heating network (DH) to supply the area with heat for domestic hot water and space heating. For this purpose, a heating central with three natural gas-fired boilers, a solar energy system with a collector area of 16,613 sq ft (1,545 m^2), and a solar storage tank capacity of 1 x 23,780 gal (90 kL) are provided.

The district heating network, the solar system, and the heat transfer stations (HTS) in the houses have been dimensioned by the Steinbeis Transfer Center in Stuttgart. The manufacturer of the HTS was selected by Neckarwerke. The house owners were obliged to buy this type of HTS to achieve a constant technical performance. Purchase and maintenance lies within the responsibility of the house owners. The power supply and the DH advance temperature (in the range between 158 and 181 °F (70 and 83 °C) depending on ambient temperature) are guaranteed by the Neckarwerke. The DH advance temperature ranges between 158 °F (70 °C) in summer and 181 °F (83 °C) in the winter. The DH return temperature should not be more than 113.0 °F (45 °C) in summer and 116.6 °F (47 °C) in the winter.

The collector area is installed as an in-roof-system on three blocks, which are connected to the common solar storage tank by underground piping. The collector area on each block operates as a separate collector loop, which is partly controlled by a central control installation in the heating central.

The arrangement of solar system, boilers, and district heating network works well. No severe problems were found in the concept. Severe problems appeared in the in-roof collector areas. Several leaks occurred in the connection elements between the flat collector plates brought the whole system to a standstill. There were so many defects that a repair was not possible. The last information received from the owner was that he considers a complete exchange of the collector areas.

Site

Location:	Residential area "Burgholzhof"
Town:	Stuttgart
Country:	Germany
Latitude:	48° 49' North
Longitude:	9° 11' East

Project description

A collector area of 16,613 sq ft (1,545 m^2) is installed as an in-roof flat collector on top of three blocks. To make it easier, the blocks are indicated here as block 1, 2, and 3. Each collector area on one block has an autonomous collector loop, a collector loop pump, a heat exchanger and a charge pump. These three collector areas are connected to an underground pipe, which works as a common advance charge loop to the solar storage tank of 23,780 gal (90 kL) capacity. There is no common return charge loop pipe because the DH return in the blocks are guided to the collector loop heat exchanger as compensation for the not existing charge loop return pipe. This system is called a three pipe system (two-pipes DH, one pipe charge loop to solar storage tank). The advantage of this three-pipe-system is, that one underground pipe and the costs of this pipe can be saved. Disadvantage is, that the system is more difficult to understand and to design as a four-pipe system (two-pipes DH, two-pipes charge loop to solar storage tank). Because of such a close connection between DH system and charge loop to the solar storage tank the hydraulic requirements of such an arrangement have to be designed very accurately to avoid not wanted misguided hydraulic volume flows.

Depending on the DH return temperature, the volume flow can be guided by shutoff valves to discharge the solar storage tank or bypass them. Three natural gas-fired boilers can feed in energy to guarantee the DH advance temperature in periods with low irradiation and less output of solar energy. The correct DH advance temperature is adjusted by the firing of the boilers. An additional control valve in the DH advance to reduce the temperature peaks caused by the stop-and-go cycles of the boilers is not installed. The boilers can be bypassed by switching a valve when sufficient energy from the solar storage tank is supplied.

Table A-24. Expected data during planning

Number of apartments supplied by DH:	1,360
Number of people supplied by DH:	2,800
Maximum energy demand for space heating supplied from DH:	19,450*10^6 Btu/yr (56,950*10^6 kWh/yr)
Maximum energy demand for domestic hot water supplied from DH:	7,510*10^6 Btu/yr (21,989*10^6 kWh/yr)
Maximum energy losses in DH underground piping:	2,390*10^6 Btu/yr (6,998*10^6 kWh/yr)
Total energy demand from heating central:	29,350*10^6 Btu/yr (85,937*10^6 kWh/yr)
Total energy supplied from solar system to DH:	2,180*10^6 Btu/yr (6,383*10^6 kWh/yr)
Solar fraction of total energy demand DH:	7.4%
Solar system efficiency (energy output from solar storage tanks/irradiation energy):	34.6%
Max. DH advance temperature (depends on ambient temperature):	158.0–181.4 °F (70–83 °C)
Max. DH return temperature	113.0–116.6 °F (45–47 °C)

Data in Table A-24 were calculated using the following equation:

Solar system efficiency = Solar energy output from solar storage tanks/Irradiation in collector area

Table A-25 lists the measured data in the period 3 July 2002 to 2 July 2003, in the following indicated as period 2002/2003. Because of severe problems concerning tightness of the collector areas and collector loops only the period 2002/2003 could be measured. Irradiation in the collector area with $412.7*10^3$ Btu/sq ft ($1.30*10^3$ kWh/m^2) is normal for this location. Solar output from the collector loops was $1672*10^6$ Btu/yr ($4.9*10^6$ MWh/yr). Considering the estimated thermal loss through the underground piping and solar storage tank (circa 4%) the solar energy output of the solar storage tank is estimated at $1607*10^6$ Btu/yr ($4.9*10^6$ MWh/yr). The output of the solar storage tank could not be measured because of the hydraulic difficulties that occur in a three-pipe arrangement. The solar system efficiency is estimated at 23.4%.

Table A-25. Measured data from 3rd July 2002 to 2nd July 2003,

	2002/2003
Irradiation in horizontal area	365.8 Btu/(sq ft*yr) (1.15 kWh/m^2*yr)
Irradiation in collector area	6,855 Btu/yr (20,071 kWh/yr)
Irradiation in collector area, specific	412.7 Btu/(sq ft*yr) (1.3 kWh/m^2*yr)
Solar energy output collector loops	1,672 Btu/yr (4,896 kWh/yr)
Solar energy output solar storage tank	1,607 Btu/yr (4,705 kWh/yr) estimated
Energy demand of DH	Not measured
Collector loop efficiency	24.4%
Solar system efficiency	23.4% estimated
Solar fraction of total energy demand DH	Not measured
DH advance temperature, yearly average	171.1 °F (77 °C)
DH return temperature, yearly average	121.8 °F (-6 °C)

The DH advance with 171.1 °F (77 °C) on a yearly average was found in the anticipated range, the DH return temperature with 121.8 °F (50 °C) on a yearly average differs from the planned temperature (not more than 113.0 to 116.6 °F [45 to 47 °C]). The result can be evaluated nevertheless as "just good."

Because the energy demand for DH could not be measured, the solar fraction of total energy demand DH could not determined.

Figure A-83 shows the 2002 measured daily data of irradiation, solar energy output from the collector loops, and the calculated efficiency of the collector loops. The irradiation shown is typical for Germany, with explicit differences between summer and winter. The solar energy output follows a similar pattern, the main harvest of energy is found from February to October, the energy harvest in the months from November to January is almost negligible. In more northern regions, this effect is more noticeable than in the southern regions where the effect declines. In the summer, the efficiency of the solar system can reach up to 40%, but the poor winter data reduce the yearly average to only 24.4%.

System details

Connection of SWH to DH

The solar system is connected to the DH in a three-pipe arrangement (Figure A-84).

Solar system

The solar system contains three collector areas on three blocks, working as in-roof collectors. The piping from these collector areas guides the solar energy to flat plate heat exchangers. In a three-pipe arrangement the charge loop transfers the solar energy to the solar storage tank in the heating central.

Collector manufacturer:	Solar Diamant
Type of collector:	SKS 2.1s flat plate collector
Orientation:	circa 30° from south deviated to east
Tilt angle:	15°
Collector area:	5162 sq ft + 5115 sq ft + 6336 sq ft = 16,613 sq ft (aperture area) 480 m^2 + 476 m^2 + 589 m^2 = 1,545 m^2 (aperture area)
Solar storage tank:	1 x 23,780 gal (90,007 L) upright cylindrical steel tank
Freeze protection:	Collector loop filled with 40% glycol, 60% water
Heat exchanger:	3 brazed flat plate type, area 188.4 sq ft + 188.4 sq ft + 220.7 sq ft (18 m^2 + 18 m^2 + 21 m^2)

Description of control strategy

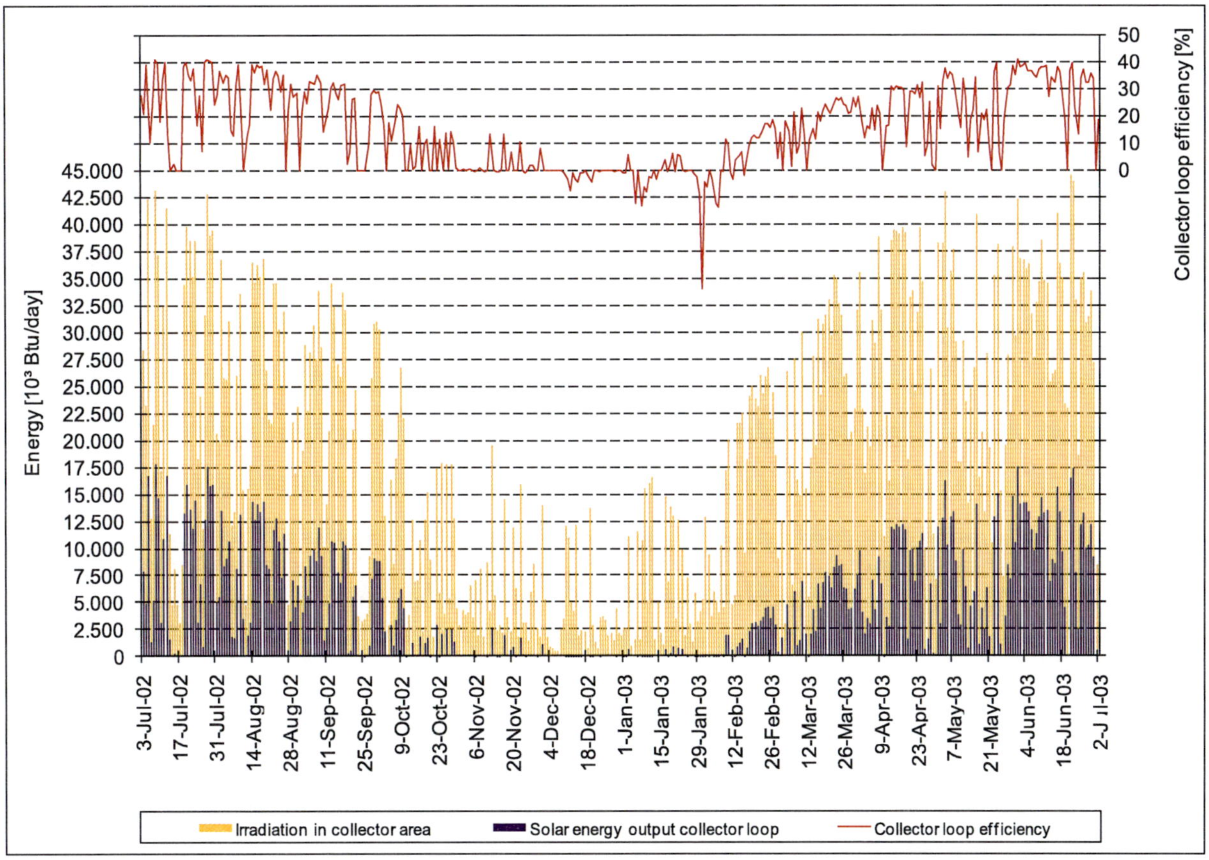

Figure A-83. Irradiation, energy output, and efficiency of collector loop in 2002/2003, daily data solution.

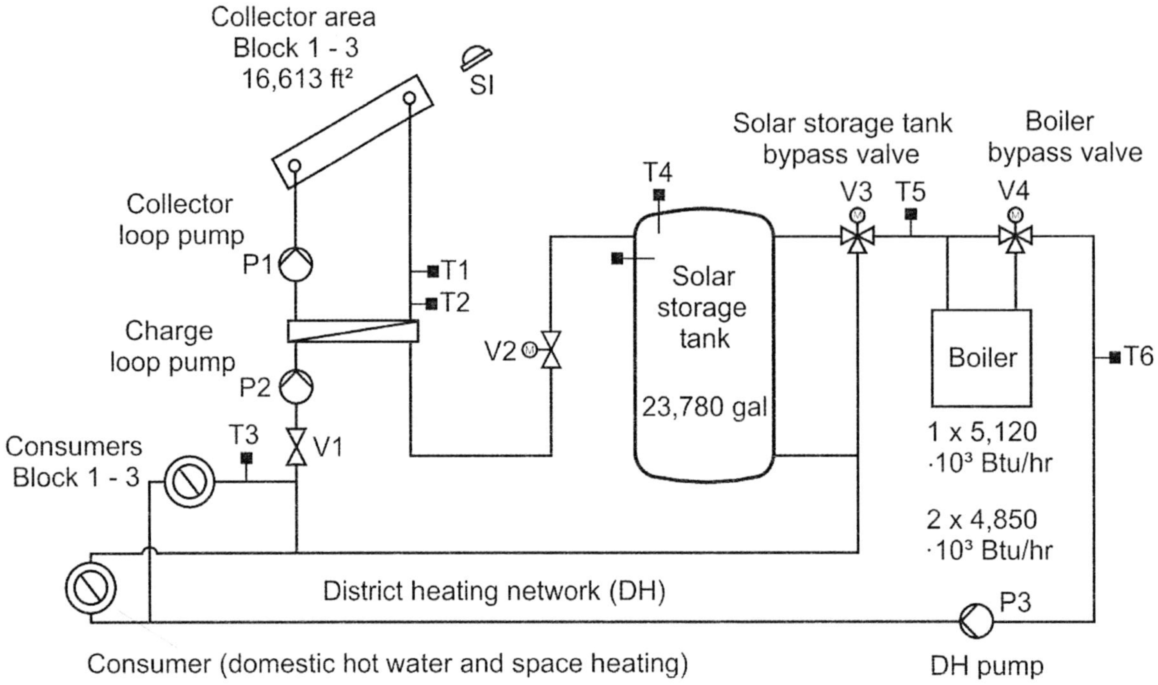

Figure A-84. Highly simplified schematics showing the positioning of the control sensors.

The control strategy described below is carried out here for the controlling of collector loop in block 1. The control strategy for the collector loops in block 2 and 3 works identically. The collector loop is fitted with a temperature sensor (T1) to prevent the underground piping from overheating and boiling. When temperatures (T1) are higher than 221 °F (105.00 °C) the pump control functions are halted and the pump P1 generally stops. When temperatures at sensor T1 fall below 221 °F (105.00 °C), and 8 hours elapse, the halted control functions clear.

The tank is fitted with a temperature sensor (T4) to prevent the solar storage tank from overheating and boiling. When by temperatures (T4) exceed 203 °F (95.00 °C), the pump P1 control functions and the valve V2 are halted and pump P1 generally stops. When temperatures at sensor T4 fall below 203 °F (95.00 °C) and a time lapse to 06: 00 h the next day passes, the halted control functions clear.

The collector loop pump (P1) and valve V2 is controlled by the measured irradiation (SI). When irradiation SI is more than 63.4 Btu/(hr * sq ft), the collector loop pump P1 runs and opens valve V2; when irradiation is less than 57.1 Btu/(hr * sq ft) and a time lapse of 30 min passes, pump P1 stops and valve V2 shuts.

The charge loop pump P2 and valve V1 is controlled by the difference in temperature between the temperature in the collector loop advance (T2) and the DH return temperature (T3). A difference in temperature (T2 – T3) of more than 12.6 °F (7 °C) makes the pump P2 run and opens valve V2; a difference less than 9.9 °F (5.5 °C) stops the pump P2 and shuts valve V2.

To prevent collector loop heat exchanger from freezing, pump P2 runs and valve V1 opens when temperature at sensor T1 is lower than 41 °F (5 °C). When the temperature at T1 is higher than 41 °F (5 °C), pump P2 stops and valve 1 shuts, unless these conditions are overridden by other conditions (see above).

Economics

The solar system costs include costs for solar collectors, piping, solar storage tanks and controls, and do not include costs for the district heating network, boiler, or the heating central building (Table A-27).

Table A-26. Summary of control activities and control conditions.

Control activity	Control Conditions
Collector loop – Pump P1	Clearance: T1 < 221 °F (105 °C) and 8 hours time lapse, T4 < 203 °F (95 °C) and a time lapse to 06: 00 h next day Bolted: T1 > 221 °F (105 °C), T4 > 203 °F (95 °C) On: SI > 63.4 Btu/(hr * sq ft), Off: SI < 57.1 Btu/(hr * sq ft) and time lapse of 30 min (On: SI > 0.20 kWh/[hr * m^2), Off: SI < 0.18 kWh/[hr * m^2] and time lapse of 30 min)
Charge loop Pump P2 Valve V1 Valve V2	 On: T2 – T3 > 12.6 °F (-10 °C), Off: T2 – T3 < 9.9 °F (-12 °C) Open: T2 – T3 > 12.6 °F (-11 °C), Shut: T2 – T3 < 9.9 °F (-12°C) Clearance: T4 < 203 °F (95 °C) Bolted T4: > 203 °F (95 °C) and time lapse to 06:00 h next day Open: SI > 63.4 Btu/(hr * sq ft), Shut: SI < 57.1 Btu/(hr * sq ft) and time lapse of 30 min (Open: SI > 0.20 kWh/[hr * m^2] , Shut: SI < 0.18 kWh/[hr * m^2] and time lapse of 30 min)
Freeze protection collector loop heat exchanger Pump P2 Valve V1	 On: T1 < 41 °F (5 °C); Clearance: T1 > 41 °F (5 °C) Open: T1 < 41 °F (5 °C), Clearance: T1 > 41 °F (5 °C)

Table A-27. Economics.

Costs solar system Costs solar system, including statics Costs planning solar system. Costs solar system, statics and planning Costs solar system, statics and planning including 15% tax	 633,333 € 10,859 € 693,729 € **797,788 €**	
Annual costs for loan Living period 20 years, 6% rate, → annuity: 8.72%	69,570 €	
	Per year	Sum total in 1 year
Solar energy output Planned solar energy output from solar storage tank Estimated solar energy output from storage tanks	 2184*10^6 Btu 1607*10^6 Btu	 2184*10^6 Btu 1607*10^6 Btu
Relation measured solar energy output/ planned energy output		73.6%
Savings of gas und CO_2 calculated with measured solar energy from solar storage tank with following assumptions: Boiler efficiency: 90%; Energy of natural gas: 27.4*10^3 Btu/cu yd_{Gas} (0.10 kWh/L_{Gas}) Emission factor: 0.129 lbs $CO_2/10^3$ Btu_{Gas} (0.06 kg CO_2/2.9 MWh_{Gas}) Saving amount of natural gas Avoidable amount of CO_2		 65,200 cu yd 115 (short) ton
Costs of solar energy from solar storage tanks with 8.72% annuity, including solar system, statics, planning and tax Costs and planned solar energy output Costs and measured solar energy output, without Costs for maintenance and repairs	 0.0319 €/10^3 Btu 0.0433 €/10^3 Btu	

District heating network

Table A-28. Expected data during planning.

Number of apartments supplied by DH:	1,360
Number of people supplied by DH:	2,800
Maximum energy demand for space heating supplied from DH:	19,450*10^6 Btu/yr (56,950*10^6 kWh/yr)
Maximum energy demand for domestic hot water supplied from DH:	7,510*10^6 Btu/yr (21,989*10^6 kWh/yr)
Maximum energy losses in DH underground piping:	2,390*10^6 Btu/yr (6,998*10^6 kWh/yr)
Total energy demand from heating central:	29,350*10^6 Btu/yr (85,937*10^6 kWh/yr)
Total energy supplied from solar system to DH:	2,180*10^6 Btu/yr (6,383*10^6 kWh/yr)
Solar fraction of total energy demand DH:	7.4%
Solar system efficiency (energy output from solar storage tank/irradiation energy):	34.6%
Max. DH advance temperature (depends on ambient temperature):	158.0–181.4 °F (14–83 °C)
Max. DH return temperature	113.0–116.6 °F (-11–47 °C)

Table A-29. Measured data from 2002/2003.

	2002/2003
Energy demand of DH	(10^6 Btu/yr) Not measured
Solar energy output solar storage tank	1,607 10^6 Btu/yr (estimated)
Solar fraction of total energy demand DH	Not measured
DH advance temperature, yearly average	171.1 °F (77 °C)
DH return temperature, yearly average	121.8 °F (50 °C)

Table A-29 lists that the solar energy output from the solar storage tank to DH reached 1607*10^6 Btu/yr and failed 2180*10^6 Btu/yr (4.7 MWh/yr and failed 6.4 MWh/yr) as planned. Investigations found out, that the performance of the flat plate collectors did not match the ensured performance by 22 to 25%. The producer of the collectors accepted a penalty of 14% of the collector area costs.

DH advance temperature (171.4 °F [77 °C] yearly average) is in the range as planned, but not the DH return temperature. The DH return temperature (121.8 °F [50 °C] yearly average) failed to match the planned temperature of not more than maximal 113.0–116.6 °F (45–47 °C).

Figure A-85 shows the DH advance temperature, which is nearly constant over the year. The DH return temperature rises in the summer because of the disappearance of energy needed for space heating in the buildings. The space heating produces a lower return temperature in the transfer stations to the DH than the domestic hot water supply. To prevent Legionella bacteria, hot water must be at least 140 °F (60 °C) at the hot water storage tank outlet, and the hot water circulation return from the building has to have not less than 131 °F (55 °C).

Experiences/Lessons learned

Energy use reduction

The only period measured data was gained was in 2002/2003, when we measured an energy output from the solar storage tanks of 1607*10^6 Btu/yr. Assuming a boiler efficiency of 90%, an energy content of natural gas of 27.4*10^3 Btu/cu yd$_{Gas}$ there is a saving of natural gas of 65,200 cu yd in this year.

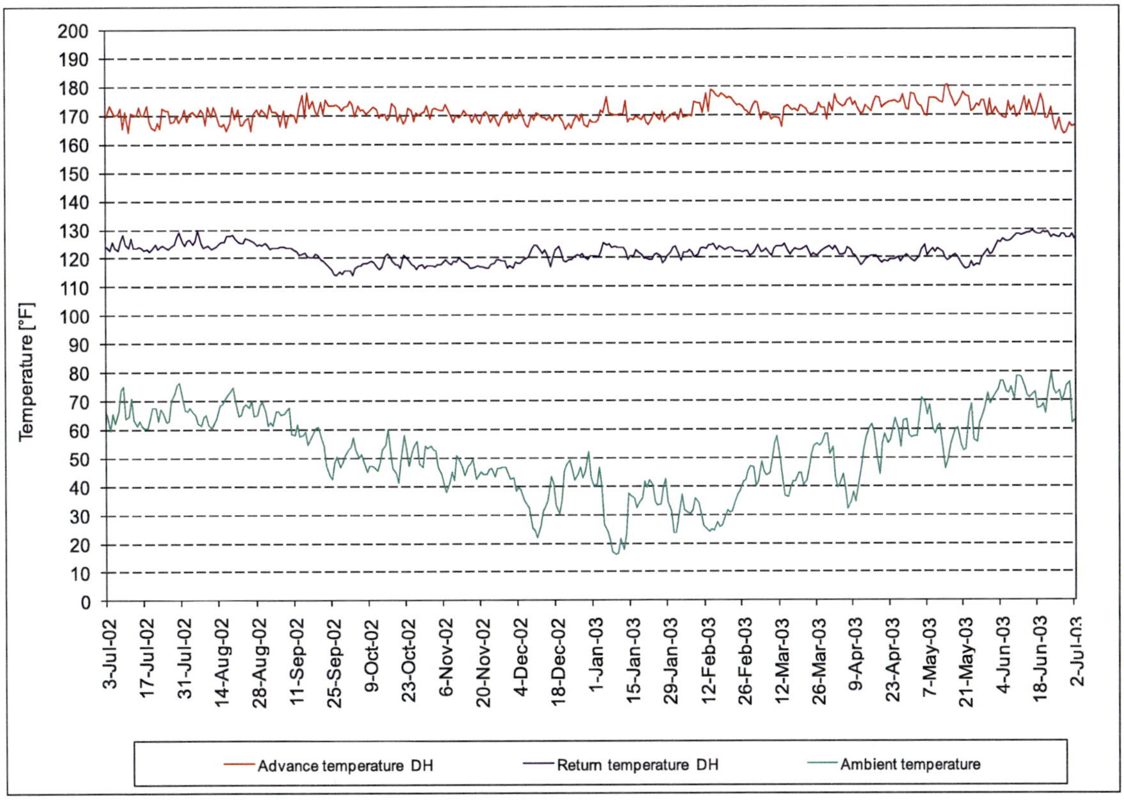

Figure A-85. Volume flow DH, volume flow through solar storage tank, and temperatures DH.

Lessons learned

From the data we measured and the monitoring of the solar system we learned:

- The principle of this solar system in combination with the district heating network worked without severe problems. In particular the here executed three-pipe-arrangement (two-pipes DH, one pipe advance charge loop) operates satisfactorily. We recommend to use three-pipe-systems only if this system and its hydraulic challenge it fully understood by the designer.
- The performance of the flat plate solar collector failed the ensured performance by 22 to 25%. Because the solar energy output from the solar storage tank was guaranteed by the builder of the solar system, a penalty was paid.
- Massive leakages in the collector areas brought the solar system to a standstill. Latest information from the owner was that a complete exchange of the collector areas is planned.

General data

Address of the project

Solar water heating connected to a district heating network
Residential area "Burgholzhof"
Stuttgart, Germany

Date of report

Measured data period: 3rd July 2002 to 2nd July 2003
Date of report: June 2010

Acknowledgements

Promoting department

Bundesministerium für Umwelt, Naturschutz und Reaktorsicherheit (BMU)
(Federal department of environment, nature conservation and nuclear reactor safety)
Alexanderstr. 3
10178 Berlin, Germany

Operating company (owner)

EnBW Energie-Vertriebsgesellschaft mBH (former Neckarwerke Stuttgart AG)
Postfach 101213
70011 Stuttgart, Germany

Planning company

Steinbeis-Transferzentrum
Heßbrühlstr. 15
70565 Stuttgart, Germany

Eproplan GmbH
Schöttlestraße 34a
70565 Stuttgart, Germany

Mounting company

Buderus Heiztechnik Esslingen (Collector area)
Rud. Otto Meyer Stuttgart (Piping)
Möhrlin GmbH Stuttgart (Heating central)

Measuring company

ZfS-Rationelle Energietechnik GmbH
Verbindungsstr. 19
40723 Hilden, Germany

References

Peuser, F. A.; Mies, M.; Rehrmann, U. Abschlussbericht für das Projekt Wohnsiedlung Burgholzhof Stuttgart. June 2007,
http: //www.zfs-energietechnik.de/main.php?RND=27410&SESS=&LANG=de&ID=169
(more and detailed information about the solar system in Stuttgart are outlined in this report)

Peuser, Felix A.; Remmers, Karl-Heinz; Schnauss, Martin. Solar Thermal Systems. Successful Planning and Construction. Solarpraxis AG Berlin Germany in association with James & James London UK, 2002. ISBN: 3-934595-24-3

VDI 6002, part 1, September 2004 (technical guideline). Solar heating for domestic water. General principles, system technology and use in residential buildings. Distributer: Beuth Verlag, 10722 Berlin, Germany

DVGW W551, April 2004 (technical guideline). Trinkwassererwärmungs- und Leitungsanlagen; Technische Maßnahmen zur Verminderung des Legionellenwachstums (Hot water systems, technical arrangements to reduce developing of legionella bacteria), no English translation available. Distributer: Wirtschafts- und Verlagsgesellschaft Gas und Wasser mbH. Josef-Wimmer-Str. 1 - 3, 53123 Bonn, Germany.

FPC – 17. Solar Thermal District Energy System at Saint Paul, MN

Title: Solar Thermal District Energy System at Saint Paul, MN

Location: Saint Paul, MN

Photo of the installation

Figure A-86. Aerial view of solar district heating installation at Saint Paul, MN.

Project summary

In Minnesota, solar energy is shifting from a novel idea to a growing energy option for utilities, businesses, and residents across the state. In 2010, District Energy St. Paul launched the largest solar project in the Midwest on the roof of the Saint Paul RiverCentre convention center. The 144 panel system became operational in March 2011 and is estimated to produce 4 MBtu/hr (1.2 MW) peak thermal energy. This large-scale, high-performance showcase will lower the carbon footprint for the system and its customers by an estimated 900,000 lb (408,233 kg) per year. Although this type of solar thermal integration is commonplace in Europe, this installation is the first solar district energy system in the United States. This groundbreaking installation fulfills the company's historic solar aspirations and moves the utility one step closer to becoming a 100% renewable fuel-based production. Table A-30 lists general system information.

Table A-30. General information for the Solar Thermal District Energy System at Saint Paul, MN.

Parameter	**Measure/Detail**
Location	Saint Paul, MN, USA
Latitude	44.95°N
Longitude	93.1°W
Total panel area	21,034 sq ft (1,956 m^2)
Solar irradiation	532,879 Btu/sq ft (1680 kWh/m^2)
Application	Solar hot water serves the host site first, including domestic hot water and space heating. Excess heat produces is exported into the district heating system.
Year of operation start	2011 (9 months of operation and collected data)

Project description

District Energy St. Paul is the largest hot water district heating system in North America and is recognized throughout the United States as a community model and a leader in renewable energy. (Figure A-87 shows the locations of district energy systems in the United States.) It currently provides heating service to more than 191 buildings and 300 single-family homes, representing over 31.7 million sq ft (2.9 million m^2) of building space, or 80% of Saint Paul's central business district and adjacent areas.

A district energy system derives energy from one or more sources and uses a system of underground pipes to aggregate and serve the thermal energy needs of proximate users. This method of networking users and aggregating their thermal energy load offers advantages over single source generation and consumption, including efficiencies and fuel flexibility. Systems can use steam, hot water, or chilled water to distribute thermal energy via piping networks to connected buildings.

Today, the system serves twice the square footage of building space with the same amount of input energy as when the system transitioned from steam to hot water in 1983, when District Energy St. Paul was created to serve the growing need to cool downtown. In 2003, the company integrated renewable energy sources through the development of the wood-fired combined heat and power plant (CHP).

Over the years, District Energy has achieved much of this success through a strong public-private partnership with the city of Saint Paul. In 2010, this partnership presented District Energy with the opportunity to add solar energy to its growing list of integrated energy solutions. The Minneapolis-Saint Paul Solar America Communities, made possible through the United States Department of Energy, had the specific goal of transforming the market to enable large-scale solar energy investment within the cities by 2015. In an effort to reach this goal, a project was developed to showcase large-scale, solar thermal energy and solar integration to serve multiple users, i.e., a "solar thermal district."

Objectives

This project was undertaken to demonstrate the value of using large-scale solar thermal energy, integrated into a distributed heating system. The Solar Thermal District Energy project was designed to address common market barriers to large scale solar thermal commercialization:

- *Low effective market penetration off solar thermal systems.* This project will serve as a high visibility reminder of the utility and economic efficiency of solar thermal energy. Although there have been local programs to encourage solar thermal energy in small, independent residential applications, this region has not been exposed to a large-scale practical application or the potential to connect and/or store the thermal energy created.
- *Lack of viable and tested business models for using solar thermal energy to connect energy needs in multiple buildings.* Both district energy and solar thermal energy are widely integrated in Europe because of their high efficiency and economic benefits. Despite the proven technical feasibility, there were no current examples of US solar thermal installation being used to service multiple buildings. This project provides exposure to planners, developers, and energy consultants to the viability of these types of installations.
- *Lack of funding for implementing large scale solar projects.* Given the financial constraints for today's city managers and planners, it is important to work with private parties to identify ways to continue to expand the solar market. Utilities play an important role in energy planning and delivery and are ideal partners, particularly when there is a financial commitment to expand solar energy projects within their portfolios.

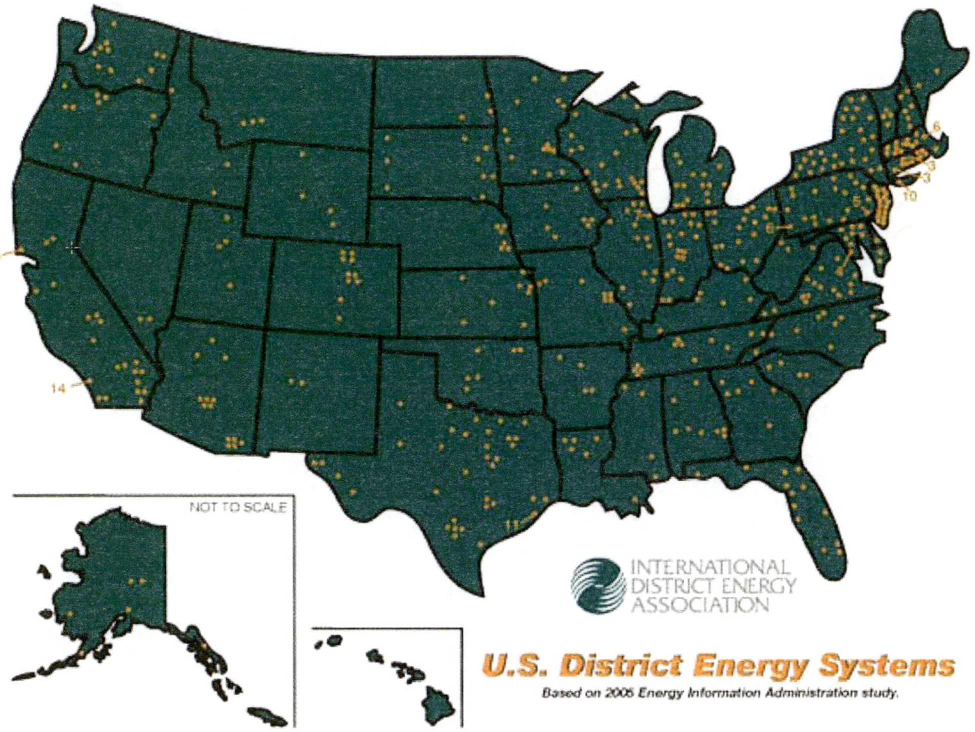

Figure A-87. US district energy systems.

Market transformation

District energy systems, which can be found in many urban areas and college and university campuses, are almost always located in the densest, most job-intensive areas of the urban core. They sell and buy energy, make significant infrastructural investment in energy delivery systems, and make large-scale delivery of energy services within a retail market. Solar thermal district energy systems can therefore play a significant role as solar energy investors by:

- capturing the long-term value of investment in solar thermal energy infrastructure,
- providing a vehicle for customer-focused solar investment, and
- diversifying the fuel source for some type of energy products. Few, if any, US district energy systems have entered into the solar energy market, either to secure energy supply or to diversify energy services.

In addition to supplementing existing district systems, the creation of solar thermal district systems may offer superior replicability. These systems could use the existing model of production and distribution of a district system, with solar as the primary generation source. A solar installation on one or more buildings could be connected to proximate buildings using hot water generated by the solar installation to deliver thermal energy to the buildings. These networks can be as small as a single building or implemented city-wide with multiple solar installments contributing to the needs of the system. A solar thermal district system might be implemented in any city with adjacent buildings that use hot water systems, have adequate roof support and roof or ground space, and have a need for heating and/or hot water production.

By demonstrating that district energy systems allow for viable solar investment opportunities, this project (Figure A-88) could be replicated in cities across the nation. Promoting district energy systems that incorporate solar energy opens the door to large-scale urban investment in solar infrastructure, in locations where energy use and job density are greatest.

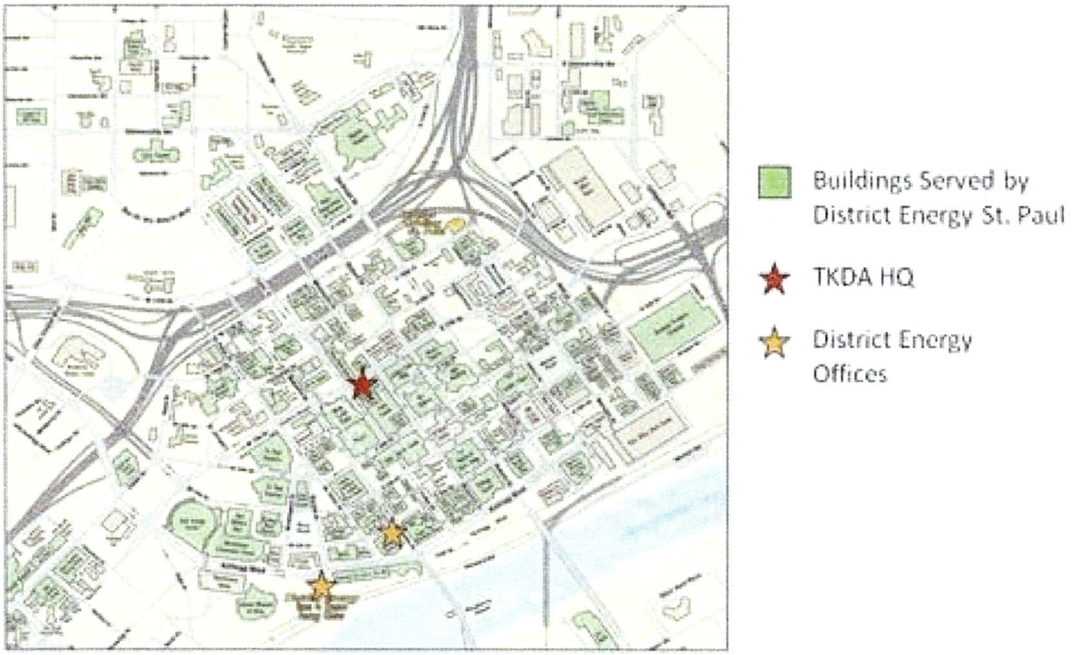

Figure A-88. District Energy St. Paul service area map.

Key project characteristics

District Energy St. Paul has been researching solar thermal integration for almost a decade, adding solar collectors to its plant for testing in 2008. This led to preliminary analysis in 2009 to assess other integrated systems, high-performance thermal collector technology, and combined thermal/photovoltaic (PV) technology.

In 2009, District Energy completed the market assessment of commercially available thermal and thermal/PV panels. Due to project and licensing requirements, and emphasis was placed on products certified through the Solar Rating and Certification Corporation (SRCC). Verifying the performance rating to the conditions of operation was a key criterion in the design process. A formal Request for Qualifications (RFQ) for the District Energy's distribution was prepared; this process was completed in March 2009. The RFQ information was then used to complete a preliminary system design, which was necessary to formally bid the collectors.

Once preliminary engineering exceeded minimum performance thresholds, District Energy pursued the grant process outlined in the project background. Final system design was initiated in January 2010 coupling internal engineering resources with the mechanical, structural, electrical, architectural and project management expertise of Tolz, King, Duvall, Anderson, and Associates (TKDA).

The formal bidding process began in May 2010. Panel performance (and related energy production) is the most crucial aspect of project feasibility although the panels account for only 16% of the project budget. Other very important aspects for project feasibility include characteristics of the building targeted for installation. Since construction was only recently completed on this solar thermal district energy system, the opportunities to learn from the project are still being defined.

Site-specific characteristics

Site characteristics will always play a major role in system design. At the selected site (Figure A-89), the building roof material was approaching the end of its useful life, making it necessary to replace the roof before starting construction of the solar project. The structural engineering review of the building determined that the panels could not be directly placed on the roof until a support structure was installed to transfer the additional weight to the building columns.

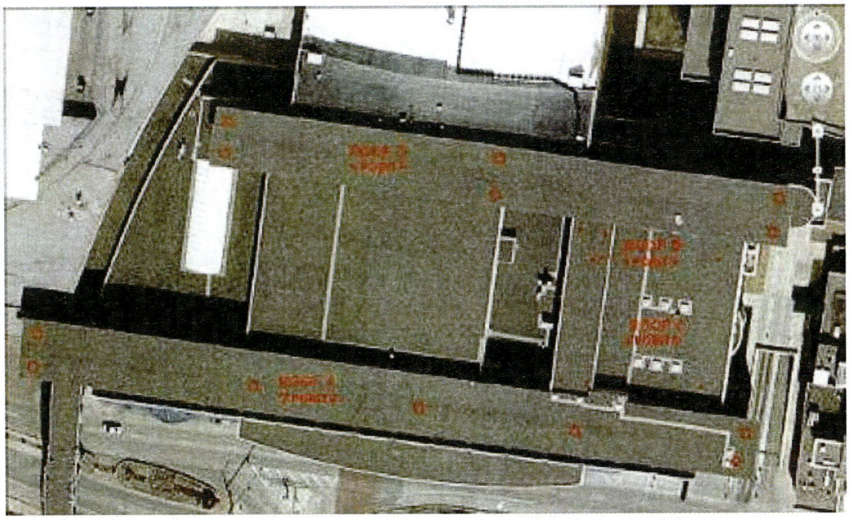

Figure A-89. Saint Paul RiverCentre pre-installation aerial view.

A key engineering design component of a thermal solar installation is that, once the solar collectors are filled with heat-transfer fluid, there must be somewhere to put the energy to avoid damaging system components. (The sun cannot be turned off for system maintenance!) This makes the mechanical configuration of the building and the fate of the energy very important to project feasibility. If the sustained load is less than solar heat produced a storage tank (or, in this case, a district heating distribution system) must be used as a heat sink to optimize energy production.

This project is unique in a number of technological ways:

1. It is integrated with district heating system.
2. It illustrates industry-leading solar collector performance.
3. It is the largest solar thermal installation in the Midwest.
4. It is both a high-temperature and high-pressure application.

Most solar thermal projects depend on the ability to store solar energy in large tanks for use at a later time. This project was designed without a conventional storage system. Instead, the excess energy that is not immediately needed by the building for space heating or hot water will be pumped into the district heating network for use by other outside customers on the district heating system.

The energy transfer fluid in the solar collector field is a 50% solution of propylene glycol and water. Minnesota winters can reach temperatures below –20 °F (–29 °C) making strong freeze protection essential. The system uses three heat exchangers to transfer the thermal energy from the glycol loop to where it is needed (Figure A-90). Domestic hot water is prioritized depending on its own setpoint up to a temperature of 140 °F (60 °C). When the heat is needed in the building, the solar collectors also feed the building heat exchanger depending on the building setpoint.

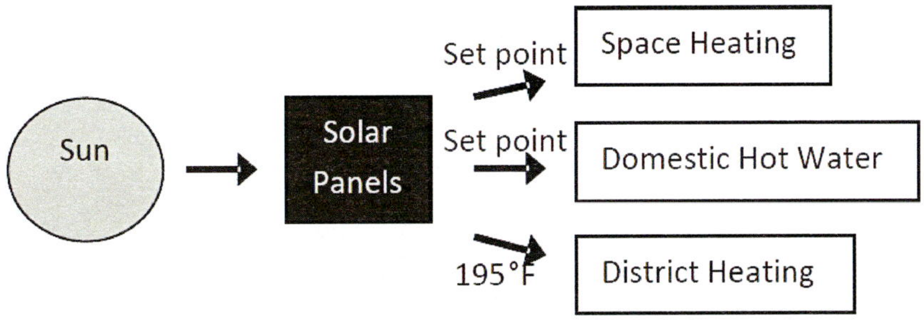

Figure A-90. Solar energy flow.

District heating integration

When both building systems are satisfied, the solar system prepares to export the energy to the district heating network (Figure A-91). The solar collectors heat the transfer fluid up to 200 °F (93 °C) or more for export. The district heating return water is extracted from the return line, heated by solar energy, and pumped into the supply line at that temperature. These high export temperatures are difficult for most flat-plate solar collectors to maintain due to limitations in their efficiency and heat retention. Few have been proven to operate consistently in this temperature range. Given the needs of the system, a specification (Table A-31) was written to solicit bids from solar manufacturers.

Collector Technology

The bid specification was developed as follows. The performance requirements are based on energy production at given ambient outdoor temperatures. The panels must produce high output temperatures and withstand high system pressures and flow rates. Panel design is rated for a maximum operating temperature of 250 °F (121 °C) and shall be 150 psig (1,034 kPa) at maximum operating temperature.

- Stagnation temperature. Panels are designed to withstand stagnation temperature conditions associated with a Saint Paul, MN extreme ambient temperature of 104 °F (40 °C) on a clear day. Panels shall be designed to withstand 150 psi/10 bar (1,034 kPa) internal temperature at this condition.
- Panels weighing 5 lb/sq ft (14.4 kg/m^2) of gross collector area (full) or less are preferred.
- Insulating materials will not outgas or break down at, or under, stagnant temperature.
- Acceptable heat loss shall be 0.7 Btu/hr/°F/sq ft (4 W/°C/m^2) or less.
- Panel starting efficiency shall be a minimum of 0.70 for flat panel collector.

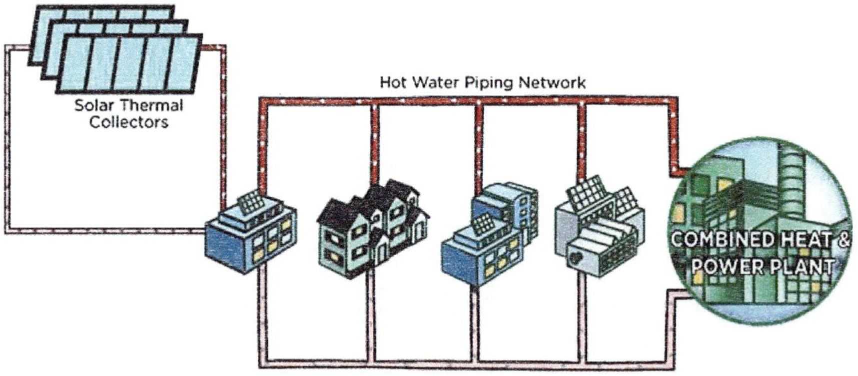

Figure A-91. Solar district energy system.

Table A-31. District heating specifications.

Parameter	**Measure**
Design peak of entire system	986,104,992 Btu/hr (289 MW)
Total volume	105,200 ft (32,065 m) (supply and return)
Design supply temperature (winter)	250 °F (121 °C)
Design return temperature (winter)	160 °F (71 °C)
Design supply temperature (summer)	180 °F (82 °C)
Supply pressure	180 psi (1,241 kPa)
Minimum pressure differential	20 psi (138 kPa)
Reliability rate	99.997%
Customer buildings	191 customer buildings, 298 single family homes

After completing the panel solicitation and review:

- Eight (8) bids were received including six (6) manufacturers located outside the United States.
- One of the US manufacturers failed to meet the technical requirements, certification, and quantity requested.
- The other US manufacturer's product would produce 41% less energy per year that the Arcon HT-SA 28/10 panel; it also appeared that that manufacturer could not supply adequate quantities. (A full installation comprised of these panels would result in a 600 kW peak system vs. the 1 MW expected from the product.)
- Arcon could meet the peak expected annual output using the available roof space; the Arcon HT-SA panel (Figures A-93 and A-94) was found to meet the technical requirements and to satisfy District Energy's requirements for project feasibility (Table A-3).

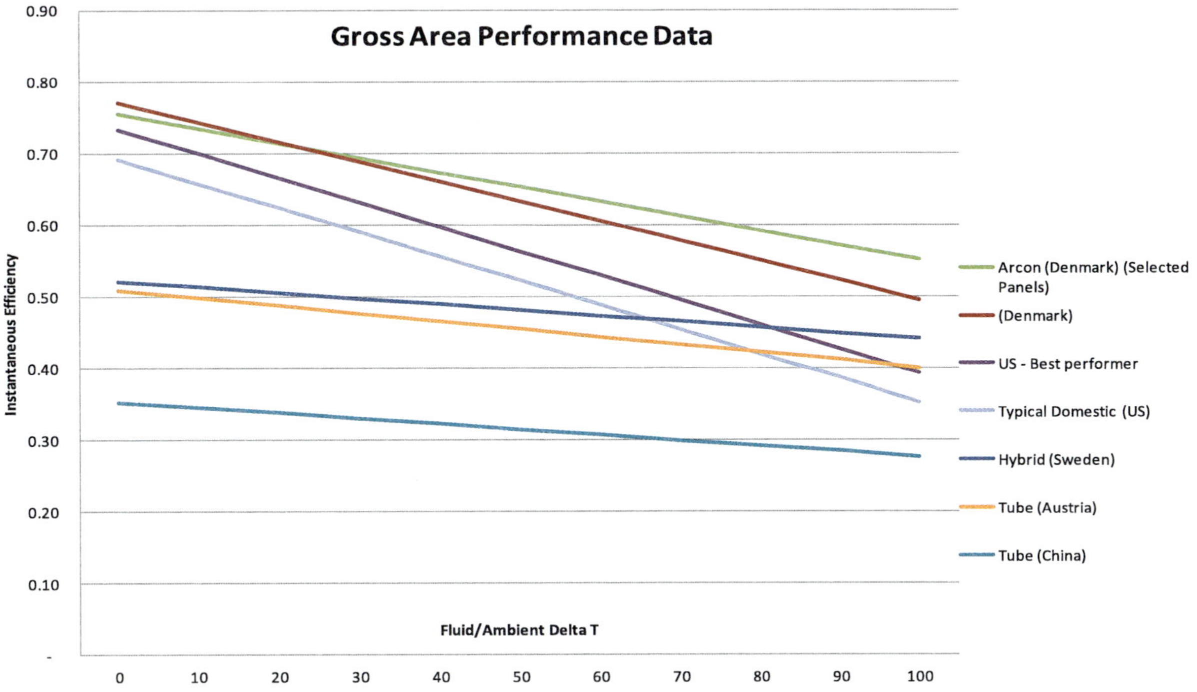

Figure A-92. Gross area performance data for competitive bidders.

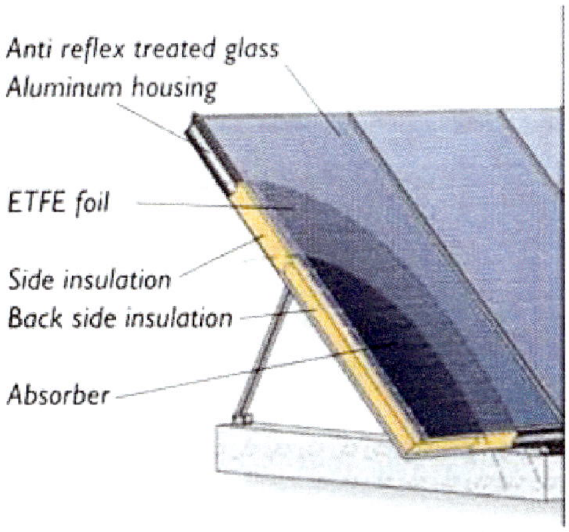

Figure A-93. Arcon HT-SA 28/10.

Figure A-94. Sheehy/Amerect installation of Arcon panel (2010).

Table A-32. Solar specifications.

Parameter	Measure/Detail
Brand of collector	Arcon
Collector specification	Standard flat-plate collector
Angle	45°
Orientation	186° south
η_o	0.817 aperture area/0.754 gross area
a_1	2.205 W/(m^{2}K) aperture area/ 2.2035 W/(m^{2}K) gross area (12.52 Btu/hr-sq ft °F aperture area/ 12.51 Btu/hr-sq ft °F gross area)
a_2	0.077 Btu/hr-sq ft °F (0.0135 W/m^{2}K)
Total panel area	21,034 sq ft (1,956 m^2)
Total roof area	30,720 sq ft (2,857 m^2)
North roof length	420 ft (128 m)
South roof length	540 ft (165 m)
Panel weight dry	550 lb (249 kg)
Panel weight wet	571 lb (259 kg)
Propylene glycol capacity	1200 gal (4,542 L)
Energy through October 2011 (total)	306,562 Btu (1047 MWh)
Energy through October 2011 (export)	167,628 Btu (572.5 MWh)
Design peak capacity	3,412 MBtu/hr (1 MW)
Measured peak capacity	4,094 MBtu/hr (1.2 MW)
Design annual solar yield	4,303 MMBtu/hr (1261 MW/hr
Designed solar fraction (per building)	42.7%
Designed solar fraction (per system)	0.351%
Measured solar fraction (per building)	29.2% on premises
Measured solar fraction (per system)	0.12%

Structural design

TKDA designed a structural "exoskeleton" (Figures A-93 and A-94) to meet local building codes that specify structures' performance under ice, wind, and snow loading. This resulted in the selection of W24 steel beams for the primary structural members. This approach would have resulted in the exoskeleton accounting for over 50% of the project projected project cost. TKDA conducted value engineering focused on reducing the systems weight and cost, and evaluated alternative materials and construction methods. The outcome resulted in the use of primary W16 steel beams, which reduced the structural costs to below 50%. This will play a more significant role in site selection for additional projects. However, this site was still exceedingly beneficial for its high profile location, public ownership, management commitment to sustainability, public exposure through conventions, and its block proximity from the Design Energy plant.

Based on panel selection, District Energy St. Paul worked with its engineering design partners, TKDA and Ramboll, to model expected performance from the Arcon panels. Initial estimates were based on monthly peaks both in building usage (comprised of space heating and DHW) and in solar production. After 9 months of operation, more refined data were available for solar output and solar fraction of the building and the overall District Energy system (Figures A-95 to A-103).

Figure A-104 shows the District Energy St. Paul "Integrated Energy Diagram™," which represents a key component of their approach to system design.

Figure A-95. Steel "Exoskeleton."

Figure A-96. Steel bracing.

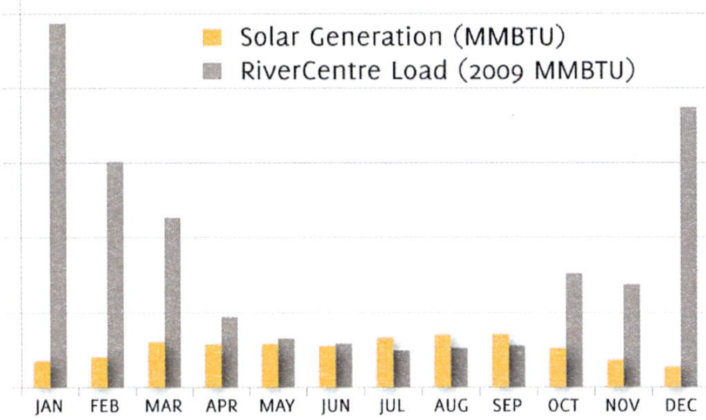

Figure A-97. Energy generation ratio – estimated pre-installation (2010).

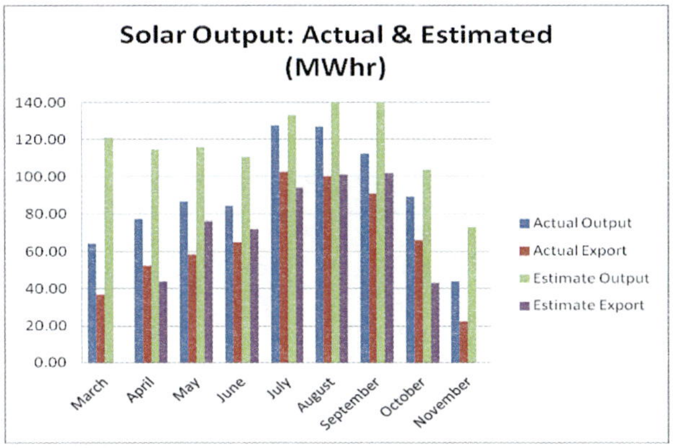

Figure A-98. Actual output (2011) vs. estimates (2010).

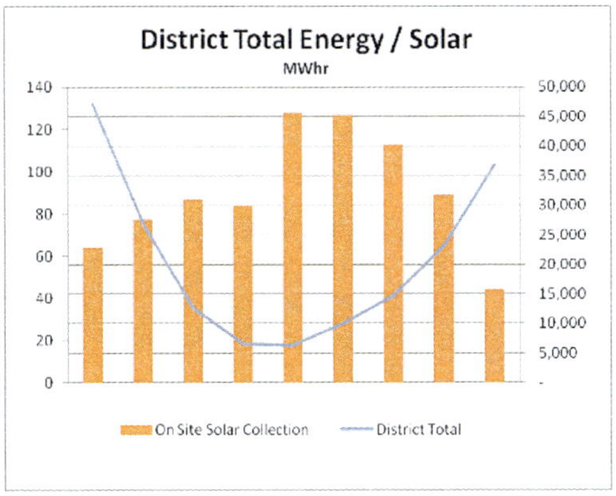

Figure A-99. District energy total solar contribution – RiverCentre installation.

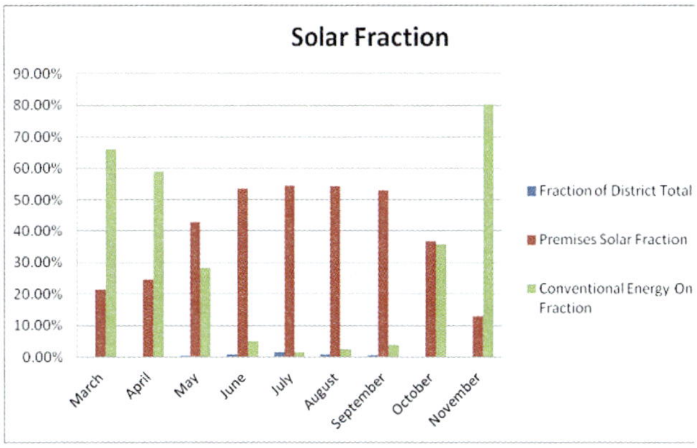

Figure A-100. Solar fraction (2011).

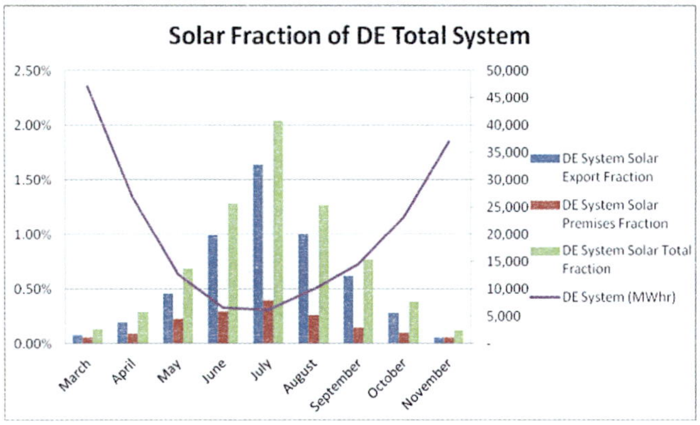

Figure A-101. Solar fraction of DE total system (2011).

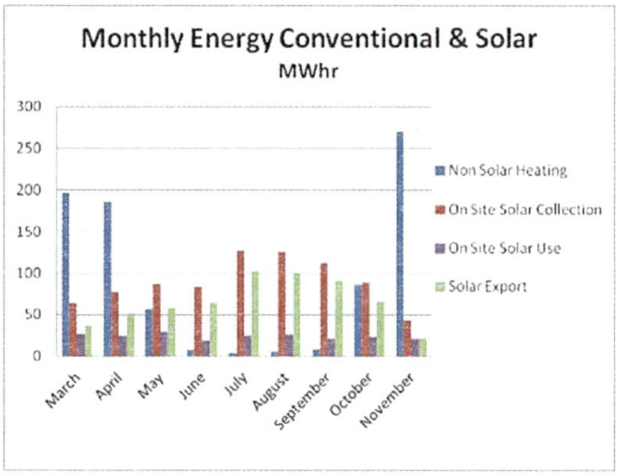

Figure A-102. Monthly energy conventional and solar (2011).

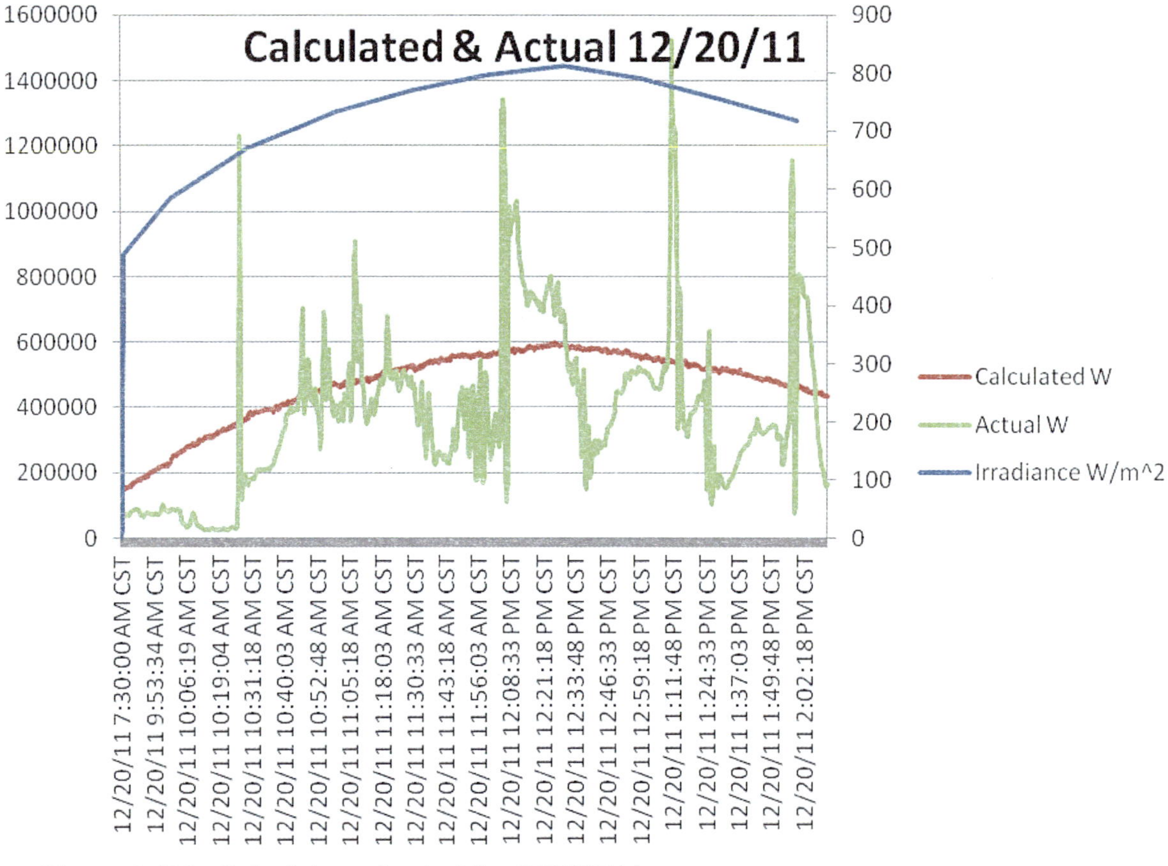

Figure A-103. Calculate and actual for 12/20/2011.

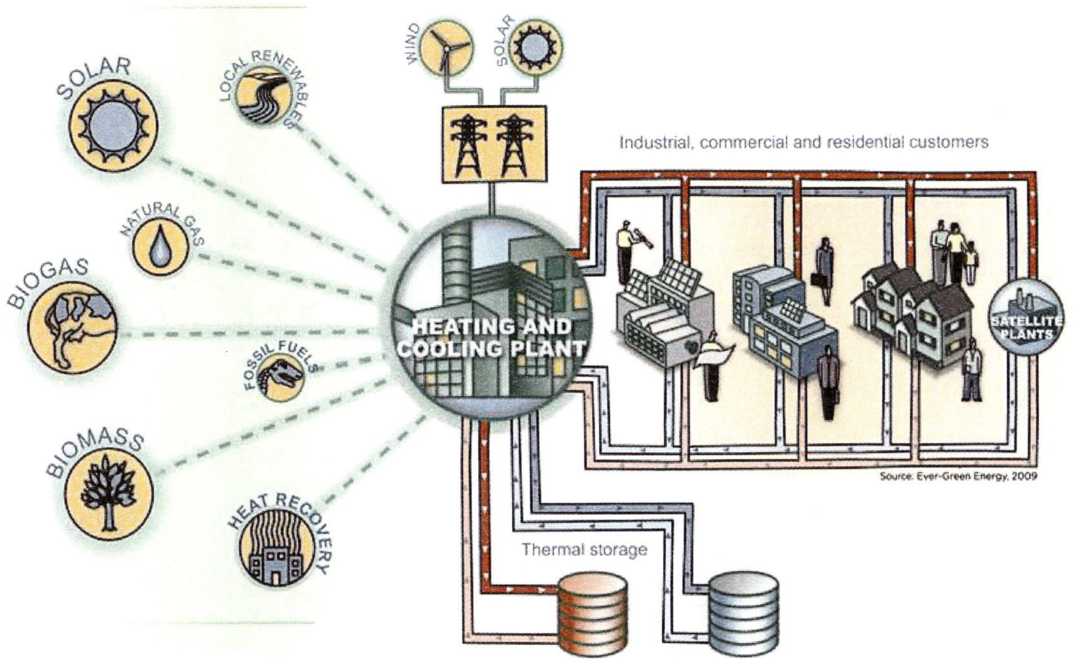

Figure A-104. District Energy St. Paul Integrated Energy Diagram™.

Acknowledgements

The Solar Thermal District Energy Project Saint Paul, Minnesota was made possible through partnerships with the City of Saint Paul, State of Minnesota, the US Department of Energy (USDOE), TKDA, Sheehy Construction, Pioneer Power, and Johnson Controls, Acknowledgement is also owed to The Department of Energy Solar America communities Program; Minnesota Department of Commerce, Division of Energy Resources (DER); Ramboll, Amerect, Inc., Linco Fab, Inc., and Arcon.

Special appreciation is expressed to District Energy St. Paul for granting permission to publish the information in this case study.

For further information, contact:

Nina Axelson
District Energy St. Paul
76 Kellogg Blvd. West
Saint Paul, MN 55102
Nina.axelson@districtenergy.com

Ray Watts
District Energy St. Paul
76 Kellogg Blvd. West
Saint Paul, MN 55102
Ray.watts@districtenergy.com

Charles Lederer
TKDA
444 Cedar Street, Suite 1500
Saint Paul, MN 55101
Charles.Lederer@tkda.com

Lon Fiedler P.E.
TKDA
444 Cedar Street, Suite 1500
Saint Paul, MN 55101
Lon.Fiedler@tkda.com

Evacuated tube collectors

ETC – 1

Title: High-temperature solar hot water system — Building 209, US Environmental Protection Agency (USEPA) Lab, Edison NJ

The Environmental Protection Agency installed three closed-loop systems with evacuated tube solar collectors for heating water at its Edison, NJ, laboratory (Building 209). The systems use a heat exchanger and food-grade propylene glycol solution for freeze protection. The collector area for all three systems is 150 sq ft that use two 80-gal preheat tanks and one 120-gal preheat tank. The technology avoids an estimated 3572 kg of carbon dioxide emissions, 23 kg of sulfur dioxide emissions, and 17 kg of nitrogen oxides emissions.

Location: Edison NJ

Project summary

Measured output: 50,000 Btu/day

Input and output heat transfer fluid temp = 180/60 °F (82/16 °C)

Annual heat collected = 121.7 MBtu/sq ft/yr (383.57 kWh/m^2/yr)

Figure A-105. Evacuated tube collectors installed on the EPA lab in New Jersey.

Economics

Total cost: $26,000

Payback period: 15 years

ETC – 2

High-temperature solar hot water system - Social Security Admin., Philadelphia PA

In 2004, the Mid-Atlantic Social Security Center in Philadelphia installed a solar hot water heating system that pre-heats domestic hot water before it reaches the boiler. The 576 sq ft (53.57 m^2) system includes insulated, evacuated tube collectors arrayed into two roof panels that provide 124,000 Btu (363 MWh) of heating for 1100 gal (4.1 kL) of water per day. The system will save $5000 per year, for a 15-year payback, and a reduction equivalent to 42,000 barrels (6,678 kL) of oil and 37,000 cu ft (1,036 m^3) of natural gas. The Center is the first Federal building in the Philadelphia region to use solar energy for heating.

Location: Philadelphia PA

Economics

Total cost: $58,000

Payback period: 15 years

Project summary

Measured output: 143 MBtu/year
Input and output heat transfer fluid temp = 140/60 °F (60/16 °C)
Storage tank size = n/A

Figure A-106. The solar water heating system installed on the SSA building is a re-circulation loop system using 360 evacuated tube collectors that cover 54 m^2 gross area.

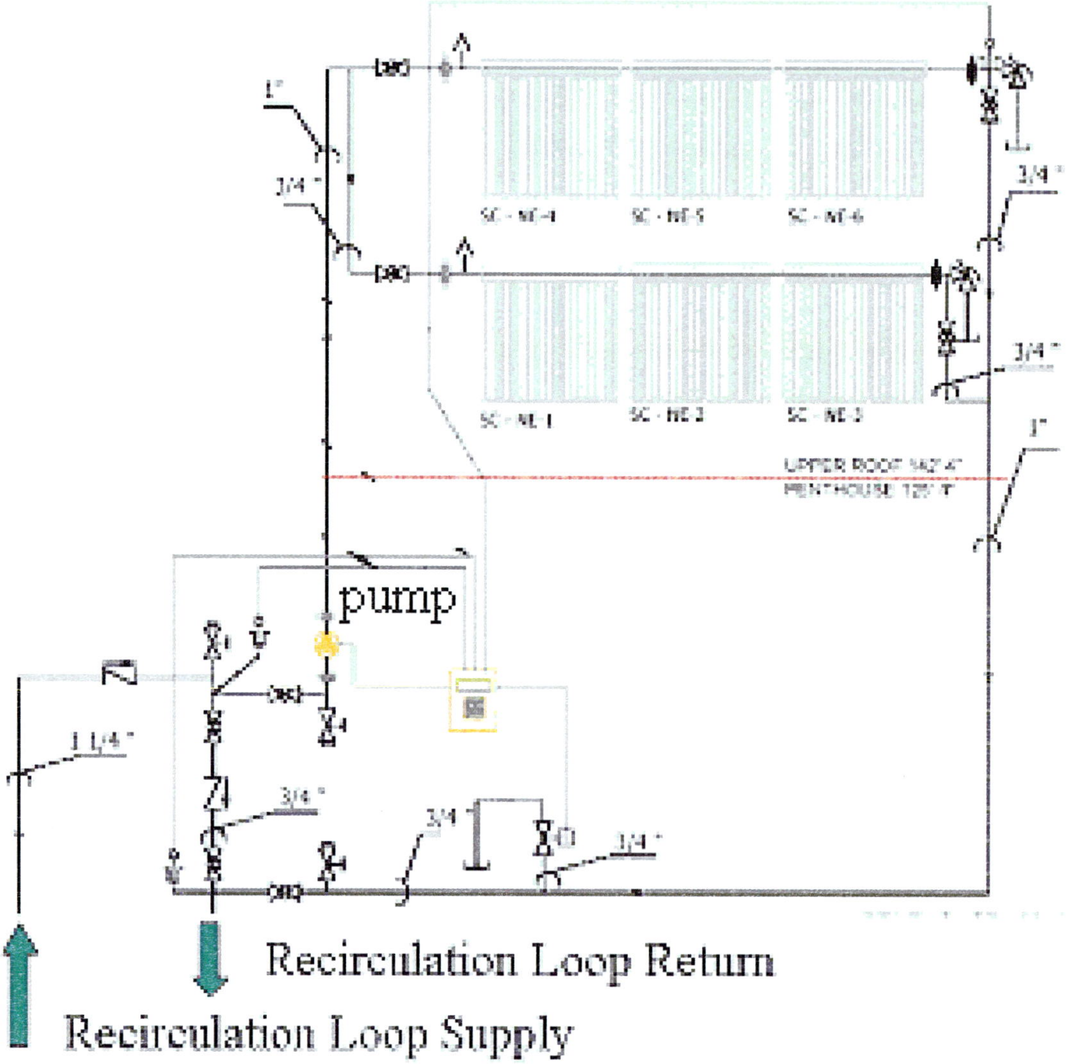

Figure A-107. Schematic diagram of solar water heating system applied to commercial building recirculation loop. This Fig. shows one of two identical systems on the Mid Atlantic Social Security Center.

ETC – 3

Title: Trade Park, Housing Estate Ritter, Karlsbad, Germany

Location: Karlsbad, Germany

Project summary

This project installed a First Paradigma DH network (an innovative DH system) for 12 new single-family passive houses:

- Gross collector area: 667 sq ft (62 m^2)
- Yearly global irradiation: 3.42 therms/sq ft (1,078 kWh/m^2) and annum

Figure A-108. Paradigma DH network.

Site

Ettlinger Str.30-64
76307 Karlsbad
Germany

Project description

Installation date:	September 2001
Design continual power:	113 MBH
Design Solar peak power:	205 MBH
Solar system yield:	112.6 MBtu (1126 therms) per annum
Specific system yield:	1.688 therms/sq ft (532 kWh/m^2)
Max. electric energy:	5 therms (14.64 MWhr) per annum
Solar share:	40%
System efficiency:	0.49 (system energy yield / radiation input).

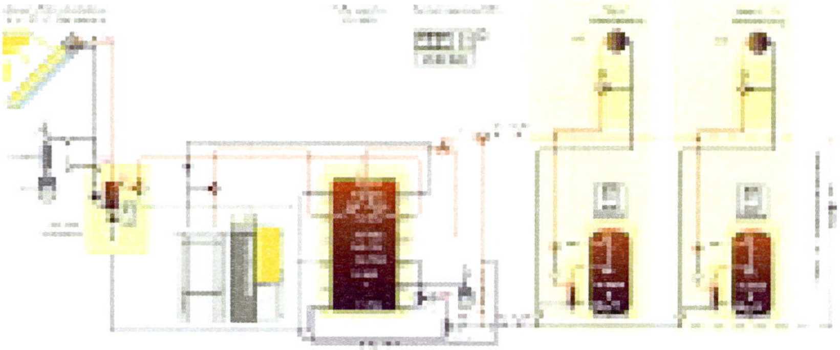

Figure A-109. Schematic view of First Paradigma DH network.

System details

DH system
Connection SWH to DH
Solar system: Paradigma XL-Solar AquaSystem
Collectors: High performance vacuum tube collectors with CPC mirror (CPC-VTC technology)
Slope 45 degree
Storage volume: 1585 gal/sq ft (66.7 kL/m^2)
Freeze protection: Active with low temperature heat from the system, passive (glycol) till 2005, without driver pump via buffer tank (see references passive and minimized).

Control strategy

AquaSystem:

- Water as heat transfer medium, active freeze protection
- Permanent automatic function control and failure diagnostic
- High (target) temperature controller (on-off controlling with the bucket principle)

Standard features:

- Two-tank-systems
- frost protection pump
- hydraulic separation (with solar counter-flow heat exchanger)
- one tank temperature solar switch
- outlet steam blockade
- solar hot-start.

Economics

Local Energy Cost $0.0286/therm (0.01 €/MWhr)
System first cost $52,500 (US $)
Operation Costs $75 (US $) per annum
Savings first year $3218 (US $) per annum
Specific saving 1st year $4.71/sq ft (35.50 €/m^2)

District heating net

Solar water temperatures
Annual energy demand
Heat transfer stations
Control strategy (hts)
140 –194 °F / 77–140 °F (4–90 °C / 25–60 °C) (hot outlet / cold inlet temperatures)
2696 therms (7,893 MWhr) per annum plate heat exchanger target temperature.

Experiences

Overall, the system ran well and provided very comfortable conditions for about 9 years.
The system provided new kind of generation of DH.
The importance of the stratification devices cannot be understated; the system was one of the first water pilot systems.

ETC – 4

Title: Festo, Esslingen, Germany

Location: Esslingen, Germany
Kastellstraße 12-14
73734 Esslingen
Germany
48°43'16.41"N; 9°18'25.02"E

Project summary

This project retrofit a plant to provide solar cooling in the summer and heating in the winter The plant is the world's largest CPC vacuum tube collector system:

- Gross collector area: 14,310 sq ft
- yearly global irradiation: 3.44 therms/sq ft (1,084 kWh/m^2) and annum.

Figure A-110. Solar cooling plant at Festo, Esslingen, Germany.

Project description

Installation date: October 2007
Design continual power: 2220 MBH
Design Solar peak power:4098 MBH
Solar system yield: 17,743 therms (51,952 MWhr) per annum
Specific system yield: 1.24 therms/sq ft (391 kWh/m^2) and ann.
Max. electric energy: 85 therms (249 MWhr) per annum
Solar share 15%
System efficiency: 0.36 (system energy yield / radiation input).

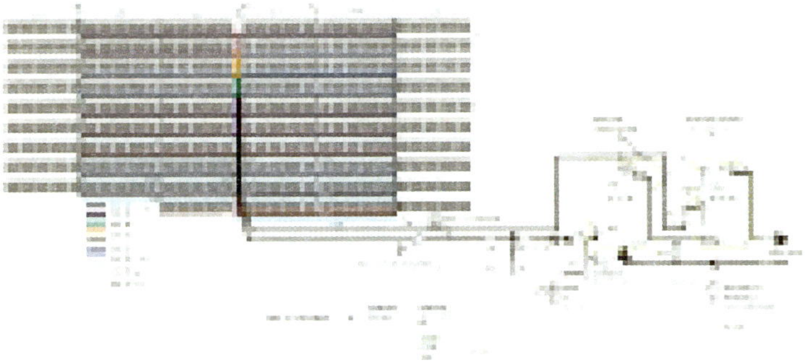

Figure A-111. Schematic view of solar cooling plant at Festo, Esslingen, Germany.

*System details**

DH system: Paradigma XL-Solar AquaSystem
Connection SWH to DH: DH network via buffer tank
Solar system:
Collectors: High performance vacuum tube collectors with CPC mirror (CPC-VTC technology)
Slope: 30 degrees
Storage volume: 4491 gal (16 m^3)
Freeze protection: Active with low temperature heat from the system.

Control strategy

AquaSystem:

- Water as heat transfer medium
- Active freeze protection
- Permanent automatic function control and failure diagnostic
- High (target) temperature controller (on-off controlling with the bucket principle)

Standard features:

- Two-tank-systems
- Frost protection pump
- Hydraulic separation (with solar counter-flow heat exchanger)
- One tank temperature solar switch
- Outlet steam blockade
- Solar hot-start.

Economics

Local Energy Cost $0.022/therm (0.01 €/MWhr)
System first cost $825,000 US
Operation Costs $750 US per annum
Savings first year $39,000 US per annum
Specific saving 1st year $2.67/sq ft (20 €/$m^2$)

District heating net

Solar water temperatures
Annual energy demand
Heat transfer stations d. control strategy (hts) 176...203/167...185 °F (80...95/75...85 °C) (hot outlet / cold inlet) unknown no.

Experiences

The system has run well from its inception. This first, large scale solar thermal system has provided a larger energy yield than promised.

Acknowledgement

LEW Automotive GmbH provided a flawless installation.

* See company internal references.

ETC – 5

Location: Coney Island, NY

Project summary

Washing station for trains First AquaSystem Project in the United States
Gross collector area yearly global irradiation: 1761 sq ft 4.5 therms per sq ft and annum

Site

Coney Island
New York, NY, USA
40°35'16.99"N; 73°58'39.84"W

Project description

Installation date planned January 2010
Design continual power 290 MBH
Design Solar peak power 512 MBH
Solar system yield 3583 therms (10,491 MWhr) per annum
Specific system yield 2.034 therms/sq ft (641 kWh/m^2) and ann.
Max. electric energy 14 therms (41 MWhr) per annum
Solar share 50%
System efficiency 0.45 (system energy yield / radiation input)

System details

DH system: unknown
Connection SWH to DH: no solar DH integration
Solar system: Paradigma XL-Solar AquaSystem 3963
Collectors: High performance vacuum tube collectors with CPC mirror (CPC-VTC technology)
Slope: 45 degree
Storage volume: 2 gal/sq ft (84 L/m^2)
Freeze protection: Active with low temperature heat from the system.

Control strategy

- AquaSystem:
  - Water as heat transfer medium, active freeze protection
  - Permanent automatic function control and failure diagnostic
  - High (target) temperature controller (on-off controlling with the bucket principle)
- Standard features:
  - Two-tank-systems
  - Frost protection pump
  - Hydraulic separation (with solar counter-flow heat exchanger)
  - One tank temperature solar switch
  - Outlet steam blockade
  - Solar hot-start.

Economics

Savings: $120 (US) per annum

District heating net

Solar water temperatures: 140–194 °F / 77–140 °F (60–90 °C / 35–70 °C) (hot outlet / cold inlet temperatures)
Annual energy demand: 716,500 kBtu (7165 therms) per annum.

ETC – 6

Title: Alta Leipziger, Oberunsel, Germany

Project summary

This project renovated the central kitchen of a large insurance company. This was one of the first AquaSystem XL-Solar Projects:

- Gross collector area yearly global irradiation: 1268 sq ft (118 m^2), 3.3 therms/sq ft (1,040 kWh/m^2) and annum

Figure A-112. Solar system at Alta Leipziger, Oberunsel, Germany.

Site: Alter Leipziger Platz 1, Oberursel, Germany

Project description

Installation date: August 2007
Design continual power: 201 MBH
Design Solar peak power: 342 MBH
Solar system yield: 1843 therms (5,396 MWhr) per annum
Specific system yield: 1.453 therms/sq ft (458 kWh/m^2) and ann.
Max. electric energy: 10 therms (29.3 MWhr) per annum
Solar share: 25%
System efficiency: 0.44 (system energy yield / radiation input).

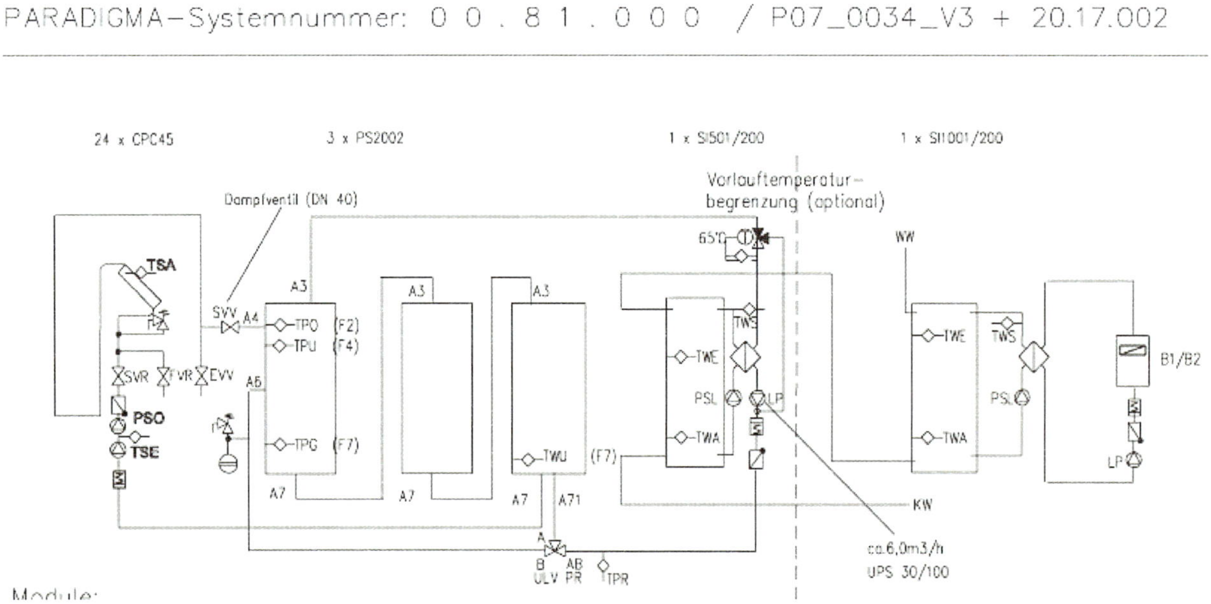

Figure A-113. Schematic view of solar system at Alta Leipziger, Oberunsel, Germany.

System details

DH system: Company internal DH network
Connection SWH to DH: No direct solar DH integration
Solar system: Paradigma XL-Solar AquaSystem
Collectors: High performance vacuum tube collectors with CPC mirror (CPC-VTC technology)
Slope: 45 degree
Storage volume: 1849 gal (7 m^3)
Freeze protection: Active with low temperature heat from the system.

Control strategy

- AquaSystem:
  - Water as heat transfer medium
  - Active freeze protection
  - Permanent automatic function control and failure diagnostic
  - high (target) temperature controller (on-off controlling with the bucket principle)
- Standard features:
  - Two-tank-systems
  - Frost protection pump
  - Hydraulic separation (with solar counter-flow heat exchanger)
  - One tank temperature solar switch
  - Outlet steam blockade
  - Solar hot-start.

Economics

Local Energy Cost: $2.64/therm (0.62 €/MWhr)
System first cost: 135,000 US $
Operation Costs: 90 US $ per annum
savings first year: 4860 US $ per annum (specific saving 1st year): $3.76/sq ft (28.34 €/$m^2$)).

District heating net

Solar water temperatures: 149–194 °F / 95–158 °F (60–90 °C /35-70 °C) (hot outlet / cold inlet temperatures).

Experiences

The system has run well from its inception. Also, the AquaSystem manages very long (>200 m) solar lines.

Acknowledgement

Paradigma partners provided a flawless installation.

ETC – 7

Title: Panoramasauna, Holzweiler, Germany

Project summary

This project provided support for network heating of a recreational swimming pool using the world's most innovative solar system, the first "Zero storage" system, with a gross collector area yearly global irradiation of 1057 sq ft (98.3 m^2), 3.06 therms/sq ft (964.5 kWh/m^2) and annum.

Site: Panoramaweg 2, 53501 Grafschaft Holzweiler, Germany

Figure A-114. Solar system at Panoramasauna, Holzweiler, Germany.

Project description

Installation date: February 2008
Design continual power: 164 MBH
Design Solar peak power: 239 MBH
Solar system yield: 1877 therms (5,496 MWhr) per annum
Specific system yield: 1.934 therms/sq ft (610 kWh/m^2) and per annum.
Max. electric energy: 10 therms (29.3 MWhr) per annum.
Solar share: 2%
System efficiency: 0.63 (system energy yield / radiation input).

Figure A-115. Schematic view of solar system at Panoramasauna, Holzweiler, Germany.

System details

DH system (local DH for about five big buildings):
Connection SWH to DH: Direct, like an additional boiler
Solar system: Paradigma XL-Solar AquaSystem
Collectors: High performance vacuum tube collectors with CPC mirror (CPC-VTC technology)

Slope: 35 / 40 degree
Storage volume: not specified
Freeze protection: Active with low temperature heat from the system.

Control strategy

- AquaSystem
  - Water as heat transfer medium
  - Active freeze protection
- Permanent automatic function control and failure diagnostic
- High (target) temperature controller (on-off controlling with the bucket principle)
- Standard features:
  - Two-tank-systems
  - Frost protection pump
  - Hydraulic separation (with solar counter-flow heat exchanger)
  - One tank temperature solar switch
  - Outlet steam blockade
  - Solar hot-start.

Economics

- Local Energy Cost $0.022/therm (0.01 €/MWhr)
- System first cost 84,000 US $
- Operation Costs 90 US $ per annum
- savings
  - First year 4125 US $ per annum
  - Specific saving 1st year $3.82/sq ft (28.79 €/m^2).

District heating net

Solar water temperatures: 158–194 °F / 149–176 °F (70–90 °C / 65– 80 °C) (hot outlet / cold inlet temperatures)
Annual energy demand: 10,236,400 kBtu (102,364 therms) per annum
Heat transfer stations control strategy (hts).

Experiences

The system has run well from its inception. The idea that "A solar thermal system has to be as easy as an additional boiler" is also exactly transferable to large scale soar thermal systems

Acknowledgement

Paradigma partners provided a flawless installation.

ETC – 8

Title: Wohnheim Langendamm, Nienburg, Germany

Site: 31582 Nienburg, Germany

Project summary

This project provided DH support for an area containing residential homes, typically integrating the new system in to an existing old DH:

- gross collector area yearly 505 sq ft
- global irradiation: 3.01 therms/sq ft (949 kWh/m^2) and annum.

Figure A-116. Solar panel installation at Wohnheim Langendamm, Nienburg, Germany.

Project description

installation date: August 2008
Design continual power: 89 MBH
Design Solar peak power: 137 MBH
Solar system yield: 819 therms (2,398 MWhr) per annum
Specific system yield: 1.622 therms/sq ft (511.2 kWh/m^2) and ann.
Max. electric energy: 5 therms (14.6 MWhr) per annum
Solar share: 8%
System efficiency: 0.54 (system energy yield / radiation input).

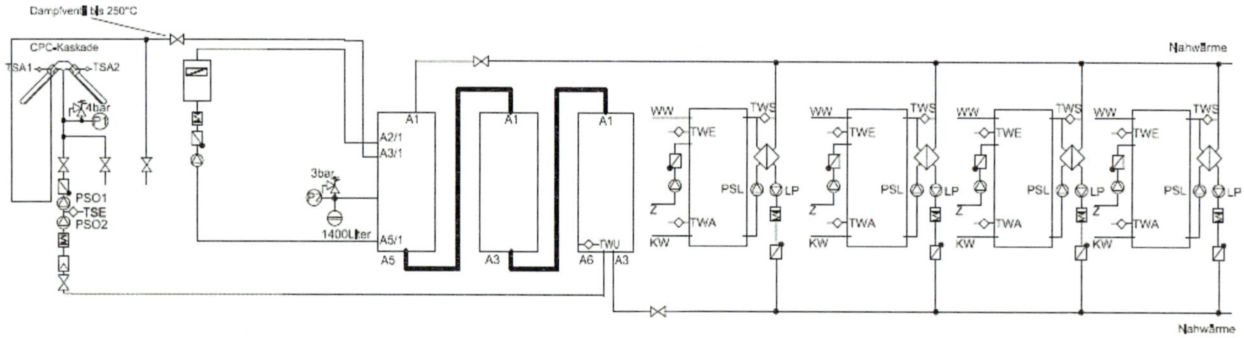

Figure A-117. Schematic view of solar system at Wohnheim Langendamm, Nienburg, Germany.

System details

DH system
Connection SWH to DH: Local DH network for four buildings via buffer tank
Solar system: Paradigma XL-Solar AquaSystem
Collectors: High performance vacuum tube collectors with CPC mirror (CPC-VTC technology)
Slope: 45 degree
Storage volume: 1321 gal (5 m^3)
Freeze protection: Active with low temperature heat from the system.

Figure A-118. Location of solar system at Wohnheim Langendamm, Nienburg, Germany.

Control strategy

- AquaSystem
  - Water as heat transfer medium
  - Active freeze protection
  - Permanent automatic function control and failure diagnostic - high (target) temperature controller (on-off controlling with the bucket principle)
- Standard features:
  - Two -tank-systems
  - Frost protection pump
  - Hydraulic separation (with solar counter-flow heat exchanger)
  - one tank temperature solar switch
  - outlet steam blockade
  - solar hot-start.

Economics

- Local Energy Cost: $0.0286/therm (0.01 €/MWhr)
- System first cost: 36,000 US $
- Operation Costs: 75 US $ per annum
- Savings
  - First year: 2340 US $ per annum
  - Specific saving 1st year: $4.49/sq ft (33.84 €/$m^2$).

District heating net

Solar water temperatures: 140–194 °F / 95–158 °F (60–90 °C / 35 – 70 °C) (hot outlet / cold inlet temperatures)
Annual energy demand: 1,023,600 kBtu (10,236 therms) per annum
Heat transfer stations control strategy (hts)

Experiences

The system has run well since its construction. It was found that old existing DH nets are no handicap for solar applications.

Acknowledgement

The system was well installed by local Paradigma partners.

ETC – 9

Title: Kraftwerk, Halle, Germany

Site: Halle, Germany

Project summary

This project has undertaken solar weekend bridging for a CHP power plant The project has won a scientific ranking procedure and is now waiting for the government subsidy declaration:

- Gross collector area: 241,024 sq ft (22,415.2 m^2)
- Yearly global irradiation: 3.16 therms/sq ft (996 kWh/m^2) and annum.

Figure A-119. View of solar system at Kraftwerk, Halle, Germany.

Project description

Installation date: Planned
Design continual power: 34,152 MBH
Design Solar peak power: 58,058 MBH
Solar system yield: 300,268 therms (879,185 MWhr) per annum
Specific system yield: 1.246 therms/sq ft (393 kWh/m^2) and ann.
Max. electric energy: 1535 therms (4,495 MWhr) per annum
Solar share: 70%
System efficiency: 0.39 (system energy yield / radiation input).

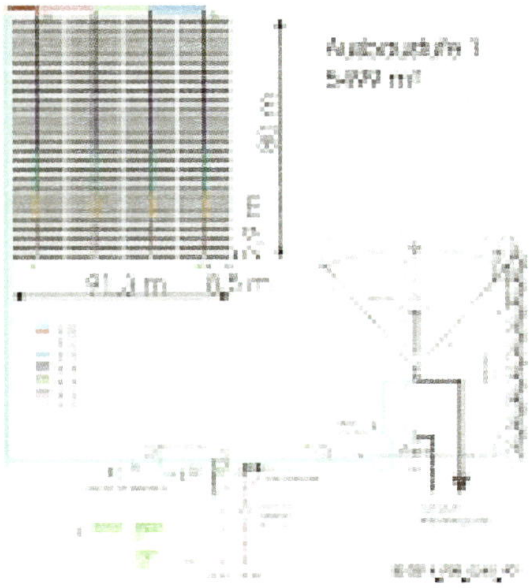

Figure A-120. Schematic view of solar system at Kraftwerk, Halle, Germany.

System details

DH system:
Connection SWH to DH: DH network for a medium-sized town via buffer tank
Solar system: Paradigma XL-Solar AquaSystem
Collectors: High performance vacuum tube collectors with CPC mirror (CPC-VTC technology)
Slope: 30 degree
Storage volume: 9511,200 gal (35 m^3)
Freeze protection: Active with low temperature heat from the system

Control strategy

- AquaSystem
    - Water as heat transfer medium
    - Active freeze protection
    - Permanent automatic function control and failure diagnostic
    - High (target) temperature controller (on-off controlling with the bucket principle)
- Standard features
    - Two-tank-systems
    - frost protection pump
    - hydraulic separation (with solar counter-flow heat exchanger)
    - One tank temperature solar switch
    - Outlet steam blockade
    - Solar hot-start.

Economics

Local Energy Cost: $0.0154 /therm (0.0036 €/MWhr)
System first cost: $12,900,000 (US)
Operation Costs: $13,500 (US) per annum
Savings first year: $462,000 (US) per annum
Specific saving 1st year: $1.86/sq ft (14 €/$m^2$).

District heating net

Solar water temperatures: 176–203 °F / 131–149 °F (80–95 °C / 55–65 °C) (hot outlet / cold inlet temperatures)
Annual energy demand
Heat transfer stations Control strategy (hts)

ETC – 10

Title: Wels, Austria

Site: Wels, Austria

Project summary

This project provided DH support for a small city from the roof of a trading show areal The project won an international bidding procedure, and is now waiting for the government subsidy declaration:

- Gross collector area yearly global irradiation: 39,629 sq ft (3,686 m^2); 2.79 therms/sq ft (879.4 kWh/m^2) and annum.

Project description

Installation date: planned 2010
Design continual power: 6830 MBH
Design Solar peak power: 10,246 MBH
Solar system yield: 58,006 therms (169,842 MWhr) per annum
Specific system yield: 1.464 therms/sq ft (461 kWh/m^2) and ann.
Max. electric energy: 208 therms (609 MWhr) per annum
Solar share : 3%
System efficiency : 0.52 (system energy yield / radiation input).

System details

DH system:

- Connection SWH to DH: DH network for a medium-sized town direct like a additional boiler
- Solar system: Paradigma XL-Solar AquaSystem
- Collectors: High performance vacuum tube collectors with CPC mirror (CPC-VTC technology)
- Slope: 30 degrees
- Storage volume: not specified
- Freeze protection: Active with low temperature heat from the system

Control strategy

- AquaSystem
  - Water as heat transfer medium
  - Active freeze protection
  - Permanent automatic function control and failure diagnostic
  - High (target) temperature controller (on-off controlling with the bucket principle)
- Standard features
  - Two-tank-systems
  - Frost protection pump
  - Hydraulic separation (with solar counter-flow heat exchanger)
  - One tank temperature solar switch
  - Outlet steam blockade
  - Solar hot-start.

Economics

- Local Energy Cost: $0.0176/therm (0.0041 €/MWhr)
- System first cost: 3,000,000 US $
- Operation Costs: 1,830 US $ per annum
- Savings
    - First year: 102,000 US $ per annum specific
    - saving 1st year: $2.53/sq ft (19.1 €/m^2).

District heating net

Solar water temperatures: 194–239 °F / 167–221 °F (90–115 °C / 75–105 °C) (hot outlet / cold inlet temperatures)
Annual energy demand: 204,728,500 kBtu (2,047,285 therms) per annum
Heat transfer stations
Control strategy (hts): unknown.

Experiences

The successful outcome of this project showed that good CPC-VTC technology can overcome substantial political obstacles and resistance.

ETC – 11

Title: AWO Rastede, Oldenburg, Germany

Site: Klingenbergstr. 73, 26133 Oldenburg, Germany

Project summary

This project provided DH support for a residential retirement community, typically by integrating new technologies with existing old DH systems:

- Gross collector area yearly global irradiation: 1054 sq ft (98.0 m2); 3.04 therms/sq ft (958 kWh/m^2) and annum.

Figure A-121. View of solar panels installed at AWO Rastede, Oldenburg, Germany.

Project description

Installation date: December 2008
Design continual power: 164 MBH
Design Solar peak power: 239 MBH
Solar system yield: 1843 therms (5,396 MWhr) per annum
Specific system yield: 1.744 therms/sq ft (550 kWh/m^2) and ann.
Max. electric energy: 10 therms (29.3 MWhr) per annum
Solar share: 10%
System efficiency: 0.57 (system energy yield / radiation input)

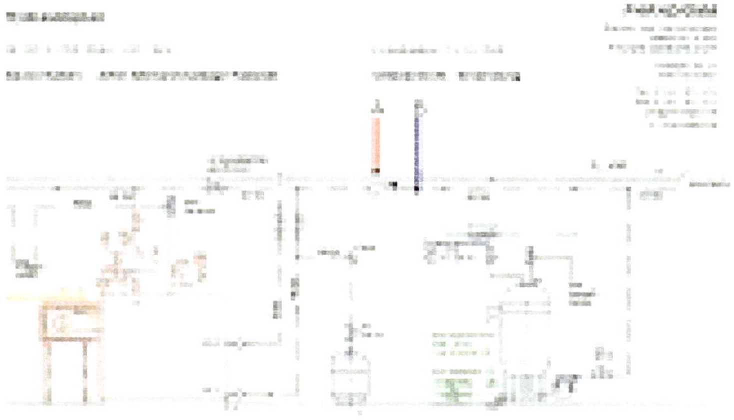

Figure A-122. Schematic view of solar system at AWO Rastede, Oldenburg, Germany.

System details

DH system

- Connection SWH to DH: Local DH network for some buildings direct like a additional boiler
- Solar system: Paradigma XL-Solar AquaSystem
- Collectors: High performance vacuum tube collectors with CPC mirror (CPC-VTC technology)
- Slope: 38 degrees.
- Storage volume: not specified
- Freeze protection: Active with low temperature heat from the system.

Control strategy

- AquaSystem
  - Water as heat transfer medium
  - Active freeze protection
  - Permanent automatic function control and failure diagnostic
  - High (target) temperature controller (on-off controlling with the bucket principle)
- Standard features:
  - Two-tank-systems
  - frost protection pump
  - Hydraulic separation (with solar counter-flow heat exchanger)
  - One tank temperature solar switch
  - Outlet steam blockade
  - Solar hot-start.

Economics

- Local Energy Cost: $0.0286/therm (0.0067 €/MWhr)
- System first cost: $112,500 US
- Operation Costs: $90 (US) per annum
- Savings
  - First year: $5265 (US) per annum
  - Specific saving 1st year: $4.91/sq ft (37 €/m^2) (US)

District heating net

Solar water temperatures: 167–185 °F / 140–158 °F (75–85 °C / 60–70 °C) (hot outlet / cold inlet temperatures)
Annual energy demand
Heat transfer stations
Control strategy (hts): unknown.

Experiences

The system has run well since its construction. The idea “A solar thermal system has to be as easy as an additional boiler” is also exactly transferable to large scale soar thermal systems

Acknowledgement

The system was well installed by local Paradigma partners.

ETC – 12

Title: METRO Istanbul, Turkey (shopping mall)

Project summary

This project renovate a shopping mall with summer solar cooling, and winter heating using the world's second largest CPC vacuum tube collector system:

- gross collector area / yearly global irradiation: 11,083 sq ft (1,031 m^2); 4.77 therms/sq ft (1,503.4 kWh/m^2) and annum.

Figure A-123. Solar system at METRO, Istanbul, Turkey

Site

Istanbul, Turkey
40.58 N; 29.05 E

Project description

- Installation date: August 2009
- Design continual power: 1708 MBH
- Design Solar peak power: 3415 MBH
- Solar system yield: 22,179 therms (64,940 MWhr) per annum
    - Specific system yield: 2.001 therms/sq ft (631 kWh/m^2) and ann.
    - Max. electric energy: 75 therms (219.6 MWhr) per annum
- Solar share: 30%
- System efficiency: 0.42 (system energy yield / radiation input).

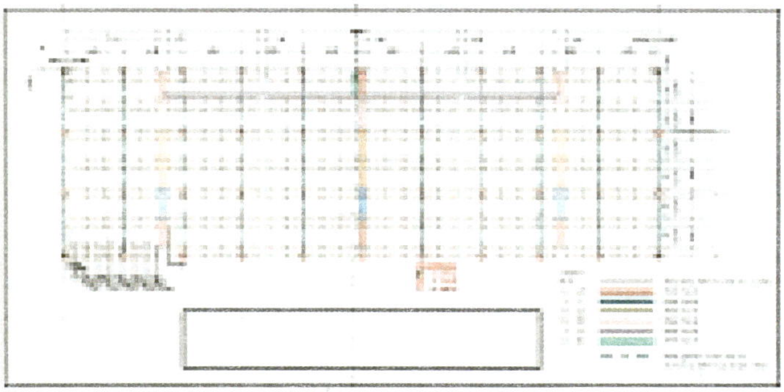

Figure A-124. Schematic view of solar system at METRO, Istanbul, Turkey.

System details

DH system: Company internal DH network via buffer tank

Connection: SWH to DH
Solar system: Paradigma XL-Solar AquaSystem
Collectors: High performance vacuum tube collectors with CPC mirror (CPC-VTC technology)
Slope: 30 degrees
Storage volume: 3963 gal (15m^3)
Freeze protection: Active with low temperature heat from the system.

Control strategy

AquaSystem
- Water as heat transfer medium
- Active freeze protection
- Permanent automatic function control and failure diagnostic
- High (target) temperature controller (on-off controlling with the bucket principle)
- Standard features:
  - Two-tank-systems
  - Frost protection pump
  - Hydraulic separation (with solar counter-flow heat exchanger)
  - One tank temperature solar switch
  - Outlet steam blockade
  - Solar hot-start.

Economics

Local Energy Cost: $0.0198/therm (0.0046 €/MWhr)
- System first cost: $750,000 (US)
- Operation Costs: $660 (US) per annum
- Savings first year: $43,875 (US) per annum; specific saving 1st year: $3.9/sq ft (29.39 €/m^2).

District heating net

Solar water temperatures: 176–203 °F / 167–185 °F (80–95 °C / 75–85 °C) (hot outlet / cold inlet temperatures)
- Annual energy demand: 7,506,700 kBtu (75,067 therms) per annum
- Heat transfer stations
- control strategy (hts).

Experiences

The solar system has run well since August 2009. The local Instrumentation and Controls [I&C] system did not use the solar energy until December 2009 so the solar system is in thermal stagnation almost daily. However, the AquaSystem is stagnation proof even under hardest conditions.

Acknowledgement

Credit is due to LEW Automotive GmbH for a flawless installation.

High Temperature Collectors

HTC – 1

Title: High-Temperature Solar Hot Water System — Phoenix Federal Correctional Institution

Location: Phoenix, AZ

A large solar water heating system installed at the Phoenix Federal Correctional Institution (FCI) in 1998 provides 70% of the facility's annual hot water needs. The solar system includes 17,040 sq ft (1,585 m^2) of parabolic trough concentrating collectors and a 23,000 gal (87,055 L) storage tank located adjacent to the collectors. The system produces up to 50,000 gal (189,250 L) of hot water daily, enough to meet the needs of 1250 inmates and staff, all of whom use kitchen, shower, and laundry facilities.

The system was financed through an Energy Savings Performance Contract (ESPC). The ESPC payments are 10% less than the energy savings so that the prison saves an average of $6700 per year, providing an immediate payback. Boiler maintenance and hot water service call costs for the facility have also been reduced. The Federal Bureau of Prisons worked with the Department of Energy (DOE) Federal Energy Management Program (FEMP) and the ESPC contractor, Industrial Solar Technology Corporation (IST), to design and install the system. Under the terms of the 20-year ESPC contract, the prison receives 10% of the total energy savings annually (an average of $6700 per year), and the other 90% goes to amortize the first costs of the system. At the end of the 20-year period, the prison will take over ownership, operation, and maintenance of the solar system and benefit from 100% of the energy savings for the remaining 10 years of its expected service life.

The solar water heating system at the Phoenix correctional facility offsets an average annual consumption of 1000 megawatt-hours (MWh) of electricity and the release of nearly 600 tons of CO_2. By comparison, conventional electricity produced in Arizona emits 1109 lb (503 kg) of CO_2 per MWh:

System cost:	$650,000 (installed)
Energy savings:	$6700/year average
Input heat transfer fluid temp =	60 °F (15.6 °C)
Annual heat collected =	14,600 MBtu/yr/sq ft (46,016 kWh/yr/m^2)

Figure A-125. Solar field of collectors at the Phoenix correctional facility (left); close-up of a row of parabolic collectors at the prison (right).

HTC – 2.

Title: Industrial Process Heating – SunChips® Manufacturing Facility, Modesto, CA

Location: Modesto, CA

The SunChips® manufacturing facility in Modesto, CA, which is owned by Frito Lay, is using solar energy to produce snacks. The solar collector field was built by American Energy Assets for Frito-Lay covering 4 acres of land and accommodating 5,391.12 m^2 (57,969 sq ft) of net collector aperture area. Before construction, the installation design was reviewed by the National Renewable Energy Laboratory. The entire system has a total annual capacity of 14,600 MMBTU/yr. Specifics on the collector field include

- 5387 m^2 (57,969 sq ft) net collector aperture area on 4 acres of ground area
- 16 N-S parallel strings, each with 24 modules in series
- One tracking drive per 16 modules
- Heat transfer fluid: Pressurized water (450 °F [232 °C] at 450 psi [3,102 kPa])
- Collector flow 600 gpm (38 kg/s)
- Storage Tank size = No additional storage
- Input heat transfer fluid temp = 788 °F (420 °C) input, 860 °F (460 °C) output
- Installed Cost = $2200,000
- Maximum Capable Delivery (Estimated at 14,790 million Btu [14,863 GJ,] by simulation).

The thermal energy demand at the manufacturing line is 2.4 MMBtu/hr (7.0 MMWh/hr). The annual thermal energy demand is ~14,600 MMBtu (42,748.8 MMWh). This is the approximate annual thermal energy output of the solar collector field at Modesto.

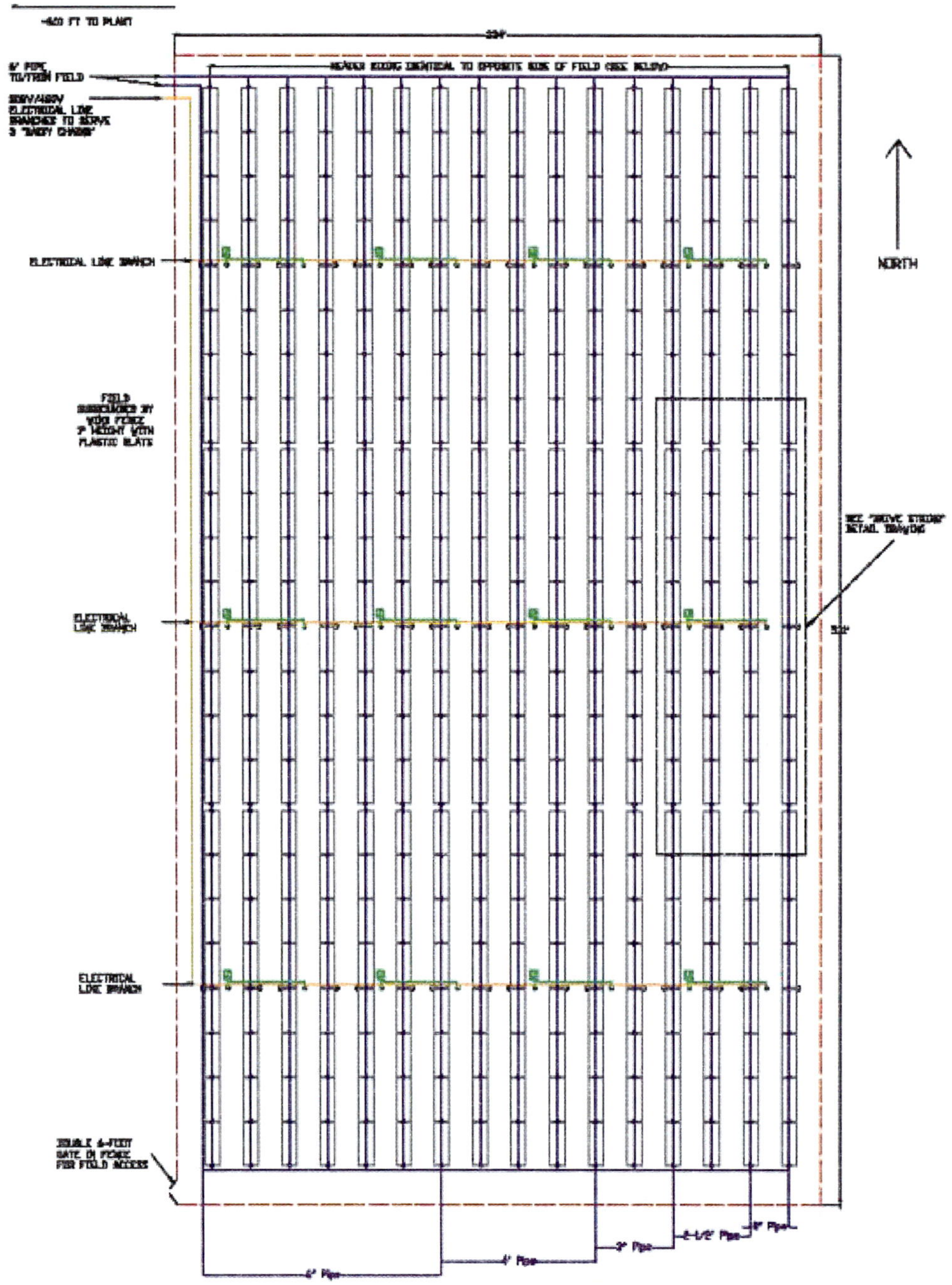

Figure A-126. Schematic of the collector field at the Modesto plant. The field comprises a huge array of concave mirrors. These mirrors track the position of the sun throughout the day, focusing the sun's energy on a black tube that runs along the focus of the array. This black tube is surrounded by a second glass tube that protects it from the air, allowing it to absorb solar energy more effectively. As super heated water passes through the black tube, the solar energy heats it up to 450 °F (232 °C). This water then runs through a boiler system that uses its heat to generate steam, which helps to cook the wheat and heat the cooking oil used in the SunChips® manufacturing process. Cooled water then flows back through the tube to the solar concentrator field to repeat the process.

Balance of System

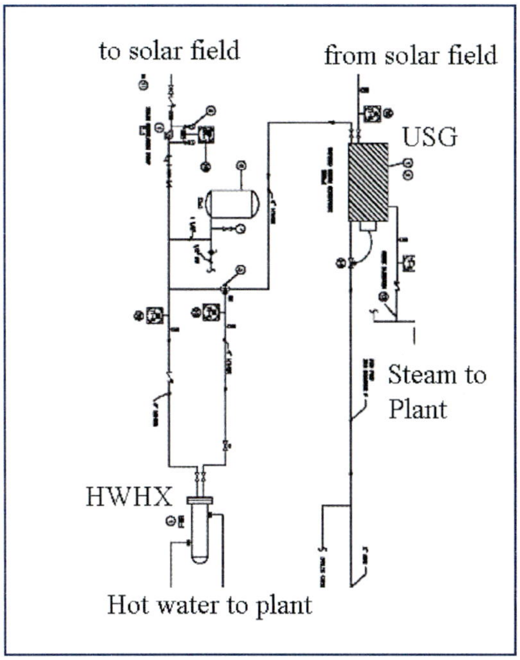

Figure A-127. The balance of systems for the SunChips manufacturing facility includes 285 m of pipe from the solar field to the plant, an unfired steam generator (see Fig. A-109), a heat exchanger, a 25 hp pump, 447 m of pipe running back to the solar field from the plant, and bypass valves and controls.

Figure A-128. Components of a 54,000 sq ft (5022 m^2) solar industrial process heat plant at Frito Lay Modesto CA, including an unfired steam generator, pump, and controls.

Figure A-129. Governor Arnold Schwarzenegger gets a close-up look at the field of parabolic solar collectors at the SunChips factory.

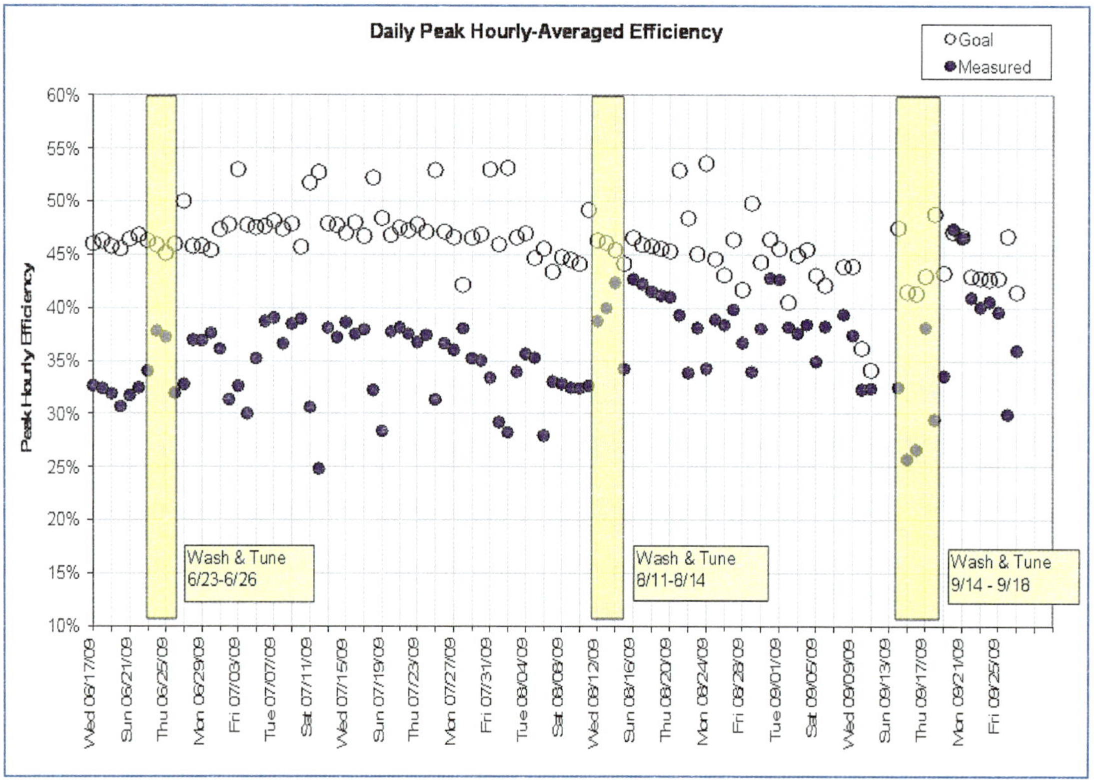

Figure A-130. Peak hourly-averaged efficiency for each day since mid-June 2009. Dark circles represent measured peak hourly efficiency for each day; open circles represent the derated-Sandia-curve goal.

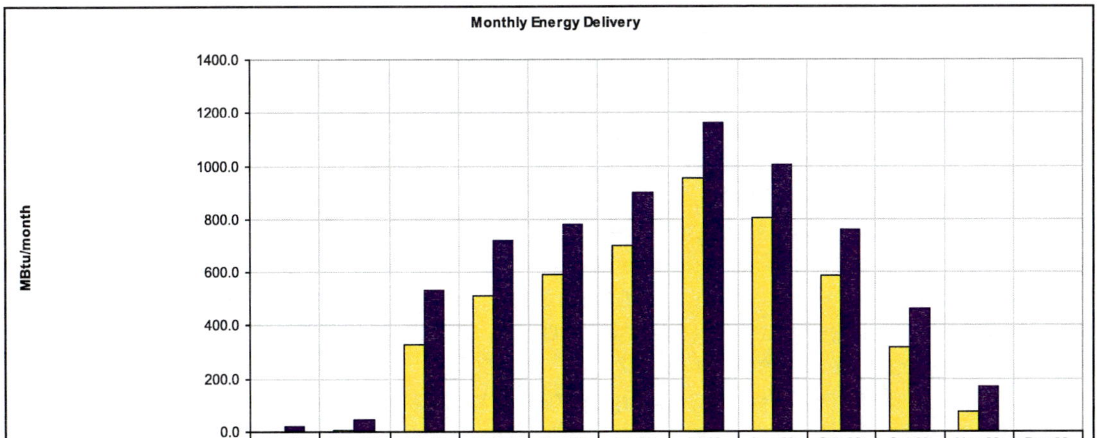

	Jan-09	Feb-09	Mar-09	Apr-09	May-09	Jun-09	Jul-09	Aug-09	Sep-09	Oct-09	Nov-09	Dec-09
Measured at Steam Generator	1.5	4.8	327.8	509.8	591.2	698.9	952.3	802.7	585.7	317.9	73.1	0.0
Measured at Field	18.2	42.4	531.4	720.2	777.8	896.7	1160.9	1000.9	758.0	463.2	167.9	0.0

Figure A-131. Chart showing the monthly energy delivery of the solar water heating system at the SunChips® factory in Modesto, CA, which is owned by Frito Lay. The annual energy collected was 14,600 MBtu/yr/sq ft in 2009.

System maintenance

General maintenance

Operation and maintenance (O&M) costs of each solar water heating system is estimated at ½ of 1% of initial cost per year. O&M is similar to that required of any hydronic heating loop and may be provided by site staff, with experts called in if something should fail. Regularly scheduled maintenance includes:

- Check the solar collectors and structure components for any damage. Note location of panel glaxing or broken evacuated tubes for replacement.
- Check tightness of mounting connectors. Repair any bent or corroded mounting components.
- Determine if any new objects, such as vegetation growth, are causing shading of the array and remove them if possible.
- Clean outer surface of collector array annually with plain water or mild dishwashing detergent. Do not use brushes, any types of solvents, abrasives, or harsh detergents.
- Check all connecting piping for leaks. Repair any damaged components.
- Check plumbing for signs of corrosion.
- Observe operational indicators of temperature and pressure to ensure proper operation of pumps and controls.
- Observe that the collector heat transfer fluid pump is running on a sunny day and not at night.
- Use insolation meter to measure incident sunlight and simultaneously observe temperature and energy output values given by system controller. Compare the values with original efficiency of system.
- Check status indicators provided by system controller. Compare indicators with measured values.
- Document all operation and maintenance activities in a workbook available to all service personnel.
- Check proper position of all valves.
- Flush system to remove mineral deposits every 10 years.

Glycol fluid care

The decomposition rate of glycol varies according to the degree of aeration, high temperature exposure and the service life of the solution. Most water/glycol solutions require periodic monitoring of the pH level and the corrosion inhibitors. The pH should be maintained between 6.5 and 8.0. Replacement of the water/glycol solution may be as often as every 12-24 months or even sooner in high temperature systems. If these solutions are used in the collector loop, the installer should specify the expected life of the solution and the amount of monitoring required. The cost of periodic fluid replacement and monitoring should be considered in the economic analysis. Since glycol-water mixtures do require a lot of maintenance (and since users can be quite negligent) it is recommended that glycols not be used in family housing solar heating and DHW systems, and that glycol-water solutions be reserved for use in large-scale installations that have regular maintenance schedules.

Appendix B: Examples of Design Options (Fort Bliss / Fort Bragg)

Integration of solar hot water generation in district heating and district cooling systems

Introduction

The US Army Corps of Engineers — specifically, the Engineer Research and Development Center, Construction Engineering Research Laboratory (ERDC-CERL) — is currently involved in the first phase of a project to analyze the potential for the use of central solar hot water (SHW) generation at US Army installations. The goal of this work is to integrate SHW into the equipment portfolio for space heating (SH), domestic hot water (DHW) generation, and air-conditioning (AC). Since several US Army installation already operate (or have proposed installing) central District Heating (DH) and District Cooling (DC) systems, the incorporation of SHW at these Army installation must considerations the combination of both SHW and central systems.

Overall, the discussion must address both technical and economic issues. While this appendix discusses both issues, it focuses primarily on the technical aspects. i.e., how to integrate SHW into central systems. The discussion of SHW technology must differentiate between two main options:

1. The first option employs a central feed-in into the DH/DC system from a small number of solar fields.
2. The second option employs a de-central feed-in of SHW from several smaller solar fields.

Economic discussion

This section focuses on the qualitative impacts of SHW on the economics on central systems. To determine the economic benefits of a central system with a shared SHW, one must compare a central system with a shared SHW with a DH/DC system without SHW. This is the most important way to understand the impact of SHW on the economics of the entire system.

One may consider an existing or designed central SHW system simply as a piece of generation equipment, like a boiler. Thus, if SHW is added to an existing central system, the energy from the SHW system replaces fossil fuel fired capacities. In the combined system, one must consider whether the SHW is generated by a centralized or de-centralized system, and whether the capacity it replaces is a base load unit with a high efficiency and low generation costs.

Suppose one begins with a DHW system that uses natural gas and/or fuel oil fired boilers to generate hot water. The (secondary) *fuel-to-heat* efficiency of such boilers would be ~85 to 90%. Each Btu from the SHW that replaces a Btu of boiler energy does not increase the energy efficiency of the DH system since only the amount of fossil fuel input changes. The rest of the energy chain downstream to the DHW/SH use is not touched. If the SHW share is cheaper than the heat from the boilers, than the addition of SHW makes the system more cost efficient (but not necessarily more energy efficient).

A discussion of co-generation/combined heat and power generation (e.g., with a gas turbine or a co-gen motor) is more complex. In such a case, the co-gen unit increases first costs, but decreases energy generation costs per Btu. If SHW is added to a co-gen unit, the SHW will likely replace most of the summer load of the co-gen unit. This could make the co-gen unit cost inefficient since the combined system will (dramatically) reduce the full load hrs/yr generated by the co-gen unit.

The lesson learned is that one must calculate the economics basing on the share of each generation option using the duration curve. In general, the most energy efficient decision is to use either SHW or co-gen, rather than the combination of both technologies.

Technical discussion

A central system is, in most cases, a distribution system that circulates hot and cold water. In fewer cases, the transport medium for heating is steam. Such systems do have a tradition in the US Army, but they are not further considered here due to the high transport medium temperatures, which are generally incompatible with SHW systems.

Whenever SHW technology is to be integrated into a central system, the most important technical issue is that of temperature. An existing central heating system like that used in the Corps Support Command (COSCOM), Fort Bragg, TX operates with relatively fixed supply temperatures. ERDC-CERL Technical Report TR-09-5* described and analyzed the COSCOM central heating system, and recommended that the system be operated with variable flow, variable speed.

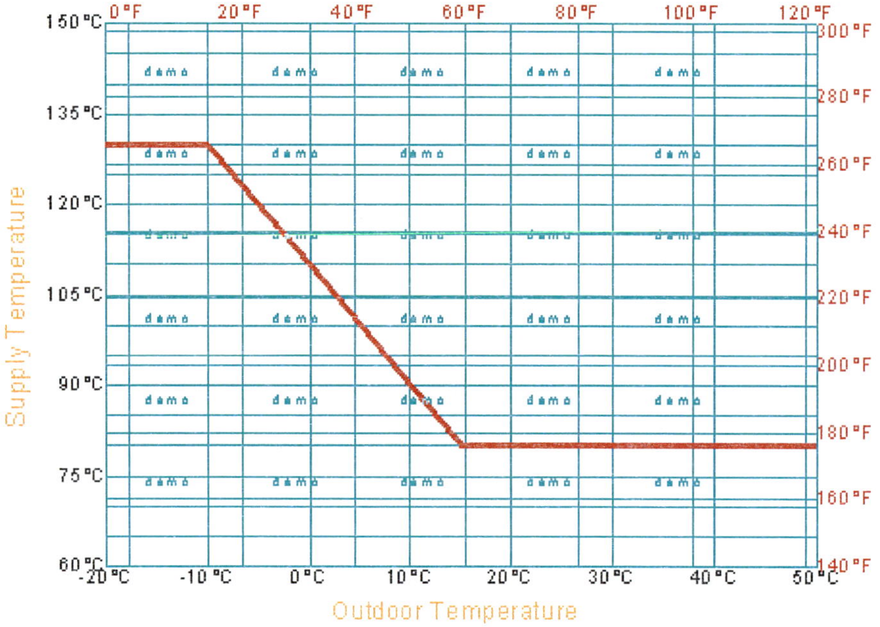

Figure B-1. Relationship between supply temperature and outdoor temperature.

Figure B-1 shows the relationship between supply water temperature and outdoor temperature; there is a strong correlation between outdoor temperature and heat load. The SH load increases as the outdoor temperature drops. If the outdoor temperature increases over 60 °F (16 °C) the SH load falls to almost zero; the DHW load represents the remaining heating load.†

The variable temperature central heating system is based on the relation:

$$\dot{Q} = \dot{m}\, c_p \Delta T$$

The supply water temperature determines how well a DH system will operate; if SHW is incorporated into the system, either the SHW generation temperature needs to be adapted to the DH supply temperature, or vice versa. When existing DH systems are adapted for SHW use, the

* William T. Brown, III, Alexander Zhivov, David M. Schwenk, David M. Underwood, Dahtzen Chu, Stephan Richter, Roland Ziegler, Harald Neuner, and Alfred W. Woody. 2009. *Heating and Cooling Master Plan for Fort Bragg, NC: Fiscal Years 2005 to 2030.* ERDC/CERL TR-09-5, 15. Champaign, IL: Engineer Research and Development Center, Construction Engineering Research Laboratory (ERDC-CERL).

† This ignores the reheat in case of dehumidification in AC operation.

SHW temperature is usually adapted to the system temperatures (Brown 2005). In the winter, when the DH system supply temperature is higher than the SHW temperature; the boiler must make up the difference in the supply temperature. In warmer weather, the SHW temperature can be sufficient when the DH system is operated in variable operation mode. i.e., when the DH supply temperature is much lower. In newly constructed DH systems, the temperature of the DH system can be adapted to the temperature that the SHW can provide (Brigade Combat Team [BCT] 4, Fort Bliss).

DH and DC systems consist of a supply pipe and a return pipe. While the supply temperature is controlled by the generation technology, the return temperature results from the system's end use (SH, DHW, AC); it is not directly influenced by the generation technology or the operator. In general, return temperatures are:

- 130 °F (54 °C) for SH only
- 150 °F (66 °C) for DHW only
- 170 °F (77 °C) for reheat in AC mode.

A single central heating system could service any (or all) of these end uses and operation modes. Thus, the annual average return temperature of any given system is unique to that system, and depends on how the temperature changes throughout the year (seasonally), on the design of the system, and on the buildings connected to the system.

In an SHW system, the temperature of the collector field is the average of the return temperature from the central system and the flow temperature of the solar collector, which also depends on the outdoor temperature.

In economic terms, the goal is to minimize the boiler fuel consumption through the year. Two approaches exist:

1. Reduce the operation hours of boilers.
2. Shift from full load boiler operation to partial load boiler operation.

Both approaches must be combined to minimize fuel consumption. Approach (1) shifts the entire central system load to the SHW system as much as possible. Approach (2) allows the boilers to be used as "boosters" in the transition time in spring and fall when the heat and/or temperature from the SHW is not sufficient to meet the load.

Another way to further reduce the DH system return temperature is through the use of a return port. In other words, the customer takes heat from the return and injects its own return water into the return of the DH system. That results in a lower return temperature and better use of the generated heat (in both SHW and boiler systems). The disadvantage of using the return port is that some customers have requirements that cannot be met by low return temperatures. Customers cannot arbitrarily be connected to the DH system return because the return temperature and temperature difference decreases as the medium passes through the return system.

The pressure in a DH/DC system is another major parameter; the pressure in the piping system determines the flow rate. The limits of the pressure are determined by the maximum pressure the weakest piece of equipment can hold, and the minimum pressure in the building with the lowest differential pressure (which must be maintained). On the other hand, the return pressure at the generation equipment needs to be higher than a certain limit to ensure that the water will not "steam-out" in DH systems. Figure B-2 shows a pressure diagram for a typical DH system.

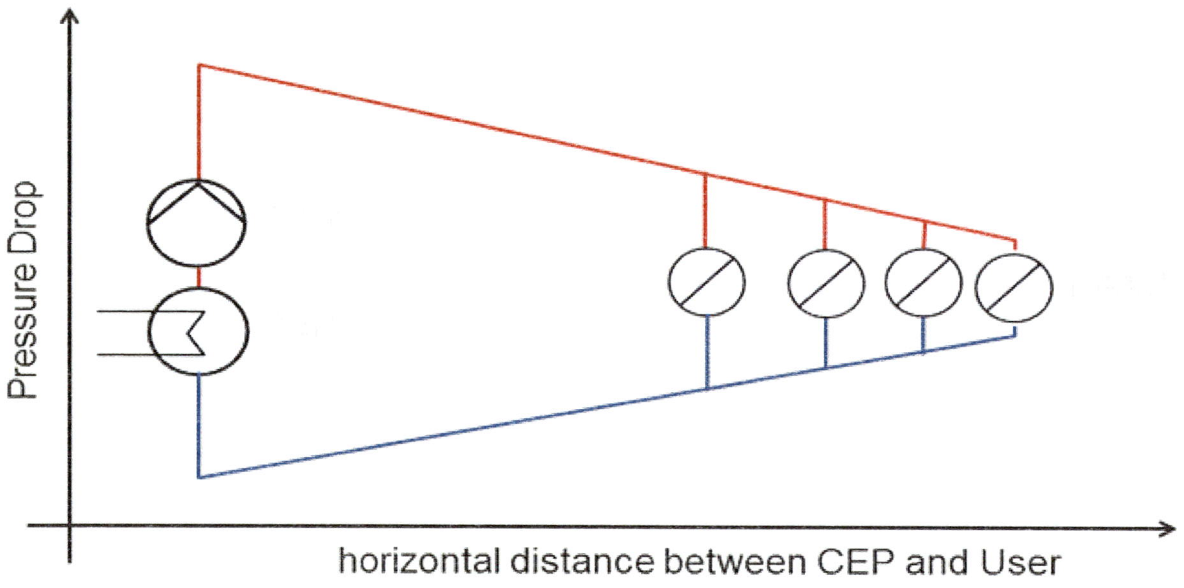

Figure B-2. Typical pressure diagram of a DH system between generation and critical building with the lowest differential pressure.

Moreover, each pump in a DH/DC system needs to operate with a defined pump head and mass flow to ensure the minimum differential pressure for each single building. If the number of pumps that feed-in into a DH system varies due to the availability of a generation site (e.g., a SHW field), the other pumps must still maintain the differential pressure.

SHW systems usually have a storage tank to buffer the load variation through a day or even a week, which creates a synergy between central systems and the SHW system. The SHW systems uses the storage tank to even out the changes between solar irradiation and load. Central systems interconnect different buildings with different load profiles, and fill pipes with water on demand. DH systems can operate at different temperatures. Thus, if there is an SHW overcapacity through the afternoon when the load is low, and a solar under capacity in the evening when the load is high, the water in the pipes can be heated through the afternoon (buffering), and the higher temperatures can be used to meet the evening heating peak.

SHW or conventional fired generation systems do not require different piping; the type of pipe depends on the supply temperature. A major advantage of the SHW system is that its efficiency increases with lower supply temperatures. Also, the complexity and first costs for the piping system decreases with lower supply water temperatures.

In a DH network, there is an important distinction between AC systems that use central DC and those that use de-central chilled water generation. The de-central AC generation (e.g., systems with absorption chillers) is effectively a DH system. Only the chiller return temperature (167 °F [75 °C]) is higher than the SH return temperature (113–149 °F [45–65 °C]) or, for DHW (149 °F [65 °C]). The supply water temperature must be higher than ~194 °F (~90 °C) at all times when the chillers operate.

The greatest energy demand for cooling characteristically correlates with the greatest SHW system supply, during the summer months. A DC system distributes chilled water generated in a Central Energy Plant (CEP). Four different DH/SHW system combinations serve these needs:

1. Existing DH systems with central SHW feed-in
2. Existing DH systems with de-central SHW feed-in

3. New designed DH systems with central SHW feed-in
4. New designed DH systems with de-central SHW feed-in.

Existing DH systems with central SHW feed-in

The Fort Bragg case study illustrates an existing DH system with central SHW feed-in. ZfS did a simulation of the solar system using T-SOL. It is assumed that the SHW is close to the existing COSCOM plant, which consists of the entire equipment to operate the DH system. ERDC/CERL TR-09-5 describes the proposed variable temperature – variable speed operation mode of the DH system for the future. Figure B-3 shows the COSCOM network model and the location of the CEP. The central SHW feed-in will be placed at the same location as the CEP.

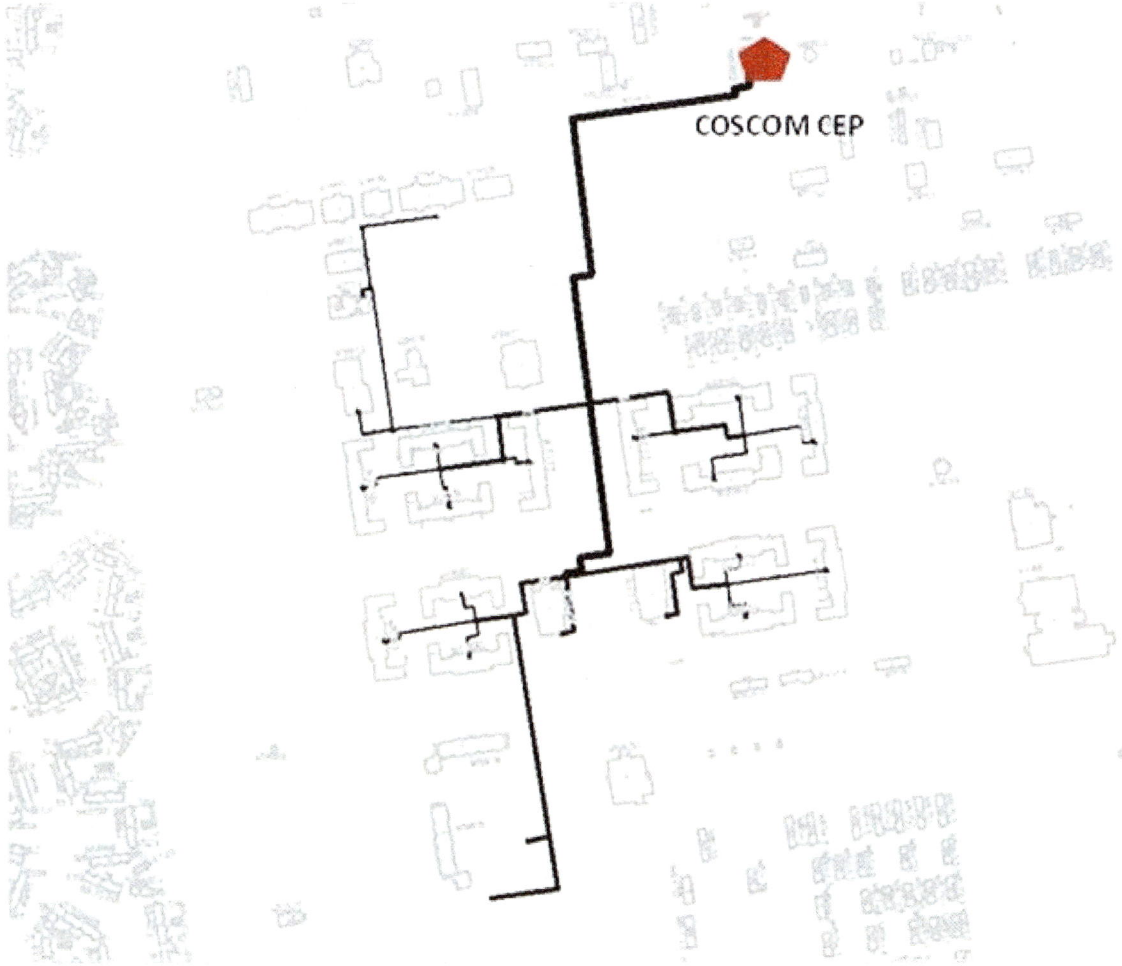

Figure B-3. COSCOM network model and the location of the CEP.

The COSCOM CEP consists of heat-only boilers; it is not economical to install a cogen system. However, a 24 hrs/day, 365 days/yr operation of the DH system is proposed, thus, SH and DHW will then be provided through the year to the connected buildings.

The SHW is designed to handle the proposed supply and return temperatures. The SHW and its storage tank provide supply water to the CEP. The SHW will replace the boiler operation hours in the CEP. The DH system itself does not feel any impact since the pumps and all the other equipment are still the same.

Assuming the solar fraction is between 10 and 25% — depending on the collector area size, solar storage tank size, and collector type — and the main contribution occurs during the summer months, the boiler operation will be reduced by approximately the same percentage. The boiler can either be partially shut down in summer, or it can be used to generate the rest of the required heat.

Finally, the SHW plant is like another boiler that has a number one priority; whenever it can provide hot water, this water is taken. Also if the flow temperature of the solar field is lower than the temperature of the central system's supply temperature, the boiler can boost the temperature.

In this case, the boiler operates 24 hrs/day, 365 days/yr without a break. The boiler has two functions, to boost the low level of the SHW temperature to the required supply temperature of the DH system, and to meet the heat demand of the DH system when the heat from SHW does not suffice.

This would be somewhat different if the SHW site were not located at the same site as the existing CEP. If this were the case, a DHW network pump would be required. The SHW pump would have to be controlled together with the other CEP pumps. This would not be too difficult if the number of CEP sites and pumps were small. However, the SHW pump takes return water (e.g., from the DH), heats is up and pumps it into the supply. Hence, the pump head and power consumption is defined by the position in the pressure diagram.

Existing DH systems with de-central SHW feed-in

In contrast with the Fort Bragg case, it is more difficult to integrate de-central, distributed SHW into a DH system. The case with a SHW at a new site in Section 3.1 (15) is the starting point for the decentralized, distributed feed-in into a DH system. However, note that the following discussion describes the situation for all kinds of DH systems with distributed feed-in (and not for SHW only).

Once the piping diameters in an existing DH system are determined for a given number of CEPs, connected buildings, and a certain temperature profile, the network topology is drawn. If a need for additional feed-in points is found, care must be taken to ensure that all pumps operate together to ensure that the minimum differential pressure is maintained throughout the system. Depending on the SHW capacity at each feed-in point, its position in the network, and the differential pressure for the current pressure, the pump must operate with a certain pump head and mass flow. However, depending on the dynamics, the current supply, return pressure, and temperature can vary dramatically.

The control of the pumps can be a great challenge in such case. The problem is even more complex if the DH system consists of loops. In a loop, the water will follow the path of the greatest pressure gradient. Unless properly coordinated, two pumps may work against each other. For example, at one moment Pump #1 may dominate and move the supply water for e.g., 1000 ft (304 m) from point A to point B. A couple of minutes later, Pump #2 may dominate and move the supply water back from B to A. If the supply water is pumped repeatedly between points A and B without re-heating at a CEP, the supply will become colder and the buildings between points A and B will not get enough heat to meet their loads. This problem is well known. No automated solution exists as yet. The problem is usually resolved by installing valves that divide a loop into two parts.

One must also consider the power of the distributed pumps, especially if the SHW is taking return water and feeding it into the supply. A solution to both these issues (the power consumption of the pumps and the operational problem discussed above) can be to use the SHW to pre-heat the return water rather than feeding the return water into the supply.

In this configuration, the de-centralized pumps must cover the internal pressure loss in the SHW. This minimizes power consumption and prevents conflicts between de-centralized pumps and a central supply pump. Disadvantages in this configuration are reduced shares of the SHW on the entire energy balance of the SHW and, depending on the boiler and co-gen equipment in the CEP, a possible decrease in the CEP's efficiency.

For this reason, the best integration point for SHW generation is at the same site as the existing CEP. In a DH system, one must always try to minimize the feed-in points.

New designed DH systems with central SHW feed-in

This ZfS simulation of Fort Bliss illustrates a newly designed DH system with a central SHW feed-in that considered de-centralized AC. Such a system increases the return temperature and increases the annual average supply water temperature, as compared to the system at Fort Bragg. Nevertheless, the DH network in this case is similar to that of the Fort Bragg system (Section 3.1 [13]), i.e., the SHW mostly replaces boiler capacity. From an economic perspective, the new construction affords an opportunity to combine co-gen with SHW since the capacities for both can be determines and configured together. Also, the pipe sizes can be determined to accommodate the available SHW supply water temperature and to meet building requirements. Thus, one can optimize the solar share (higher at lower supply temperatures), larger pipe size (align with higher first costs), and other piping materials (like pre-insulated bonded pipes with PE-medium or steel medium pipes).

Figure B-4 shows the network model of the Fort Bliss DH system. The DH system can be used to provide domestic hot water and space heating, and to satisfy cooling energy demand. The SHW feed-in will be placed at the same location as the CEP.

New designed DH systems with de-central SHW feed-in

Some of the problems described in the previous section can be anticipated and avoided in the design of new DH system that will contain a number of distributed SHW feed-ins into the DH network.

Such a system will require a non-constant hydraulic flow simulation of the DH system. That is to say, the load curve must be simulated to reflect the different possible feed-in situations from the distributed SHW sites. These simulations can be used to derive the best suited distribution system, and the optimum way to connect new buildings to the system needs.

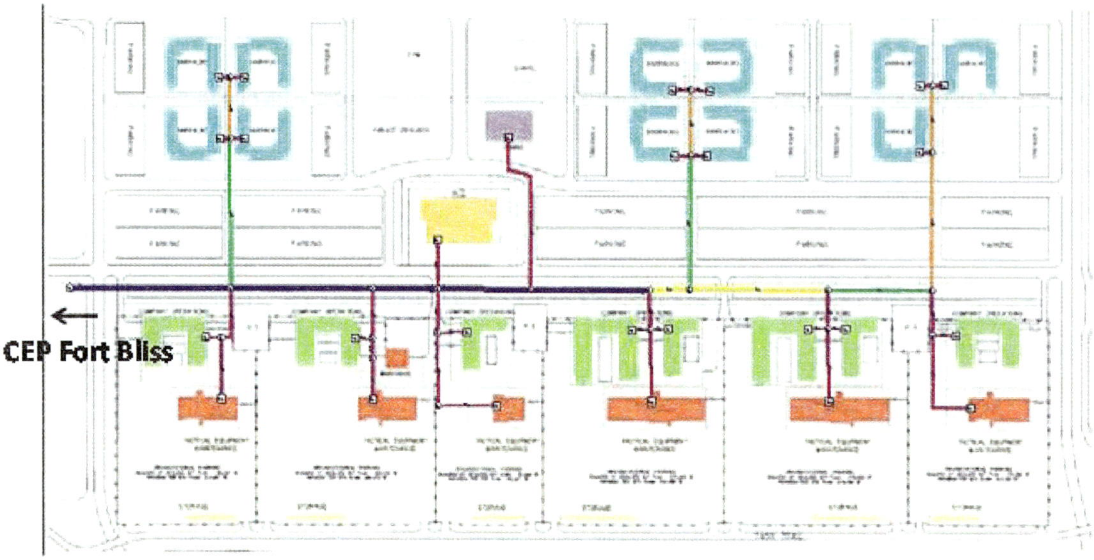

Figure B-4. Fort Bliss network model of the DH system.

Conclusions

In both centralized or de-centralized SHW systems, an SHW replaces part of the boiler heat generation. If the SHW temperature is too low, the boiler must boost the low level temperature of the SHW. Otherwise, if the SHW temperature is adequate, the boiler can be shut down several hours in the year when the heat demand does not exceed the available heat from the SHW. In a centralized system, the SHW can increase energy efficiency and lower heat generation costs relative to a de-central SHW or a fossil-fuel fired central system.

Both SHW and DH/DC systems have special characteristics that can be combined to achieve synergies. But if the characteristics of both systems are not well considered, the integration of both types of systems can result in economic and technical problems. When existing DH systems are upgraded to include SHW, the addition of SHW is, in most cases, similar to adding another boiler.

When SHW is designed as part of a newly constructed DH system, the temperature conditions of the SHW, the DH, and the connected buildings can be adjusted to achieve optimal efficiencies and economics.

The greatest challenge occurs when distributed SHW sites with distributed pumps are integrated into an existing DH system. Such an integration requires a significant effort in the design of the pipes sizes — an effort that is much easier in new construction than in retrofit or rehabilitation.

Fort Bliss

Design options for a solar system in Fort Bliss TX: District heating network for cooling energy demand retrofitted by a solar system furnished with vacuum pipe collectors

General information

- Fort Bliss TX, 31.8° N, 106.4° W
  Solar Water Heating (SWH) Connected to District Heating Networks (DH)
- First Estimation of Size Solar System: Demand of Cooling
- Collector: Vacuum pipe

Input data

- Climate
  - USA_TX_El.Paso.Intl.AP.722700_TMY3EPW.csv
  - Lowest temperature: - 42 °F (5.6 °C)
  - Highest temperature: 102 °F (38.9 °C)
  - Year average temperature: 64.40 °F (18.0 °C)
    Global irradiation (horizontal): 654,997 Btu/sq ft *yr (2065 kWh/m^2 *yr)

Energy demand DH (cooling) (demand profile received from GEF, Germany)

- Energy Demand (COOLING)
  Total energy demand: 4.16 MBtu/yr (14215 MWh)/yr
- Advance temperature DH: 203 °F (95 °C)
- Return temperature DH: 167 °F (75 °C)

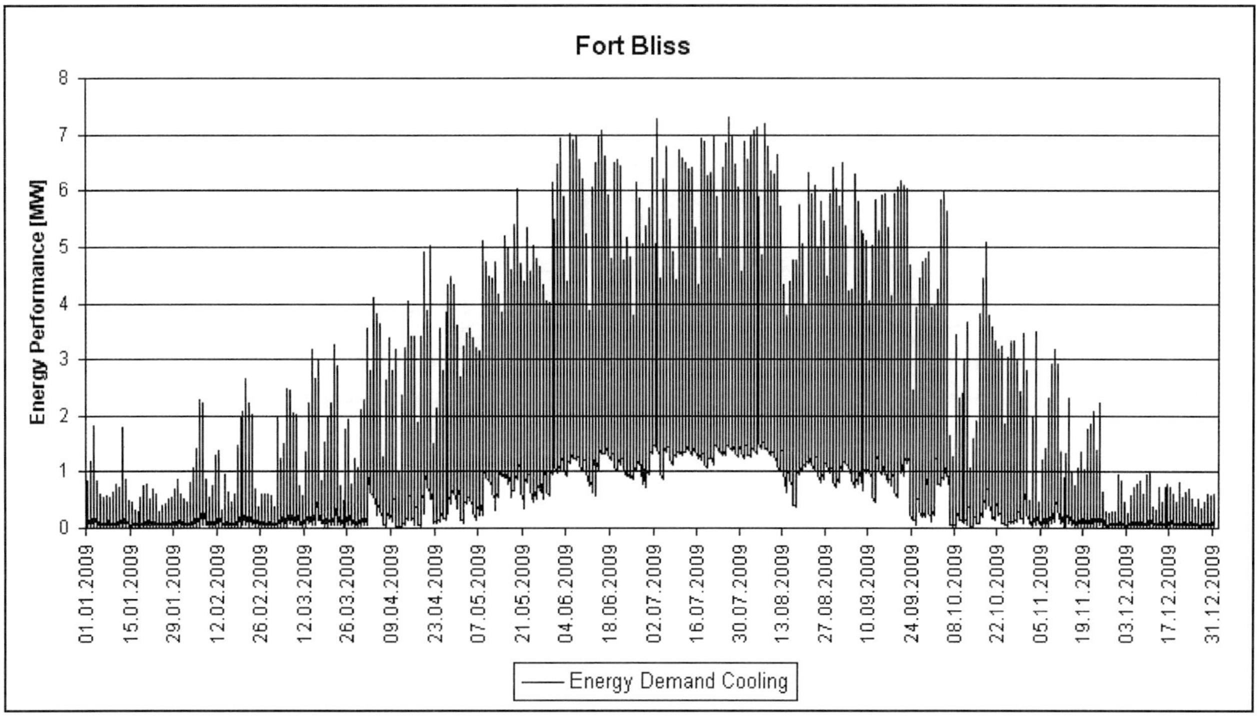

Figure B-5. Energy demand for cooling (Fort Bliss).

- o Solar Collector
  - * Type: Vacuum pipe collector
  - * Maker: Paradigma 45 Star Azzurro
  - * Collector area: 21,520 sq ft (2,000 m^2), 32,280 sq ft (3,000 m^2), 43,040 sq ft (4,000 m^2), 53,800 sq ft (5,000 m^2), 64,560 sq ft (6,000 m^2)
  - * Orientation: South
  - * Tilt angle: 30 degrees
  - * Shading: Shading, when sun lower than 10 degrees elevation
- o Collector Area Piping
  - * Length outdoor (one way): 1.64 ft (0.5 m) piping per 10.76 sq ft (1 m^2) collector area
    Length indoor (one way): 0.16 ft (0.05 m) piping per 10.76 sq ft (1 m^2) collector area
  - * Thickness isolation: Thickness isolation = diameter piping, but not more than 7.9 in. (200 mm)
  - * Heat transfer isolation: 0.09 Btu/hr-ft-°F (0.05 W/[m*K])
  - * Velocity in piping: 1.64 ft/s (0.5 m/s)
    Diameter piping: Depends on selected collector area
- o Heat Exchanger Collector Loop
  - * Temperature difference: 5 K (-450.67 °F)
- o Solar Storage Tank
  - * Type: Cylinder upright tank
  - * Number: 1
  - * Volume: 26,420 gal (100 m^3); 39,630 gal (150 m^3); 52,840 gal (200 m^3); 66,050 gal (250 m^3); 79,260 gal (300 m^3)
  - * Height to diameter: 5 to 1
  - * Thickness Isolation: 11.8 in. (300 mm)
  - * Heat transfer isolation: 0.09 Btu/hr-ft-°F (0.05 W/[m*K])

Used simulation program for calculations

The simulation program T*SOL Expert 4.5 Release 5, developed by Dr. Valentin EnergieSoftware GmbH, Berlin, Germany, was used to calculate the simulation results.

Figure B-6 shows required main components of the selected T*SOL system type A14.1: the collector area, heat exchanger, solar storage tank, and the integration of the district heating network. Note that the arrangement is highly simplified in comparison with a real solar system application and connection to a district heating network system. This design is used simply to gauge the needed size of the solar system. More detailed calculations require more sophisticated and laborious simulation programs as TRNSYS (developed by University of Wisconsin).

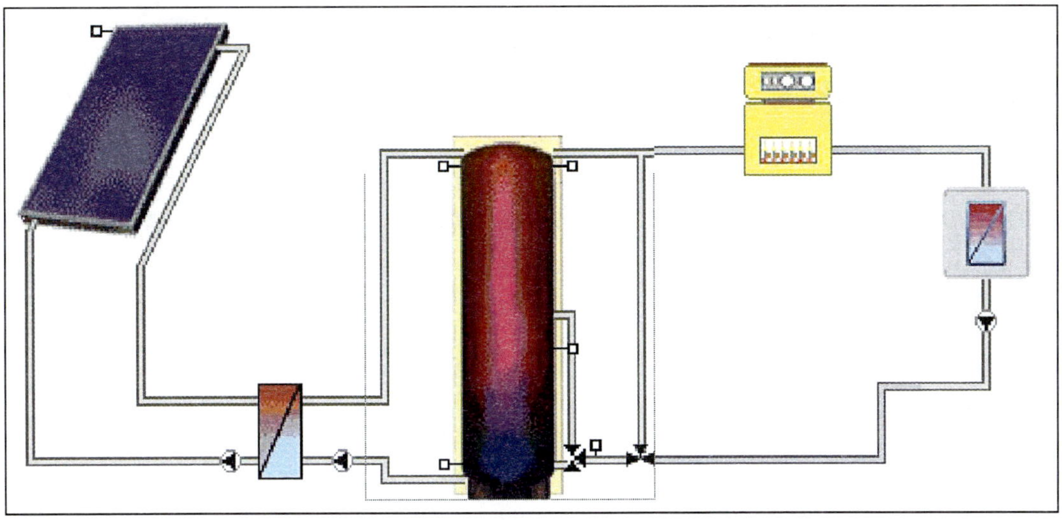

Figure B-6. System type.

Results

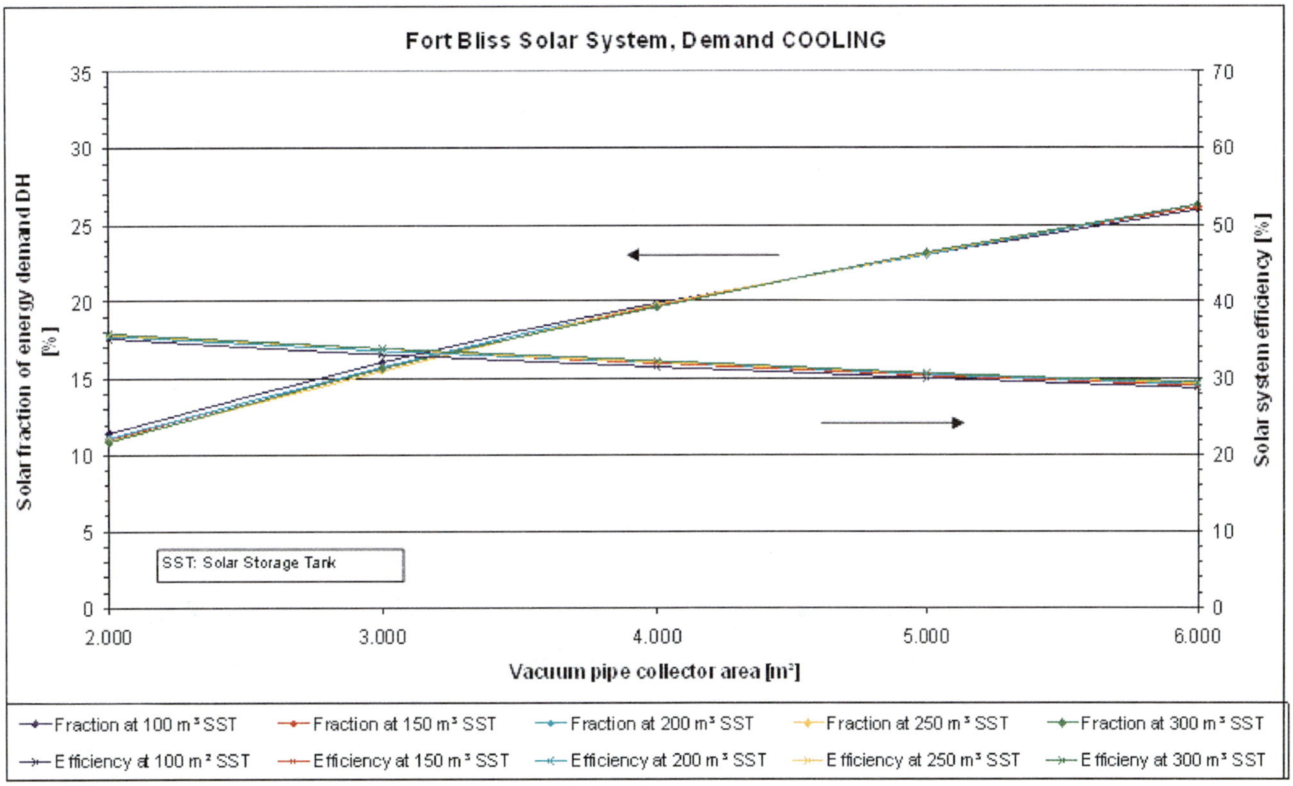

Figure B-7. Solar fraction and solar efficiency depending on vacuum pipe area, Fort Bliss, cooling.

Figure B-7 shows the collector area, solar storage tank volume, solar fraction of energy demand DH, and solar system efficiency for a given district heating network energy demand (DH Fort Bliss COOLING), and for the given climate for El Paso. Because of many assumptions made for this calculation (e.g., selected collector type, collector area arrangement, length of piping collector area, dimension of isolation, return temperature DH), the results shown here can give only a rough first impression of the size of a solar system that fits the demand of the given energy and climate:

- Because of the good correlation between cooling demand and solar radiation in summer, the solar energy production is nearly independent from the size of the solar storage tank.

- An increase of the size of the collector area causes only a slight decrease of the solar system efficiency, the solar fraction rises notably.
- A solar system design based on a solar fraction of 25% or more could be considered.

Design options for a solar system in Fort Bliss TX: District heating network for domestic hot water, space heating and cooling energy demand retrofitted by a solar system furnished with vacuum pipe collectors

- General Information
  - Fort Bliss TX, 31.8° N, 106.4° W
    Solar Water Heating (SWH) Connected to District Heating Networks (DH)
  - First Estimation of Size Solar System: Demand of DHW + SH + COOLING
  - Collector: Vacuum Pipe
- Input Data
  - Climate
    - * USA_TX_El.Paso.Intl.AP.722700_TMY3EPW.csv
    - * Lowest temperature: 22 °F (-5.6 °C)
    - * Highest temperature: 102 °F (38.9 °C)
    - * Year average temperature: 64.40 °F (18 °C)
      Global irradiation (horizontal): 654,997.36 Btu/sq ft *yr (2,065 kWh/m^2*yr)
  - **Energy Demand DH (DHW + SH + COOLING)**
    Demand profile received from GEF, Germany
    - * Total Building Cluster Energy Demand (DHW + SH + COOLING)
      Total energy demand: 7 MBtu/y5 (23,797 MWh/yr)
    - * Advance temperature DH: 203 °F (95 °C)
    - * Return temperature DH: 158 °F (70 °C)

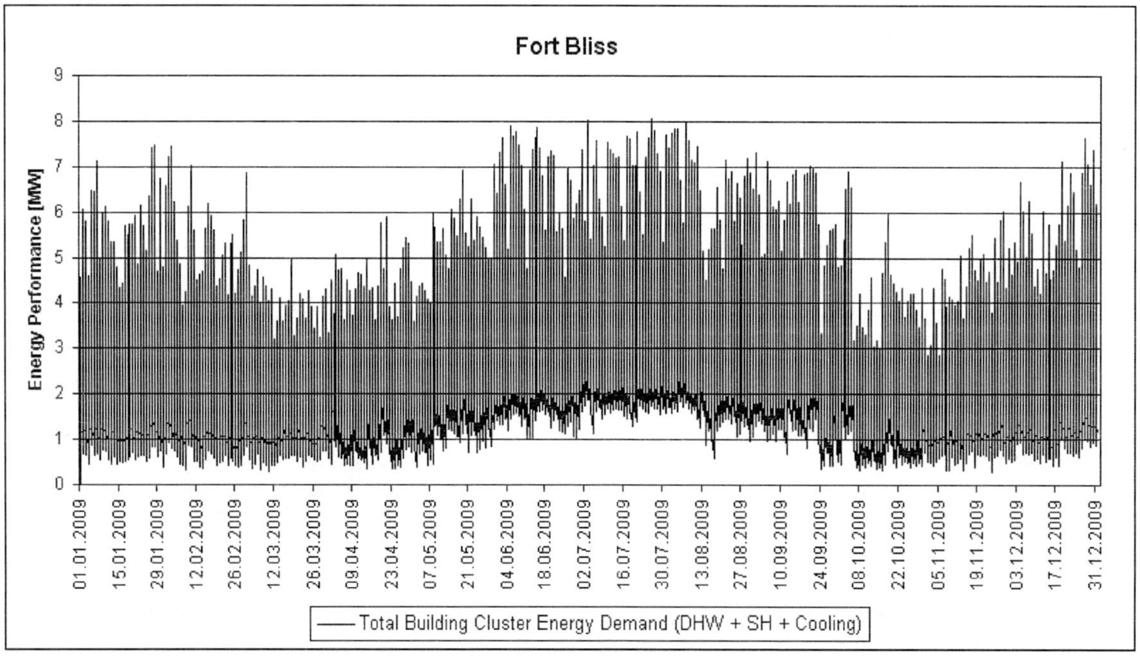

Figure B-8. Energy demand for DHW, space heating and cooling, Fort Bliss solar collector.

- Type: Vacuum pipe collector
  Maker: Paradigma CPC 45 Star azzurro

- Collector area: 43,040 sq ft (4,000 m²), 64,560 sq ft (6,000 m²); 86,080 sq ft (8,000 m²) ; 107,600 sq ft (10,000 m²); 129,120 sq ft (12,000 m²)
- Orientation: South
  Tilt angle: 30 degrees
- Shading: Shading, when sun lower than 10-degree elevation
- Collector Area Piping
  - Length outdoor (one way): 1.6 ft (0.5 m) piping per 10.8 sq ft (1 m²) collector area
    Length indoor (one way): 0.16 ft (0.05 m) piping per 10.8 sq ft (1 m²) collector area
  - Thickness isolation = diameter piping, but not more than 7.9 in. (200 mm)
  - Heat transfer isolation: 0.09 Btu/hr-ft-°F (0.05 W/[m*K])
  - Velocity in piping: 1.6 ft/s (0.5 m/s)
    Diameter piping: Depends on selected collector area
- Heat Exchanger Collector Loop
  - Temperature difference: 5 K (-450.67 °F)
- Solar Storage Tank
  - Type: Cylinder upright tank
  - Number: 1
  - Volume: 52,840 gal (200 m³) ; 79,260 gal (300 m³) ; 105,680 gal (400 m³) ; 132,100 gal (500 m³) ; 158,520 gal (600 m³)
  - Height to diameter: 5 to 1
  - Thickness Isolation: 11.8 in. (300 mm)
  - Heat transfer isolation: 0.09 Btu/hr-ft-°F (0.05 W/[m*K])

Used simulation program for calculations

The simulation program T*SOL Expert 4.5 Release 5, developed by Dr. Valentin EnergieSoftware GmbH, Berlin, Germany, was used to calculate the simulation results.

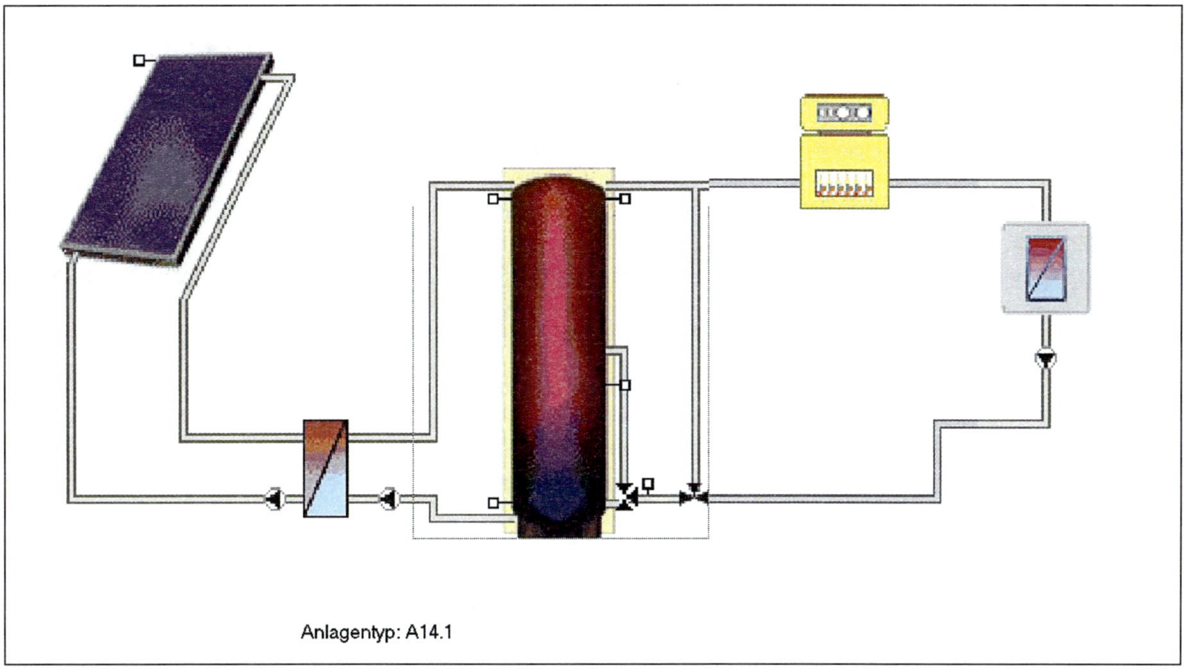

Figure B-9. System type.

Figure B-9 shows the required main components of the selected T*SOL system type: the collector area, heat exchanger, solar storage tank, and the integration of the district heating network. Note that the arrangement is highly simplified in comparison with a real solar system application and connection to a district heating network system. This design is used simply to gauge the needed size of the solar system. More detailed calculations require more sophisticated and laborious simulation programs as TRNSYS (developed by University of Wisconsin).

Results

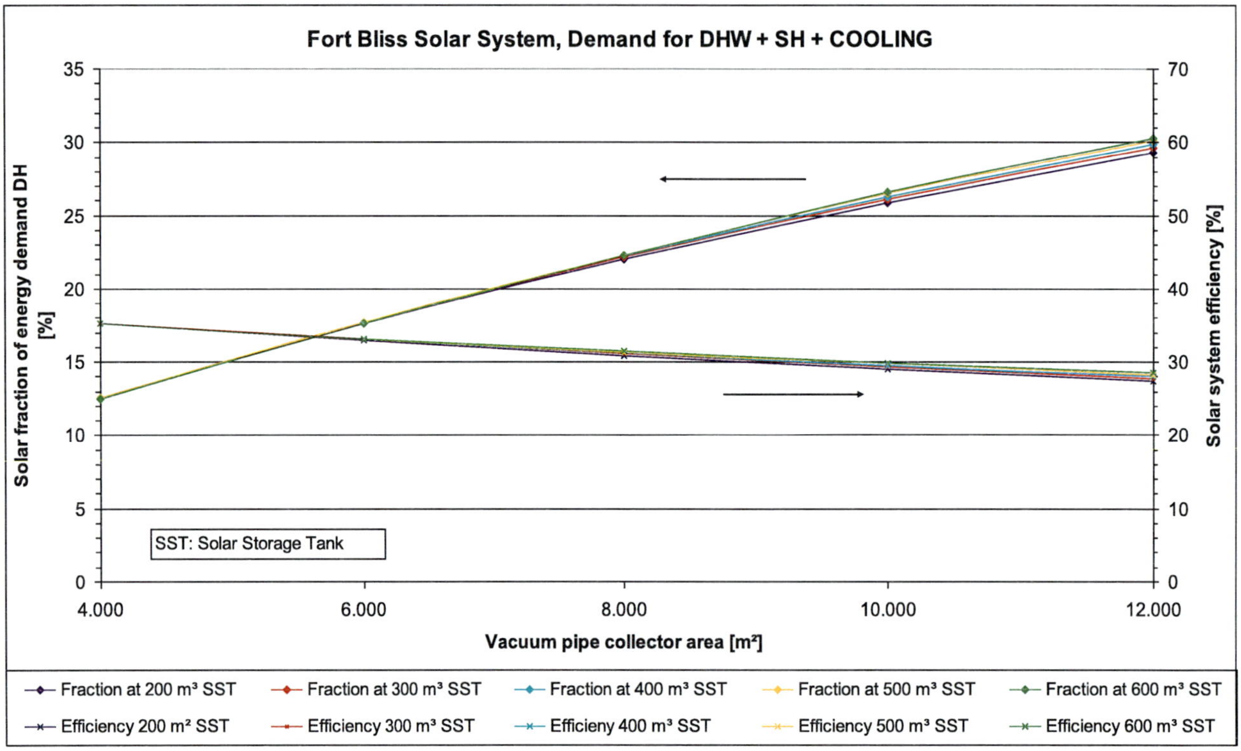

Figure B-10. Solar fraction and solar efficiency depending on vacuum pipe area, Fort Bliss, DHW, space heating and cooling.

Figure B-10 shows the collector area, solar storage tank volume, solar fraction of energy demand DH, and solar system efficiency for a given district heating network energy demand (DH Fort Bliss DHW + SH + COOLING) and the given climate for El Paso. Because of many assumptions made for this calculation (e.g., selected collector type, collector area arrangement, length of piping collector area, dimension of isolation, return temperature DH) the results shown in figure B-10 give only a rough first impression of the size of a solar system that fits the given demand of energy and the given climate:

- Because of the good correlation between cooling demand and solar radiation in summer, the solar energy production is nearly independent from the size of the solar storage tank. At a solar fraction exceeding 25%, the size of the solar storage tank may need to be slightly increased.
- An increase of the size of the collector area causes only a slight decrease of the solar system efficiency, although the solar fraction rises notably.
- A solar system design based on a solar fraction of 30% or more could be considered.

Design options for a solar system in Fort Bliss TX: District heating network for domestic hot water and space heating energy demand retrofitted by a solar system furnished with flat plate collectors

General information

- Fort Bliss TX, 31.8° N, 106.4° W
  Solar Water Heating (SWH) Connected to District Heating Networks (DH)
- First Estimation of Size Solar System: Demand of DHW + SH
- Collector: Flat Plate

Input data

Climate

- USA_TX_El.Paso.Intl.AP.722700_TMY3EPW.csv
- Lowest temperature: 22 °F (-5.6 °C)
- Highest temperature: 102 °F (39 °C)
- Year average temperature: 64 °F (18 °C)
  Global irradiation (horizontal): 654,997.36 Btu/sq ft *yr (2,065 kWh/m^2*yr)

Energy demand DH for DHW + SH (demand profile received from GEF, Germany)

- Energy Demand (DHW + SH)
  Total energy demand: 2.8 MBtu/yr (9,583 MWh/yr)
- Advance temperature DH: 203 °F (95 °C)
- Return temperature DH: 140 °F (60 °C)

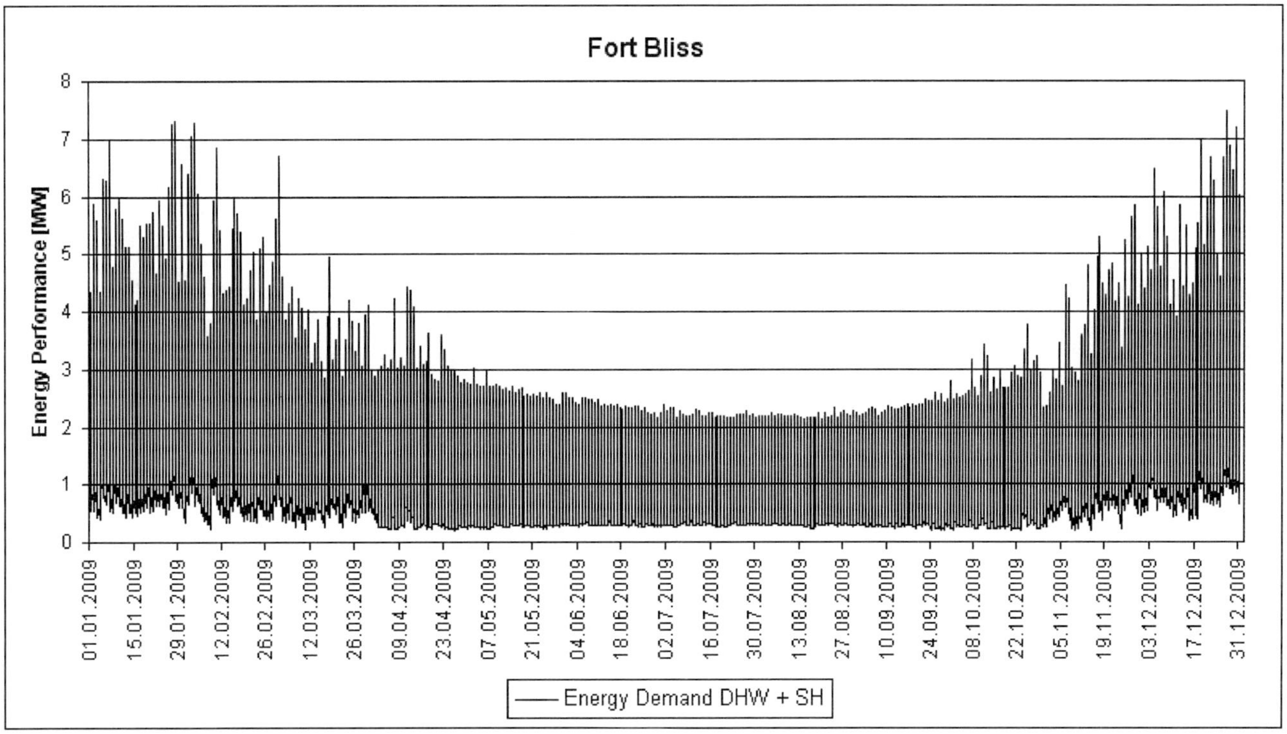

Figure B-11. Energy demand for DHW and space heating, Fort Bliss.

Solar collector

- Type: Flat plate collector
- Maker: Solvis Fera F80
- Collector area: 21,520 sq ft (2,000 m^2); 32,280 sq ft (3,000 m^2); 43,040 sq ft (4,000 m^2); 53,800 sq ft (5,000 m^2); 64,560 sq ft (6,000 m^2)
- Orientation: South
  Tilt angle: 30 degrees
- Shading: Shading, when sun lower than 10-degree elevation

Collector area piping

- Length outdoor (one way): 1.6 ft (0.5 m) piping per 10.8 sq ft (1 m^2) collector area
- Length indoor (one way): 0.16 ft (0.05 m) piping per 10.8 sq ft (1 m^2) collector area
- Thickness isolation: diameter piping, but not more than 7.9 in. (200 mm)
- Heat transfer isolation: 0.09 Btu/hr-ft-°F (0.05 W/[m*K])
- Velocity in piping: 1.6 ft/s (0.5 m/s)
  Diameter piping: Depends on selected collector area

Heat exchanger collector loop

- Temperature difference: 5 K (-450.67 °F)

Solar storage tank

- Type: Cylinder upright tank
- Number: 1
- Volume: 26,420 gal (100 m^3); 39,630 gal (150 m^3); 52,840 gal (200 m^3); 66,050 gal (250 m^3) ; 79,260 gal (300 m^3)
- Height to diameter: 5 to 1
- Thickness Isolation: 11.8 in. (300 mm)
- Heat transfer isolation: 0.09 Btu/hr-ft-°F (0.05 W/[m*K])

Used simulation program for calculations

The simulation program T*SOL Expert 4.5 Release 5, developed by Dr. Valentin EnergieSoftware GmbH, Berlin, Germany, was used to calculate the simulation results.

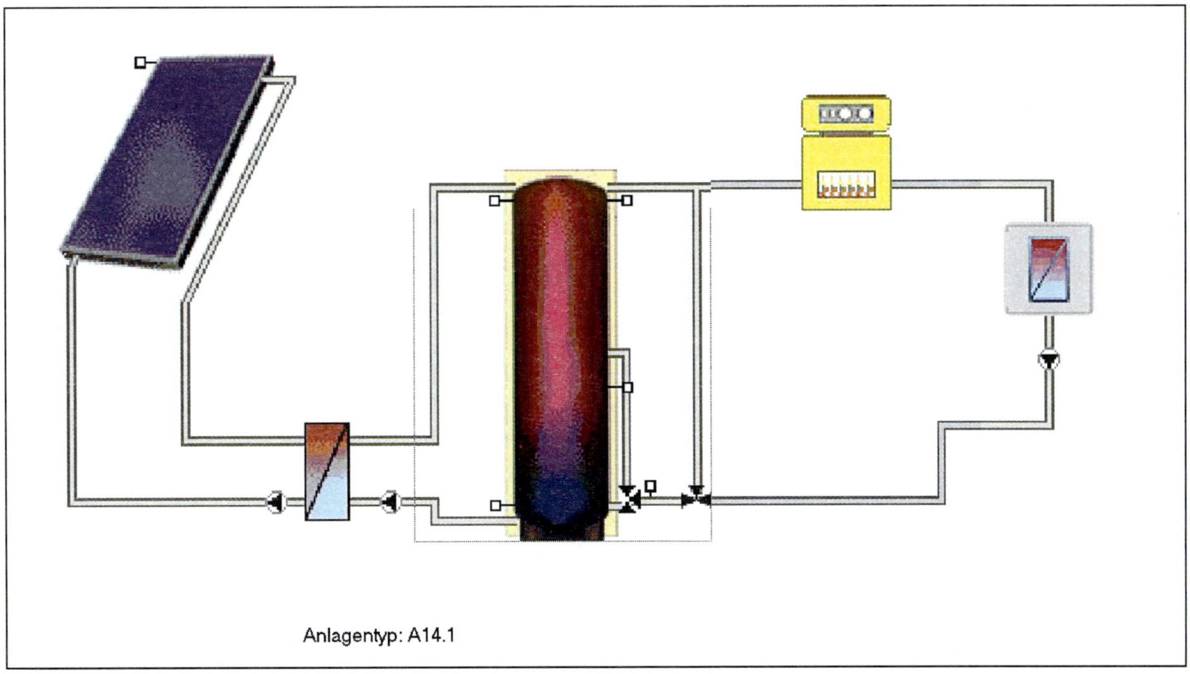

Figure B-12. System type.

Figure B-12 shows the required main components of the selected T*SOL system type A14.1: the collector area, heat exchanger, solar storage tank and the integration of the district heating network. Note that the arrangement is highly simplified in comparison with a real solar system application and connection to a district heating network system. This design is used simply to gauge the needed size of the solar system. More detailed calculations require more sophisticated and laborious simulation programs as TRNSYS (developed by University of Wisconsin).

Results

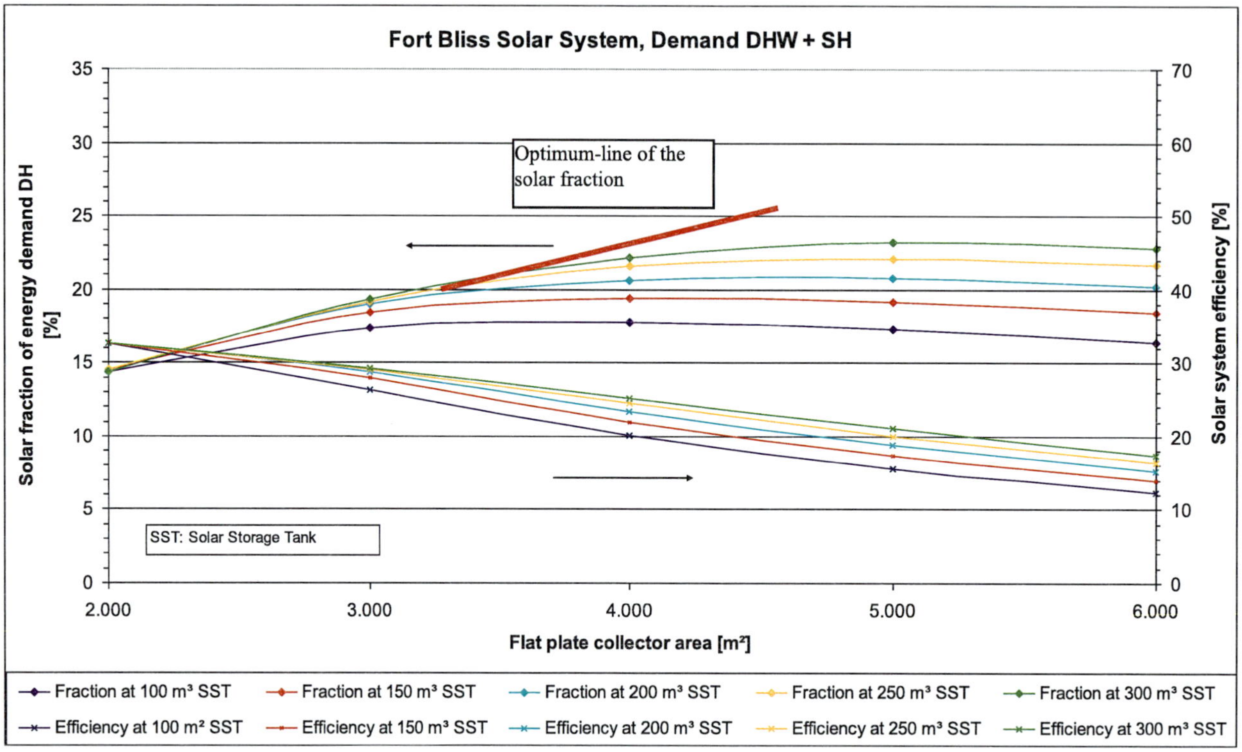

Figure B-13. Solar fraction and solar efficiency depending on flat plate area, Fort Bliss, DHW and space heating.

Diagram shows the coherence of collector area, solar storage tank volume, solar fraction of energy demand DH and solar system efficiency for a given district heating network (DH Fort Bliss DHW + SH) energy demand and a given climate for El Paso. Because of many assumptions made for this calculation (e.g., selected collector type, collector area arrangement, length of piping collector area, dimension of isolation, return temperature DH) the above shown results can give only a very first rough impression of the size of a solar system fitting to the given demand of energy and the given climate.

The optimum of the solar fraction depends on the collector area size and solar storage tank size, taken from the diagram:

- Solar fraction: 18%, storage tank: 26,420 gal (100 m^3), collector area: 37,660 sq ft (3,500 m^2)
- Solar fraction: 19%, storage tank: 39,630 gal (150 m^3), collector area: 40,888 sq ft (3,800 m^2)
- Solar fraction: 21%, storage tank: 52,840 gal (200 m^3), collector area: 44,116 sq ft (4,100 m^2)
- Solar fraction: 22%, storage tank: 66,050 gal (250 m^3), collector area: 47,344 sq ft (4,400 m^2)
- Solar fraction: 23%, storage tank: 79,260 gal (300 m^3), collector area: 50,572 sq ft (4,700 m^2)

Fort Bragg Design options for a solar system in Fort Bragg NC (COSCOM plant) — District heating network for domestic hot water and space heating energy demand retrofitted by a solar system furnished with vacuum pipe collectors

General

- Fort Bragg NC (COSCOM Plant), 35.13 N, 78.95 W
  Solar Water Heating (SWH) Connected to District Heating Networks (DH)

- First Estimation of Size Solar System
- Collector: Vacuum Pipe

Input Data

Climate

- USA_NC_Fort.Bragg-Simmons.AAF.746930_TMY3EPW.csv
- Lowest temperature: 3.9 °F (-15.6 °C)
- Highest temperature: 100 °F (37.8 °C)
- Year average temperature: 61 °F (16.2 °C)
- Global irradiation (horizontal): 480,225.7 Btu/sq ft*yr (1,514 kWh/m^{2*yr})

Energy Demand DH (DHW + SH)

- Heating and Cooling Master Plan for Fort Bragg, NC; February 2009, partly modified and corrected by ZfS
  Total energy demand (DHW + SH): 3,816 MWh/yr
- Advance temperature DH: 203 °F (95 °C)
- Return temperature DH: 140 °F (60 °C)

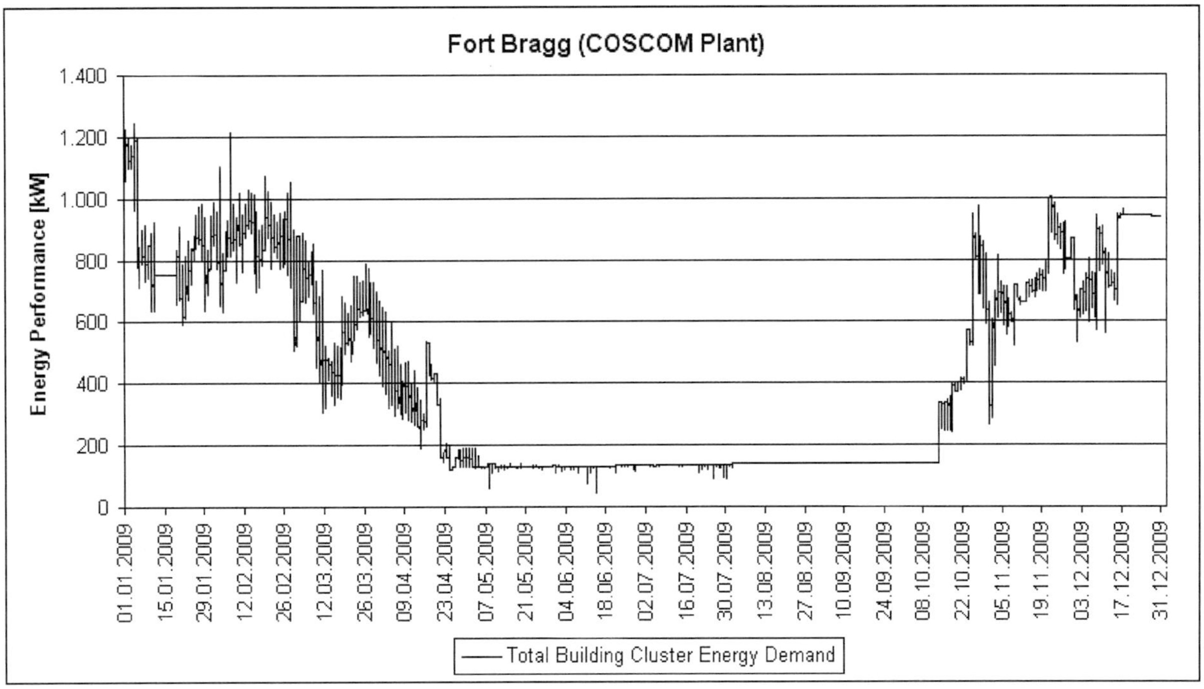

Figure B-14. Metered energy data from COSCOM Plant.

Solar Collector

- Type: Vacuum pipe collector
  Maker: Paradigma CPC45 Star azzurro
  Collector area: 5,380 sq ft (500 m^2); 10,760 sq ft (1,000 m^2); 16,140 sq ft (1,500 m^2); 21,520 sq ft (2,000 m^2); 26,900 sq ft (2,500 m^2)
- Orientation: South
  Tilt angle: 45°
- Shading: Shading, when sun lower than 10° elevation

Collector Area Piping

- Length outdoor (one way): 2 ft (0.5 m) piping per 11 sq ft (1 m^2) collector area
- Length indoor (one way): ~2 in. (0.05 m) piping per 10.76 sq ft (1 m^2) collector area
- Thickness isolation: Thickness isolation = diameter piping, but not more than ~8 in. (200 mm)
- Heat transfer isolation: 0 Btu/hr-ft-°F (0.05 W/[m*K])
- Velocity in piping: ~2 ft/s (0.5 m/s)
- Diameter piping: Depends on selected collector area

Heat Exchanger Collector Loop

- Temperature difference: 5 K (-450.67 °F)

Solar Storage Tank

- Type: Cylinder upright tank
- Number: 1
- Volume: 6,605 gal (25 m^3); 13,210 gal (50 m^3); 19,815 gal (75 m^3); 26,420 gal (100 m^3); 33,025 gal (125 m^3)
- Height to diameter: 5 to 1
- Thickness isolation: 12 in. (300 mm)
- Heat transfer isolation 0.09 Btu/hr-ft-°F (0.05 W/[m*K])

Used Simulation Program for Calculations

The simulation program T*SOL Expert 4.5 Release 5, developed by Dr. Valentin EnergieSoftware GmbH, Berlin, Germany, was used to calculate the simulation results.

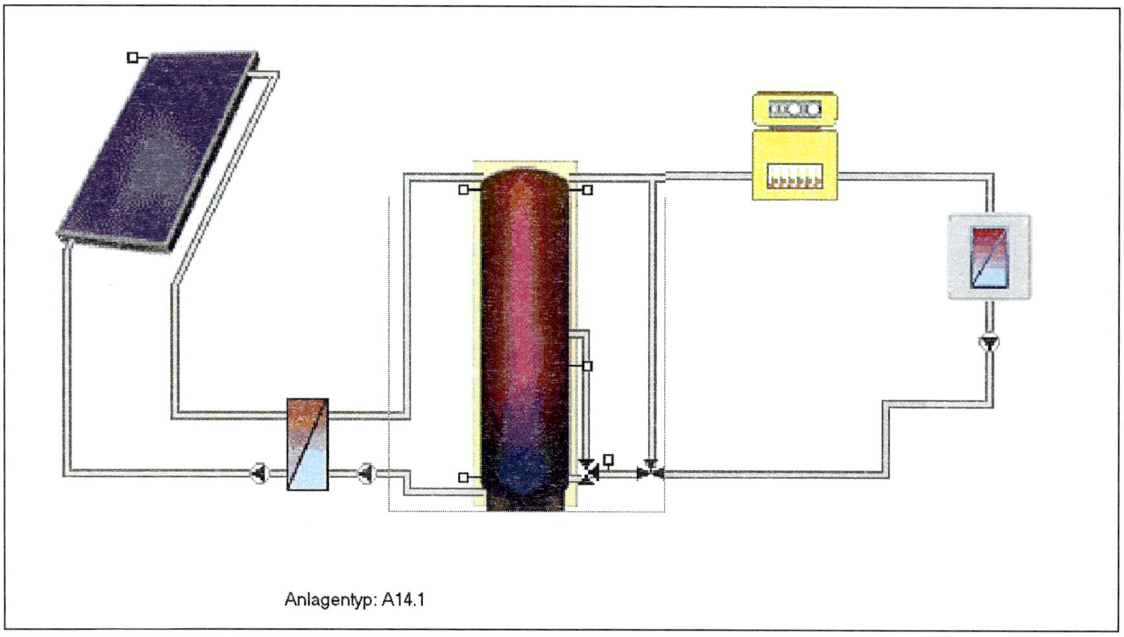

Figure B-15. System type.

Figure B-15 shows the required main components of the selected T*SOL system type A14.1: the needed as collector area, heat exchanger, solar storage tank and the integration of the district heating network. Note that the arrangement is highly simplified in comparison with a real solar system application and connection to a district heating network system. This design is used simply to gauge the needed size of the solar system. More detailed calculations require more sophisticated and laborious simulation programs as TRNSYS (developed by University of Wisconsin).

Results

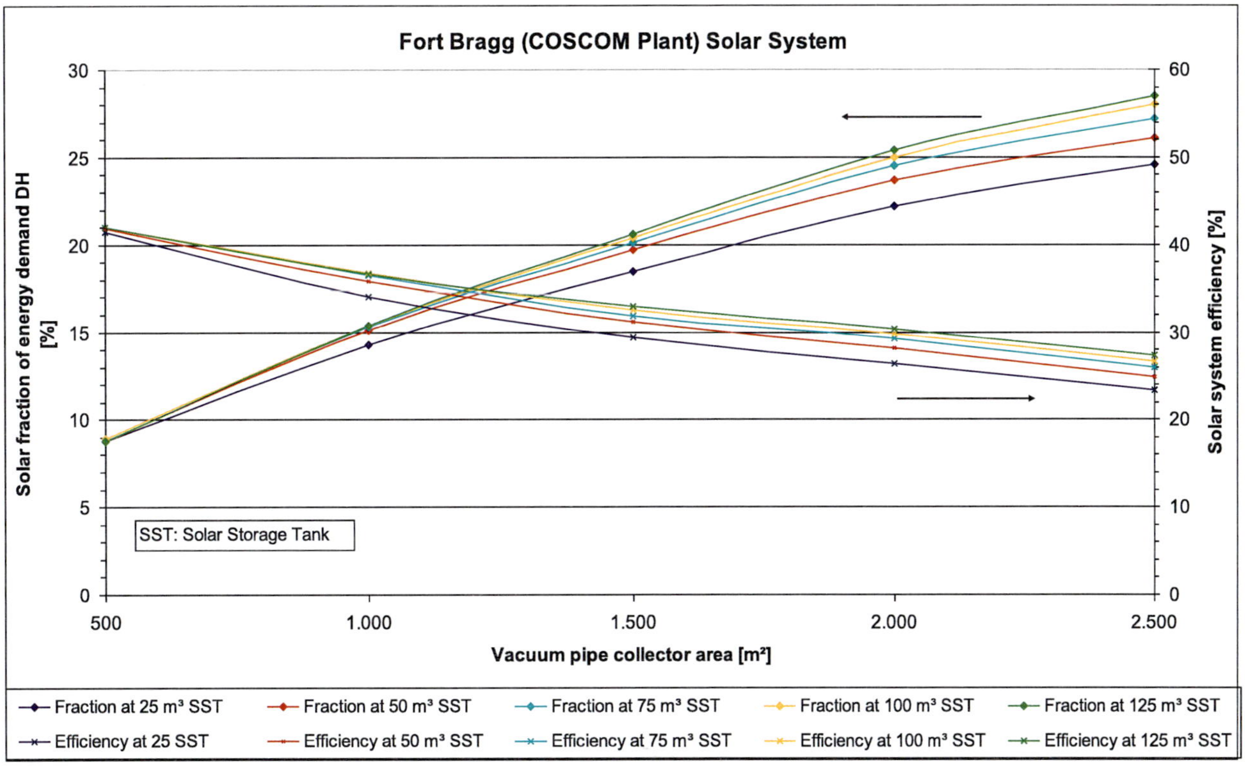

Figure B-16. Solar fraction and solar efficiency depending on vacuum pipe area, Fort Bragg, DHW and space heating.

Figure B-16 shows the collector area, solar storage tank volume, fraction of solar thermal production of energy demand DH and solar system efficiency for a given district heating network energy demand (DH Fort Bragg COSCOM Plant), and a given climate for Fort Bragg. Because of many assumptions made for this calculation (e.g., selected collector type, collector area arrangement, length of piping collector area, dimension of isolation, return temperature DH), the results shown in Figure B-16 can give only a rough first impression of the size of a solar system that fits the given energy demand and climate.

Table B-33 lists the required collector areas and tank volumes required to reach the same solar fraction either by fitting the solar system with flat plate collectors or with vacuum pipe collectors.

Table B-33. Flat plate and vacuum pipe collector area and solar storage volume at the same solar fraction.

	10% solar fraction	**15% solar fraction**	**20% solar fraction**	**25% solar fraction**
Flat plate collector Collector area Solar storage tank	9,684 sq ft (900 m^2) 13,210 gal (50 m^3)	17,216 sq ft (1,600 m^2) 26,420 gal (100 m^3)	32,280 sq ft (3,000 m^2) 39,630 gal (150 m^3)	Not possible to reach with flat plate collectors
Vacuum pipe collector Collector area Solar storage tank	6,456 sq ft (600 m^2) 6,605 gal (25 m^3)	10,760 sq ft (1,000 m^2) 13,210 gal (50 m^3)	16,140 sq ft (1,500 m^2) 19,815 gal (75 m^3)	21,520 sq ft (2,000 m^2) 26,420 gal (100 m^3)

Example

A solar fraction of 15% can be reached either by using flat plate collectors with:

- collector area: 17,216 sq ft (1,600 m^2), solar storage tank capacity: 26,420 gal (100 m^3).

or by using vacuum pipe collectors:

- collector area: 10,760 sq ft (1,000 m^2), solar storage tank capacity: 13,210 gal (50 m^3).

Fort Bragg Design options for a solar system in Fort Bragg NC (COSCOM plant) — District Heating Network for domestic hot water and space heating energy demand retrofitted by a solar system furnished with flat plate collectors

General

- Fort Bragg NC (COSCOM Plant), 35.13 N, 78.95 W
- Solar Water Heating (SWH) Connected to District Heating Networks (DH)
- First Estimation of Size Solar System
- Collector: Flat Plate

Input Data

Climate

- USA_NC_Fort.Bragg-Simmons.AAF.746930_TMY3EPW.csv
- Lowest temperature: 3.92 °F (-15.6 °C)
- Highest temperature: 100 °F (37.8 °C)
- Year average temperature: 61 °F (16.2 °C)
- Global irradiation (horizontal): 480 MBtu/sq ft*yr (1,514 kWh/(m^2*yr))

Energy Demand DH (DHW + SH)

- Heating and Cooling Master Plan for Fort Bragg, NC; February 2009, partly modified and corrected by ZfS
  Total energy demand (DHW + SH): 1.1 MBtu/yr (3,816 MWh/yr)
- Advance temperature DH: 203 °F (95 °C)
- Return temperature DH: 140 °F (60 °C)

Solar collector

- Type: Flat plate collector
  Maker: Solvis Fera F80
- Collector area: 10,760 sq ft (1,000 m^2) ; 16,140 sq ft (1,500 m^2) ; 21,520 sq ft (2,000 m^2) ; 26,900 sq ft (2,500 m^2) ; 32,280 sq ft (3,000 m^2)
- Orientation: South
  Tilt angle: 45°
- Shading: Shading, when sun lower than 10° elevation.

Collector area piping

- Length outdoor (one way): 1.6 ft (0.5 m) piping per 10.78 sq ft (1 m^2) collector area
- Length indoor (one way): 0.16 ft (0.05 m) piping per 10.78 sq ft (1 m^2)collector area
- Thickness isolation: Thickness isolation = diameter piping, but not more than 200 mm
- Heat transfer isolation: 0 Btu/hr-ft-°F (0.05 W/[m*K])
- Velocity in piping: 1.6 ft/s (0.5 m/s)

- Diameter piping: Depends on selected collector area.

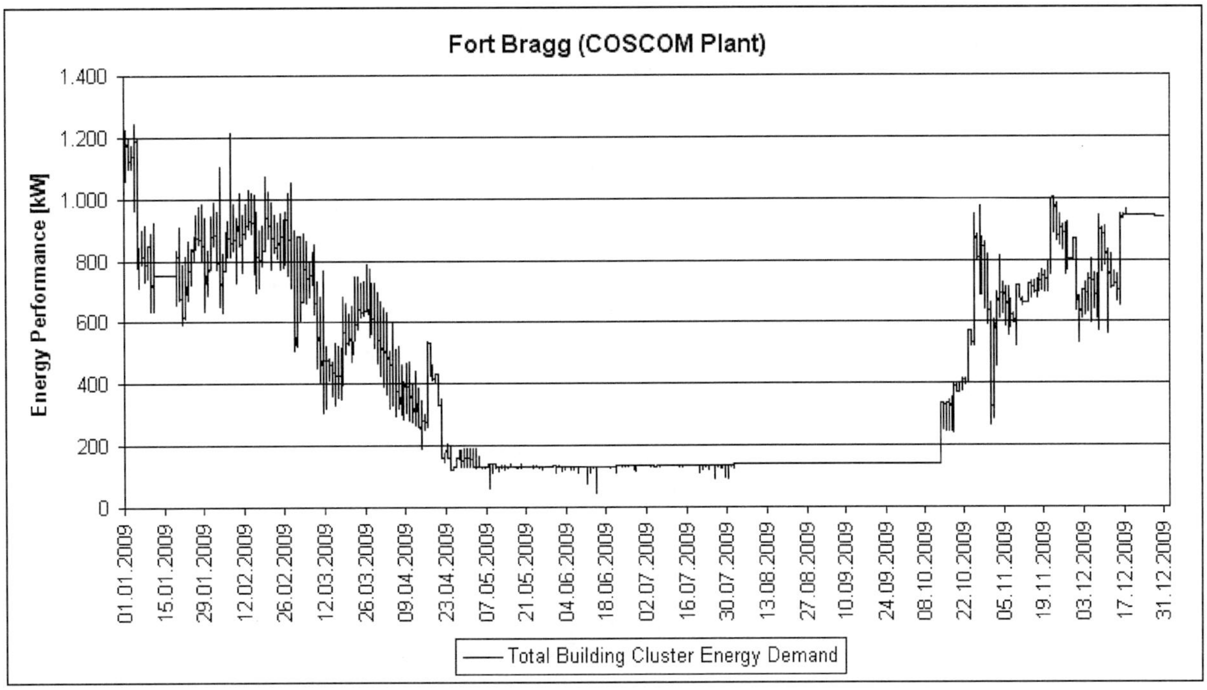

Figure B-17. Metered energy data from COSCOM Plant.

Heat Exchanger Collector Loop

- Temperature difference: (9 °F (5 K)

Solar Storage Tank

- Type: Cylinder upright tank
- Number: 1
- Volume: 13,210 gal (50 m^3); 19,815 gal (75 m^3); 26,420gal (100 m^3); 33,025gal (125 m^3); 39,630 gal (150 m^3)
- Height to diameter: 5 to 1
- Thickness isolation: 11.8 in. (300 mm)
- Heat transfer isolation: 0 Btu/hr-ft-°F (0.05 W/[m*K]).

Used Simulation Program for Calculations

The simulation program T*SOL Expert 4.5 Release 5, developed by Dr. Valentin EnergieSoftware GmbH, Berlin, Germany, was used to calculate the simulation results.

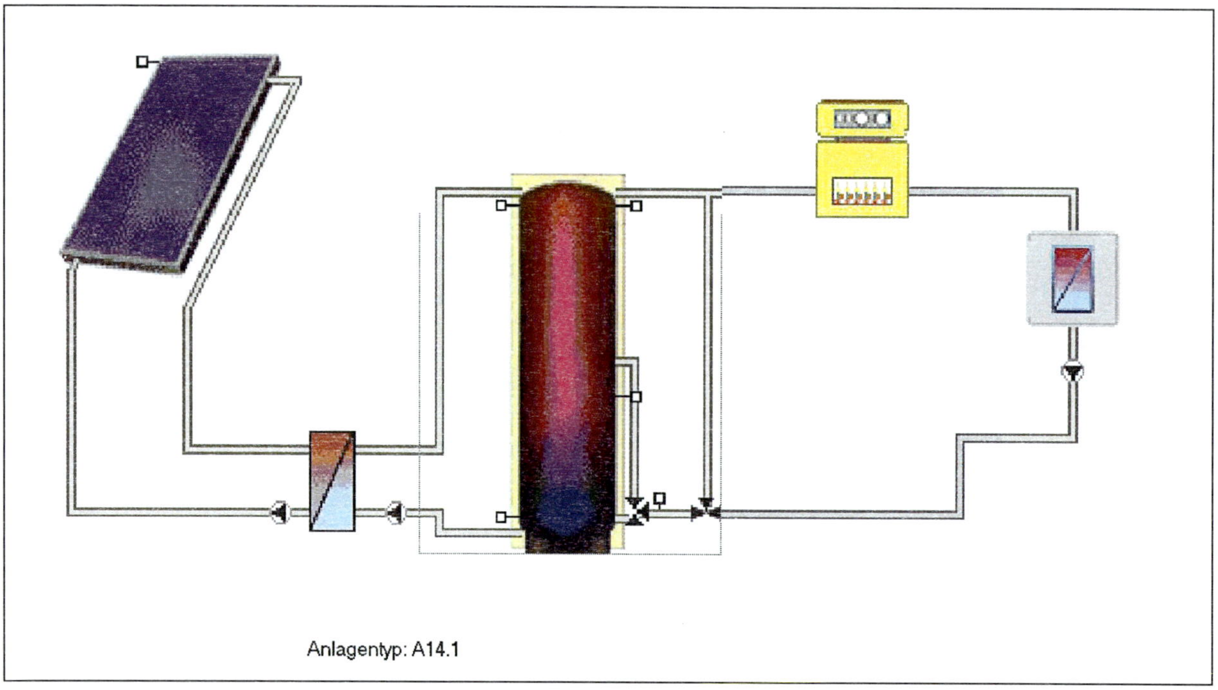

Figure B-18. System type.

Figure B-18 shows the required main components for the selected T*SOL system type A14.1: the collector area, heat exchanger, solar storage tank and the integration of the district heating network. Note that the arrangement is highly simplified in comparison with a real solar system application and connection to a district heating network system. This design is used simply to gauge the needed size of the solar system. More detailed calculations require more sophisticated and laborious simulation programs as TRNSYS (developed by University of Wisconsin).

Figure B-19 shows the collector area, solar storage tank volume, solar fraction of energy demand DH and solar system efficiency for a given district heating network energy demand (DH Fort Bragg COSCOM Plant) and a given climate for Fort Bragg. Because of many assumptions made for this calculation (e.g., selected collector type, collector area arrangement, length of piping collector area, dimension of isolation, return temperature DH), the above shown results can give only a rough first impression of the size of a solar system that fits the given demand of energy and the given climate.

The optimum of the solar fraction depends on the collector area size and solar storage tank size, Taken from the diagram is:

- Solar fraction: 16.0%, storage tank: 13,210 gal (50 m^3), collector area: 24,210 sq ft (2,250 m^2)
- Solar fraction: 17.0%, storage tank: 19,815 gal (75 m^3), collector area: 26,900 sq ft (2,500 m^2)
- Solar fraction: 18.0%, storage tank: 26,420 gal (100 m^3), collector area: 29,052 sq ft (2,700 m^2)
- Solar fraction: 19.0%, storage tank: 33,025 gal (125 m^3), collector area: 30,128 sq ft (2,800 m^2)
- Solar fraction: 19.5%, storage tank: 39,630 gal (150 m^3), collector area: 31,204 sq ft (2,900 m^2)

Fort Irwin

No simulation available

Results

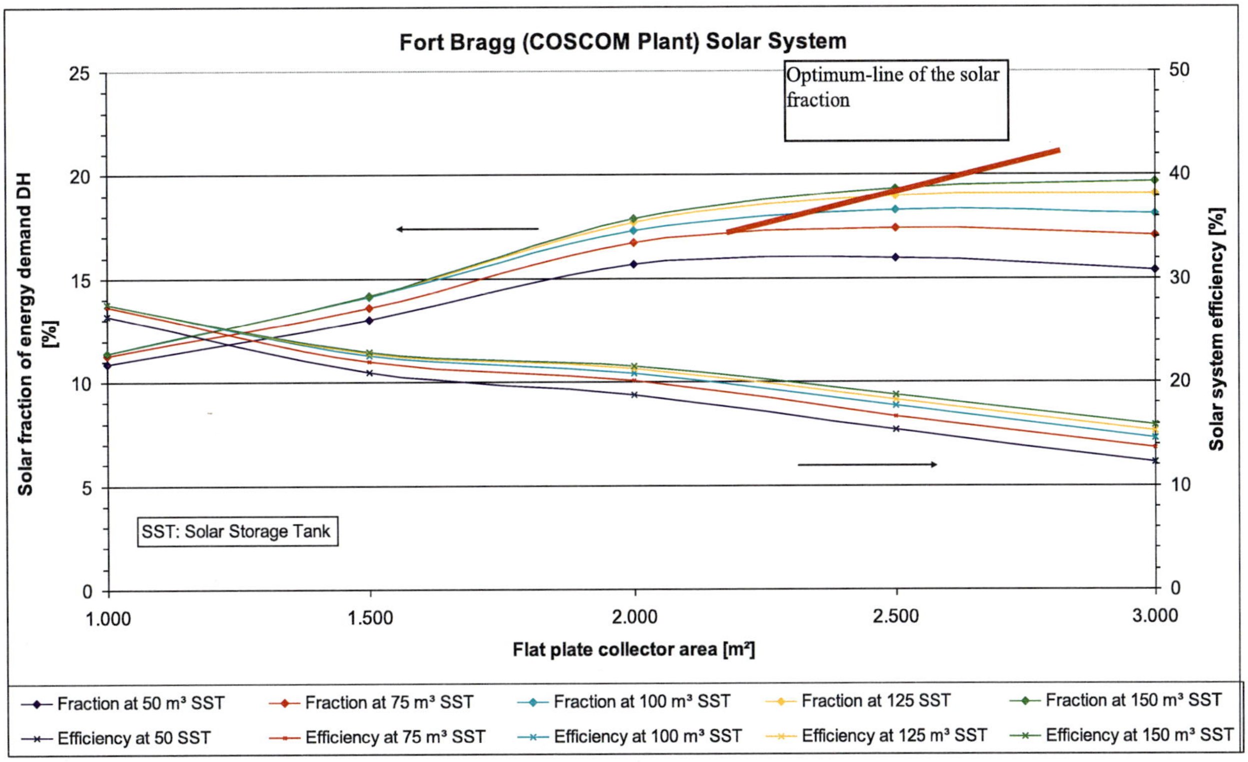

Figure B-19. Solar fraction and solar efficiency depending on flat plate area, Fort Bragg, DHW and space heating.

Comparison of economic results — Summary simulation results for Fort Bliss and Fort Bragg

Boundaries of the economical calculations and system costs

The economic calculation of simple payback is based on the submissions of the solar system companies Arcon (Denmark) and Paradigma (Germany) for a 53,800 sq ft (5,000 m^2) solar system with 132,100 gal (500 m^3) solar storage. To calculate the price for a 53,800 sq ft (5,000 m^2) system with a smaller storage (e.g., 26,420 gal [100 m^3]) it is assumed, that the specific costs of a 26,420 gal (100 m^3) storage are 10% more expensive than the 132,100 gal (500 m^3) storage. The specific price of other storage sizes between 26,420 and 132,100 gal (100 and 500 $m^{3)}$ was interpolated accordingly.

For simplicity, it is assumed that there are no economies of scale in the specific costs of the collector field (€/m^2). The specific solar system costs of a 500 – 5,000 m^2 system with the same storage size remains the same.

Additional costs for planning amounted to 20% of the submitted system costs (without cost for shipping and supervision). All costs exclude VAT.

The formula to calculate simple payback is:

(System-costs) / (saved energy costs per year)

Boundaries to calculate the saved energy cost per year

- Price of conventional energy (1 year): $0.01.2/kBtu (3 €-Cent per kWh)
- Price increase fuel / year: 5%
- Efficiency of conventional system during summer: 65% (time when solar system is mainly used)

Using these criteria, the system costs will be:

- Systems with flat plate collectors costs: ~300 - 326 €/m^2
- Systems with vacuum pipes: 400-430 €/m^2.

Table B-34. Specific costs of solar systems with flat plate collectors connected to district heating net in Fort Bliss or Fort Bragg dependent on m^2 collector area and m^3 solar storage volume.

IP	**1,000 m^2**	**1,500 m^2**	**2,000 m^2**	**2,500 m^2**	**3,000 m^2**	**6,000 m^2**
13,210 gal	$40/sq ft	$40/sq ft	$40/sq ft	$40/sq ft	$40/sq ft	$40/sq ft
19,815 gal	$40/sq ft	$40/sq ft	$40/sq ft	$40/sq ft	$40/sq ft	$40/sq ft
26,420 gal	$40/sq ft	$40/sq ft	$40/sq ft	$40/sq ft	$40/sq ft	$40/sq ft
33,025 gal	$41/sq ft	$41/sq ft	$41/sq ft	$41/sq ft	$41/sq ft	$41/sq ft
39,630 gal	$41/sq ft	$41/sq ft	$41/sq ft	$41/sq ft	$41/sq ft	$41/sq ft
79,260 gal	$43/sq ft	$43/sq ft	$43/sq ft	$43/sq ft	$43/sq ft	$43/sq ft
SI	**1,000 m^2**	**1,500 m^2**	**2,000 m^2**	**2,500 m^2**	**3,000 m^2**	**6,000 m^2**
50 m^3	299 €/m^2	299 €/m^2	299 €/m^2	299 €/m^2	299 €/m^2	299 €/m^2
75 m^3	302 €/m^2	302 €/m^2	302 €/m^2	302 €/m^2	302 €/m^2	302 €/m^2
100 m^3	305 €/m^2	305 €/m^2	305 €/m^2	305 €/m^2	305 €/m^2	305 €/m^2
125 m^3	308 €/m^2	308 €/m^2	308 €/m^2	308 €/m^2	308 €/m^2	308 €/m^2
150 m^3	310 €/m^2	310 €/m^2	310 €/m^2	310 €/m^2	310 €/m^2	310 €/m^2
300 m^3	326 €/m^2	326 €/m^2	326 €/m^2	326 €/m^2	326 €/m^2	326 €/m^2

Table B-35. Absolute costs of solar systems with flat plate collectors connected to district heating net in Fort Bliss or Fort Bragg dependent on m^2 collector area and m^3 solar storage volume.

IP	**10,760 sq ft**	**16,140 sq ft**	**21,520 sq ft**	**26,900 sq ft**	**32,280 sq ft**	**64,560 sq ft**
13,210 gal	$424,580	$637,580	$849,160	$1,062,160	$1,273,740	$2,547,480
19,815 gal	$428,840	$643,260	$857,680	$1,072,100	$1,286,520	$2,573,040
26,420 gal	$433,100	$648,940	$866,200	$1,082,040	$1,297,880	$2,598,600
33,025 gal	$437,360	$654,620	$873,300	$1,091,980	$1,310,660	$2,624,160
39,630 gal	$440,200	$661,720	$881,820	$1,101,920	$1,322,020	$2,641,200
79,260 gal	$462,920	$694,380	$925,840	$1,157,300	$1,388,760	$2,777,520
SI	**1,000 m^2**	**1,500 m^2**	**2,000 m^2**	**2,500 m^2**	**3,000 m^2**	**6,000 m^2**
50 m^3	299,000 €	449,000 €	598,000 €	748,000 €	897,000 €	1,794,000 €
75 m^3	302,000 €	453,000 €	604,000 €	755,000 €	906,000 €	1,812,000 €
100 m^3	305,000 €	457,000 €	610,000 €	762,000 €	914,000 €	1,830,000 €
125 m^3	308,000 €	461,000 €	615,000 €	769,000 €	923,000 €	1,848,000 €
150 m^3	310,000 €	466,000 €	621,000 €	776,000 €	931,000 €	1,860,000 €
300 m^3	326,000 €	489,000 €	652,000 €	815,000 €	978,000 €	1,956,000 €

Table B-36. Absolute costs of solar systems with vacuum pipe collectors connected to district heating net in Fort Bliss or Fort Bragg dependent on m² collector area and m³ solar storage volume.

IP	21,520 sq ft	43,040 sq ft	64,560 sq ft	86,080 sq ft	107,600 sq ft	129,120 sq ft
26,420 gal	$53/sq ft	$53/sq ft	$53/sq ft			
52,840 gal	$53/sq ft	$53/sq ft	$53/sq ft	$53/sq ft	$53/sq ft	$53/sq ft
79,260 gal	$54/sq ft	$54/sq ft	$54/sq ft	$54/sq ft	$54/sq ft	$54/sq ft
105,680 gal		$55/sq ft	$55/sq ft	$55/sq ft	$55/sq ft	$55/sq ft
132,100 gal		$56/sq ft	$56/sq ft	$56/sq ft	$56/sq ft	$56/sq ft
158,520 gal		$57/sq ft	$57/sq ft	$57/sq ft	$57/sq ft	$57/sq ft
SI	**2,000 m²**	**4,000 m²**	**6,000 m²**	**8,000 m²**	**10,000 m²**	**12,000 m²**
100 m³	397 €/m²	397 €/ m²	397 €/ m²			
200 m³	404 €/ m²	404 €/ m²	404 €/ m²	404 €/ m²	404 €/ m²	404 €/ m²
300 m³	411 €/ m²	411 €/ m²	411 €/ m²	411 €/ m²	411 €/ m²	411 €/ m²
400 m³		418 €/ m²	418 €/ m²	418 €/ m²	418 €/ m²	418 €/ m²
500 m³		425 €/ m²	425 €/ m²	425 €/ m²	425 €/ m²	425 €/ m²
600 m³		431 €/ m²	431 €/ m²	431 €/ m²	431 €/ m²	431 €/ m²

Table B-37. Absolute costs of solar systems with vacuum pipe collectors connected to district heating net in Fort Bliss or Fort Bragg dependent on m² collector area and m³ solar storage volume.

IP	21,520 sq ft	43,040 sq ft	64,560 sq ft	86,080 sq ft	107,600 sq ft	129,120 sq ft
26,420 gal	$1,126,060	$2,253,540	$3,379,600			
52,840 gal	$1,148,780	$2,296,140	$3,444,920	$4,592,280	$5,741,060	$6,888,420
79,260 gal	$1,168,660	$2,337,320	$3,505,980	$4,674,640	$5,843,300	$7,011,960
105,680 gal		$2,375,660	$3,564,200	$4,752,740	$5,939,860	$7,128,400
132,100 gal		$2,412,580	$3,619,580	$4,825,160	$6,032,160	$7,239,160
158,520 gal		$2,448,080	$3,672,120	$4,894,740	$6,118,780	$7,342,820
SI	**2,000 m²**	**4,000 m²**	**6,000 m²**	**8,000 m²**	**10,000 m²**	**12,000 m²**
100 m³	793,000 €	1,587,000 €	2,380,000 €			
200 m³	809,000 €	1,617,000 €	2,426,000 €	3,234,000 €	4,043,000 €	4,851,000 €
300 m³	823,000 €	1,646,000 €	2,469,000 €	3,292,000 €	4,115,000 €	4,938,000 €
400 m³		1,673,000 €	2,510,000 €	3,347,000 €	4,183,000 €	5,020,000 €
500 m³		1,699,000 €	2,549,000 €	3,398,000 €	4,248,000 €	5,098,000 €
600 m³		1,724,000 €	2,586,000 €	3,447,000 €	4,309,000 €	5,171,000 €

Discussion of calculation-results from Fort Bliss

Comparison of the energy yield and payback time of solar systems with vacuum pipe collectors in district nets with cooling demand

Tables B-38 and B-39 list the calculated specific solar energy yield (kWh/[yr·m²]) for district heating nets that meet cooling demand (Table B-38: cooling, Table B-39: DHW, space heating and cooling). The calculations were done at a yearly average net return temperature of 167 °F (75 °C) using vacuum pipe collectors for cooling and 158 °F (70 °C) for DHW, space heating and cooling. The minimum size of the collector area was configured to ensure that that the solar fraction does not fall below 10%.

In the case where the solar system is connected to a district heating net that is used solely for cooling (Table B-38), the specific solar yield per m² collector area lies in the range of about

810 - 620 kWh/(yr*m^2) (2567–197 kBtu/[yr*sq ft]). The highest solar yield per m^2 collector area can be achieved with a 2,000 m^2-collector-field. A storage volume of 100 m^3 will be sufficient. The solar fraction at this dimensioning is about 11% (see Figure B-19). In a detailed planning phase, even a solar system without solar storage can be taken into account.

Collector fields smaller than 21,520 sq ft (2,000 m^2) might have a slightly higher specific yield. They were not calculated because the solar fraction will drop to marginal values lower than 10%.

Table B-38. Specific solar yield in Btu/(yr sq ft) [kWh/(yr·m²)] Fort Bliss, vacuum pipe-collectors, demand: cooling.

IP	21,520 sq ft	32,280 sq ft	43,040 sq ft	53,800 sq ft	64,560 sq ft
26,420 gal	256,924	240,430	224,253	207,442	195,389
39,630 gal	248,043	235,989	222,033	208,394	196,023
52,840 gal	250,263	235,989	223,302	207,442	197,609
66,050 gal	245,822	232,817	223,302	208,394	197,609
79,260 gal	243,602	234,403	221,081	209,345	197,609
SI	**2,000 m^2**	**3,000 m^2**	**4,000 m^2**	**5,000 m^2**	**6,000 m^2**
100 m^3	810	758	707	654	616
150 m^3	782	744	700	657	618
200 m^3	789	744	704	654	623
250 m^3	775	734	704	657	623
300 m^3	768	739	697	660	623

The calculations were done for collector areas of 43,040 sq ft (4,000 m^2), and for larger areas in cases where the solar system is connected to a district heating net that is used for DHW, space heating, and cooling (Table B-39). The specific demand lies in the range of about 750 – 580 kWh/(yr·m^2) (238–184 kBtu/[yr*sq ft]). The highest solar yield per m^2 collector area is achieved with a 43,040 sq ft (4,000 m^2)-collector-field. A storage volume of 52,840 gal (200 m^3) will be sufficient, independent from the collector area. The solar fraction at this dimensioning is about 12% (see Figure B-19). In a detailed planning phase, even a solar system with a smaller solar storage than 52,840 gal (200 m^3) can be taken into account.

Table B-39. Specific solar yield in Btu/(yr sq ft) [kWh/(yr·m²)] Fort Bliss, vacuum pipe-collectors; demand: DHW, space heating, cooling.

IP	43,040 sq ft	64,560 sq ft	86,080 sq ft	107,600 sq ft	129,120 sq ft
52,840 gal	237,893	221,399	207,442	195,389	184,287
79,260 gal	237,893	221,399	209,345	196,975	186,191
105,680 gal	237,893	222,667	210,297	198,561	188,094
132,100 gal	237,893	222,667	210,297	200,147	189,362
158,520 gal	235,989	221,399	210,297	200,781	190,631
SI	**4,000 m^2**	**6,000 m^2**	**8,000 m^2**	**10,000 m^2**	**12,000 m^2**
200 m^3	750	698	654	616	581
300 m^3	750	698	660	621	587
400 m^3	750	702	663	626	593
500 m^3	750	702	663	631	597
600 m^3	744	698	663	633	601

A comparison of the specific energy yield on the same collector areas (e.g., at 43,040 sq ft (4,000 m^2) - red) shows that the energy yield is higher for DHW, space heating, cooling system (750 kWh/[yr·m^2]), e.g., Table B-39 compared Table B-38). This is due to the fact, that in case of additional space heating and DHW, the energy demand is higher and temporarily better distributed over the year.

Simple payback time

The payback times of both cooling systems (pure cooling demand or with cooling, DHW and space heating) lie in the range of 11 to 15 years. The influence of the solar storage is small and not shown.

Table B-40. Simple payback time; Fort Bliss, vacuum pipe-collectors; demand: cooling.

	21,520 sq ft (2,000 m^2)	32,280 sq ft (3,000 m^2)	43,040 sq ft (4,000 m^2)	53,800 sq ft (5,000 m^2)	64,560 sq ft (6,000 m^2)
26,420 gal (100 m^3)	11 yr	11 yr	12 yr	13 yr	14 yr

Table B-41. Simple payback time, Fort Bliss, vacuum pipe-collectors demand: DHW, space heating and cooling.

	43,040 sq ft (4,000 m^2)	64,560 sq ft (6,000 m^2)	86,080 sq ft (8,000 m^2)	107,600 sq ft (10,000 m^2)	129,120 sq ft (12,000 m^2)
52,840 gal (200 m^3)	12 yr	13 yr	13 yr	14 yr	15 yr

Payback time — Annuity method / economic results – Fort Bliss – cooling – vacuum tube collectors

Table B-42 lists the basic data used to calculate the payback time with the annuity method, the energy prices in the 30-year period, and the energy savings in the 30-year period.

Three dedicated solar systems are chosen for each simulation case. Table B-43 lists system data for each solar system.

Table B-42. Basic data - Fort Bliss - cooling - vacuum tube collectors.

Basic Data	€ / kWh	$ / MBTU
Life time (years)	20	
Interest Rate (%)	6.00%	
Planning cost factor (% of Investment)	115%	
Price increase Fuel / year (%)	5.00%	
Price increase Maintenance / year (%)	2.00%	
Fuel price (€/kWh) [$/kWh]	0.0300	0.0390
Boiler Efficiency (%)	70%	
Assumed full load hrs / year	4,000	
Maintenance Boiler / MWh (€) [$]	20,000	26.000
Heat Demand (kWh) [MBTU]	14,215,000	48.501.580
Boiler capacity (MW) [MBTUh]	3.55	12,125
Maintenance Cost Boiler (€) [$]	71,075	92,398
Maintenance Cost Solar (% System Invest)	2%	
Degradation Solar (%/year)	0.00%	

Calculation parameters

- Heating Cost Fuel: Cost of Conventional Energy – US$ / kWh
- Cost Heat Production Fuel: Heating Cost Fuel (Energy Requirement) + Maintenance Cost Fuel
- Cost Heat Production Fuel + Solar: Heating Cost Fuel + Solar (Energy Requirement) + Maintenance Cost (Fuel + Solar) + Interest Solar System
- Energy Cost Savings: Cost Heat Production Fuel + Solar - Cost Heat Production Fuel
- Heat Production Price / kWh: Cost Heat Production / Heating Demand
- Payback Time: Total Investment – Energy Cost Savings

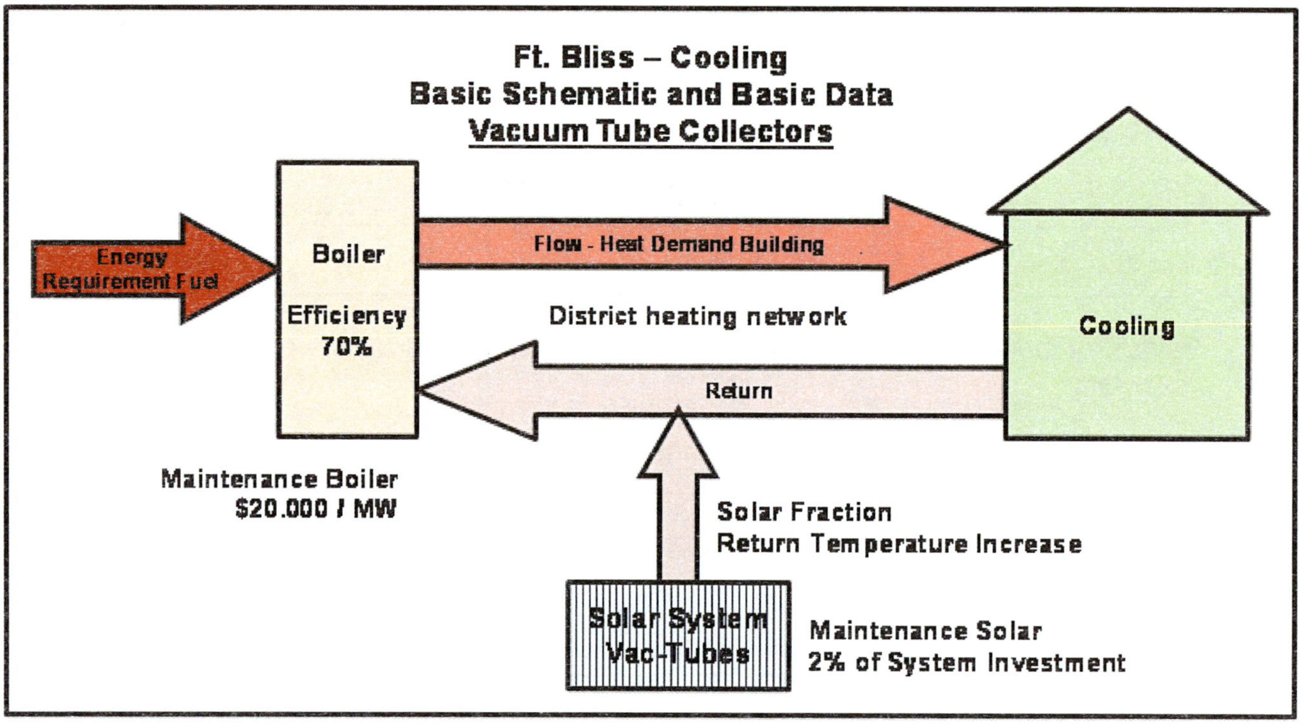

Figure B-20. Basic schematic – Fort Bliss – vacuum tube collectors.

Table B-43. Solar system data – Fort Bliss – cooling – vacuum tube collectors.

System 1		
Area Solar system (m^2)[sq ft]	2.000,00	21.528,00
Volume storage tank (m^3) [ft^3]	100,00	3.531.500,00
Yield Solar System / year (kWh)[MBTU]	1.620.510,00	5.529.180,12
Solar fraction (%)	**11,40%**	
Total Heat Demand (kWh) [MBTU]	14.215.000,00	48.501.580,00
Total Energy Requirement (kWh) [MBTU]	20.307.142,86	69.287.971,43
Investment System (€) [$]	689.947,83	896.932,17
Planning Cost (€) [$]	103.492,17	134.539,83
Total Investment (€) [$]	793.440,00	1.031.472,00

System 2		
Area Solar system (m^2)[sq ft]	6.000,00	64.584,00
Volume storage tank (m^3) [ft^3]	150,00	5.297.250,00
Yield Solar System / year (kWh)[MBTU]	3.710.115,00	12.658.912,38
Solar fraction (%)	**26,10%**	
Total Heat Demand (kWh) [MBTU]	14.215.000,00	48.501.580,00
Total Energy Requirement (kWh) [MBTU]	20.307.142,86	69.287.971,43
Investment System (€) [$]	2.089.565,22	2.716.434,78
Planning Cost (€) [$]	313.434,78	407.465,22
Total Investment (€) [$]	2.403.000,00	3.123.900,00

System 3		
Area Solar system (m^2)[sq ft]	4.000,00	43.056,00
Volume storage tank (m^3) [ft^3]	100,00	3.531.500,00
Yield Solar System / year (kWh)[MBTU]	2.828.785,00	9.651.814,42
Solar fraction (%)	**19,90%**	
Total Heat Demand (kWh) [MBTU]	14.215.000,00	48.501.580,00
Total Energy Requirement (kWh) [MBTU]	20.307.142,86	69.287.971,43
Investment System (€) [$]	1.380.000,00	1.794.000,00
Planning Cost (€) [$]	207.000,00	269.100,00
Total Investment (€) [$]	1.587.000,00	2.063.100,00

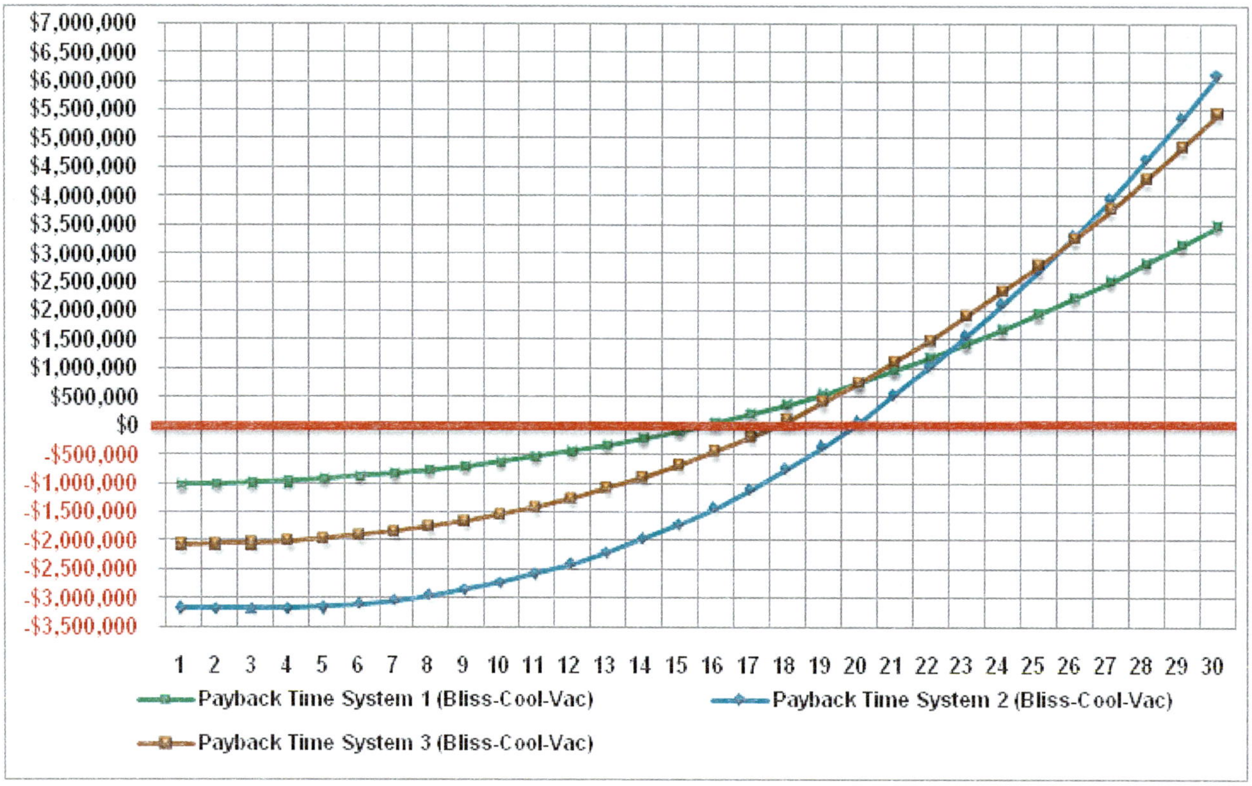

Figure B-21. Payback time - Fort Bliss – cooling – vacuum tube.

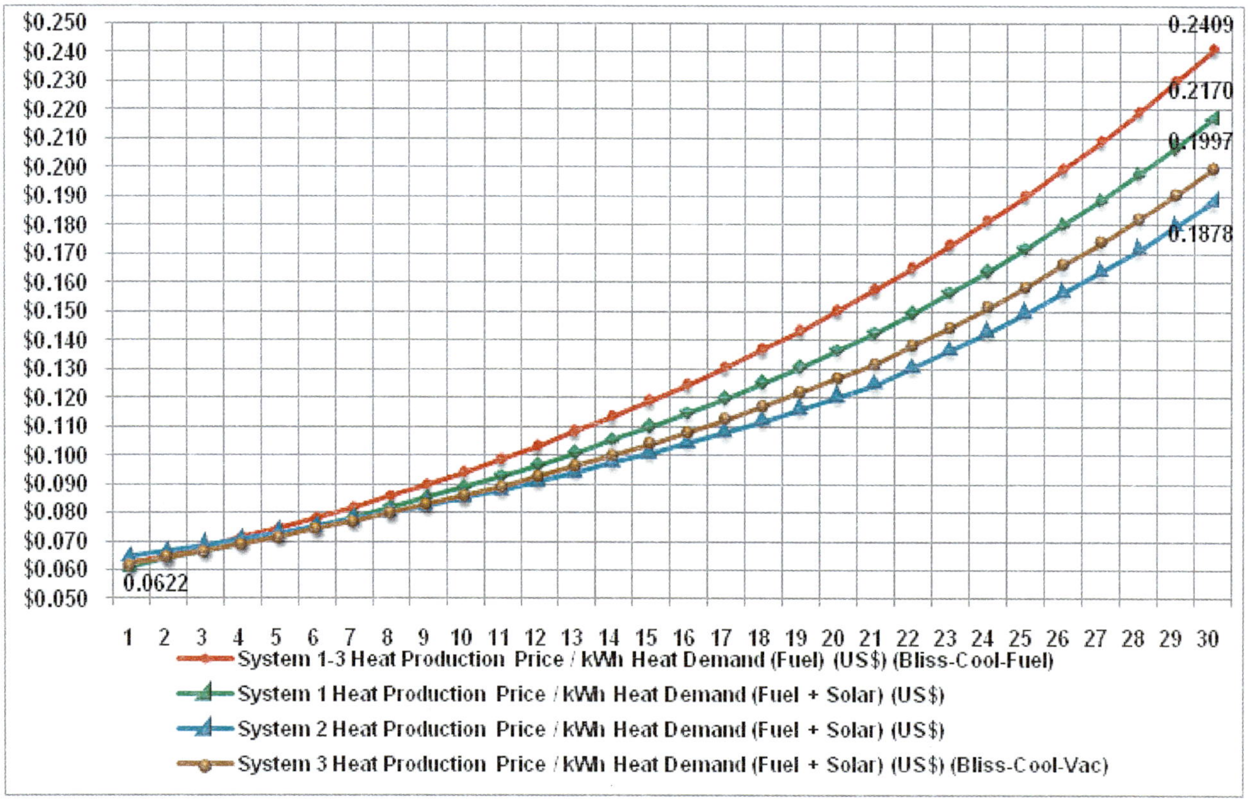

Figure B-22. Energy cost - Fort Bliss – cooling – vacuum tube.

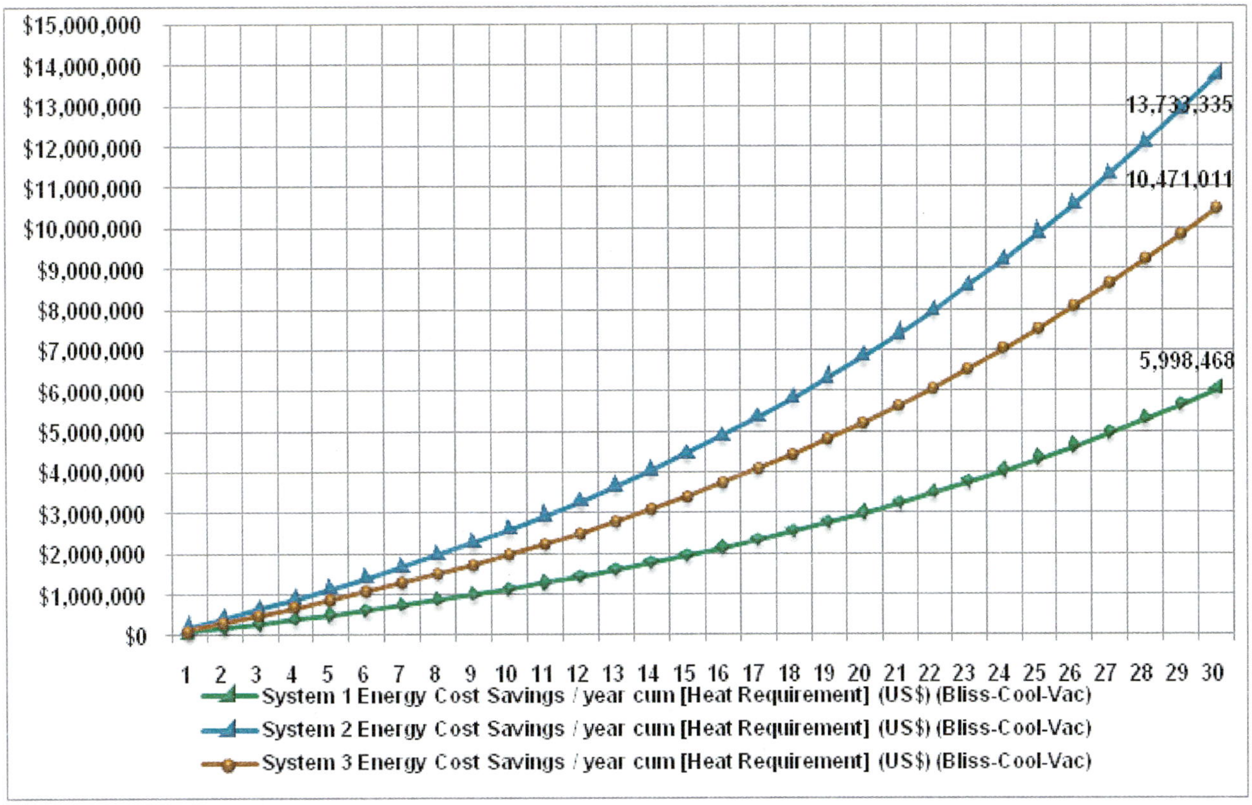

Figure B-23. Energy cost saving / year (cum) – Fort Bliss – cooling – vacuum tube.

Payback time – annuity method / economic results – Fort Bliss – DHW / SPH / Cooling – vacuum tube collectors

Table B-44 lists the basic data used to calculate the payback time with the annuity method, the energy prices in the 30-year period, and the energy savings in the 30-year period.

Three dedicated solar systems are chosen for each simulation case. Table B-45 lists the system data for each solar system.

Table B-44. Basic data - Fort Bliss - DHW / SPH / cooling - vacuum tube collectors.

Basic Data	$ / MBTU	€ / kWh
Life time (years)		20
Interest Rate (%)		6.00%
Planning cost factor (% of Investment)		115.00%
Price increase Fuel / year (%)		5.00%
Price increase Maintenance / year (%)		2.00%
Fuel price (€/kWh) [$/kWh]	0.0390	0.0300
Boiler Efficiency (%)		70.00%
Assumed full load hrs / year		5.000
Maintenance Boiler / MWh (€) [$]	26,000	20.000
Heat Demand (kWh) [MBTU]	81,195.364	23,797.000
Boiler capacity (MW) [MBTUh]	16,239	4.76
Maintenance Cost Boiler (€) [$]	123,744	95,188
Maintenance Cost Solar (% System Invest)		2.00%
Degradation Solar (%/year)		0.00%

Calculation parameters

- Heating Cost Fuel: Cost of Conventional Energy – US$ / kWh
- Cost Heat Production Fuel: Heating Cost Fuel (Energy Requirement) + Maintenance Cost Fuel
- Cost Heat Production Fuel + Solar: Heating Cost Fuel + Solar (Energy Requirement) + Maintenance Cost (Fuel + Solar) + Interest Solar System
- Energy Cost Savings: Cost Heat Production Fuel + Solar - Cost Heat Production Fuel
- Heat Production Price / kWh: Cost Heat Production / Heating Demand
- Payback Time: Total Investment – Energy Cost Savings

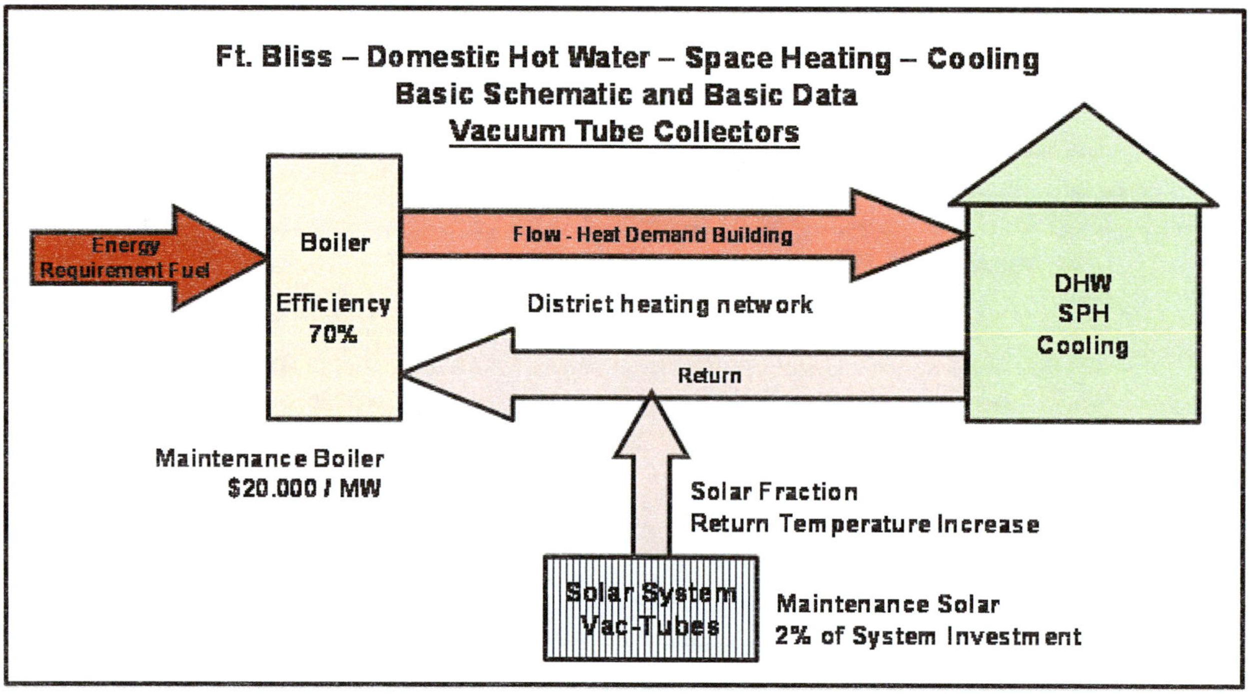

Figure B-24. Basic schematic – Fort Bliss – DHW / SPH / cooling – vacuum tube collectors.

Table B-45. Solar system data - Fort Bliss - DHW / SPH / cooling - vacuum tube collectors.

System 1		
Area Solar system (m²)[sq ft]	4.000,00	43.056,00
Volume storage tank (m³) [ft³]	200,00	7.063.000,00
Yield Solar System / year (kWh)[MBTU]	2.998.422,00	10.230.615,86
Solar fraction (%)	**12,60%**	
Total Heat Demand (kWh) [MBTU]	23.797.000,00	81.195.364,00
Total Energy Requirement (kWh) [MBTU]	33.995.714,29	115.993.377,14
Investment System (€) [$]	1.406.191,30	1.828.048,70
Planning Cost (€) [$]	210.928,70	274.207,30
Total Investment (€) [$]	1.617.120,00	2.102.256,00

System 2		
Area Solar system (m²)[sq ft]	12.000,00	129.168,00
Volume storage tank (m³) [ft³]	300,00	10.594.500,00

Yield Solar System / year (kWh)[MBTU]	7.043.912,00	24.033.827,74
Solar fraction (%)	**29,60%**	
Total Heat Demand (kWh) [MBTU]	23.797.000,00	81.195.364,00
Total Energy Requirement (kWh) [MBTU]	33.995.714,29	115.993.377,14
Investment System (€) [$]	4.293.913,04	5.582.086,96
Planning Cost (€) [$]	644.086,96	837.313,04
Total Investment (€) [$]	4.938.000,00	6.419.400,00

System 3		
Area Solar system (m^2)[sq ft]	8.000,00	86.112,00
Volume storage tank (m^3) [ft^3]	200,00	7.063.000,00
Yield Solar System / year (kWh)[MBTU]	5.235.340,00	17.862.980,08
Solar fraction (%)	**22,00%**	
Total Heat Demand (kWh) [MBTU]	23.797.000,00	81.195.364,00
Total Energy Requirement (kWh) [MBTU]	33.995.714,29	115.993.377,14
Investment System (€) [$]	2.812.173,91	3.655.826,09
Planning Cost (€) [$]	421.826,09	548.373,91
Total Investment (€) [$]	3.234.000,00	4.204.200,00

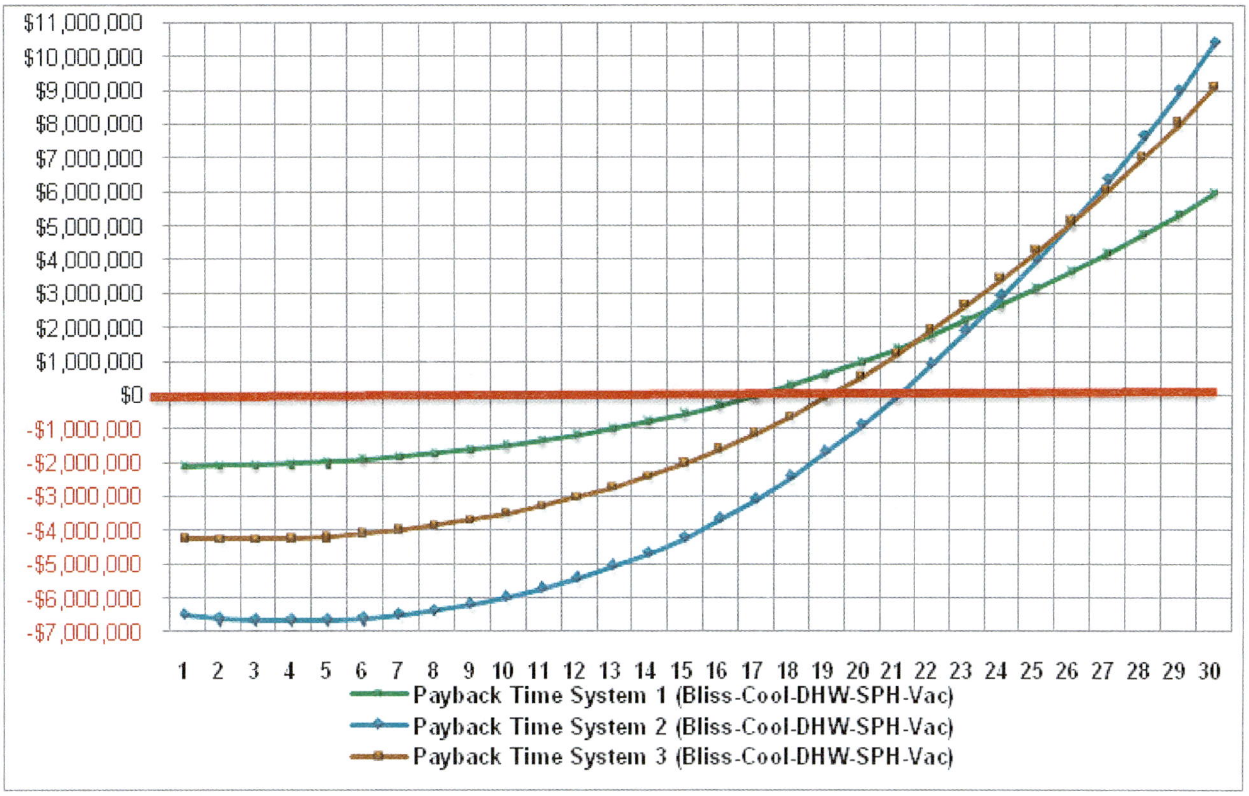

Figure B-25. Payback time - Fort Bliss – DHW / SPH / cooling – vacuum tube collectors.

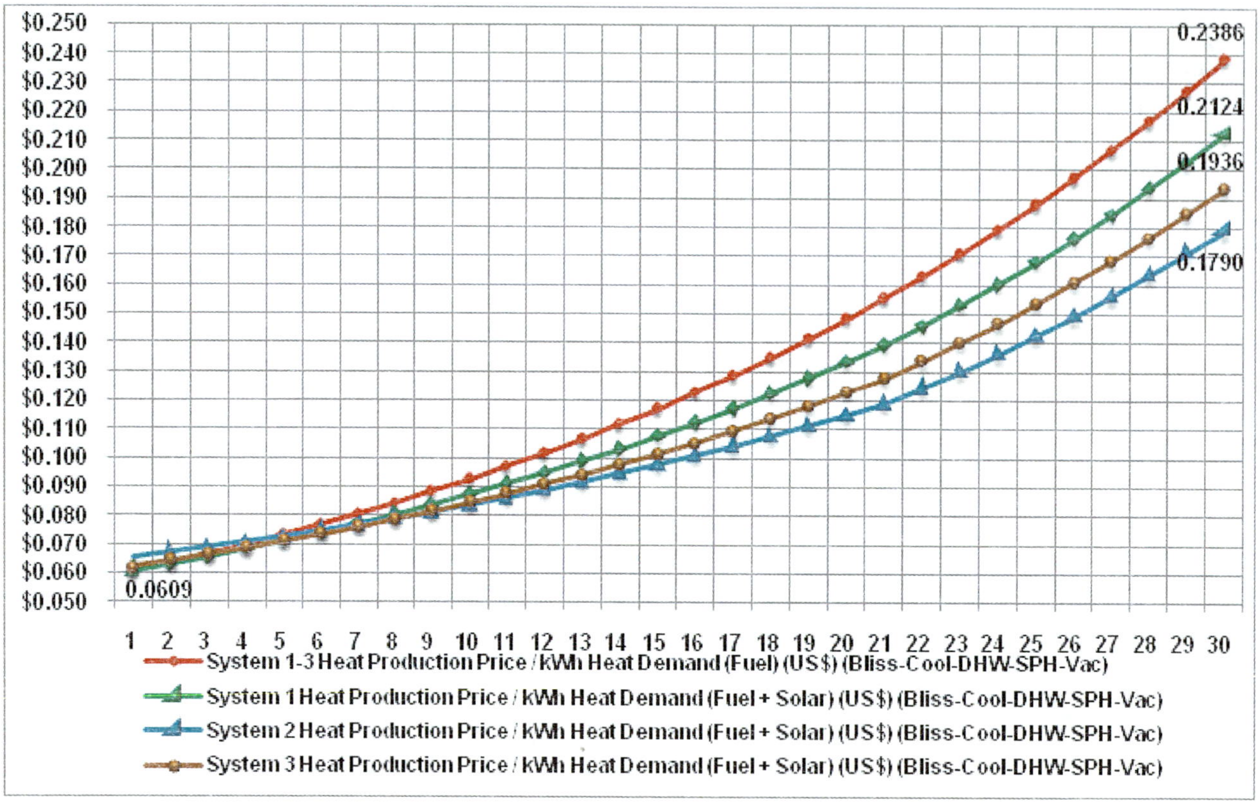

Figure B-26. Energy cost - Fort Bliss – DHW / SPH / cooling – vacuum tube collectors.

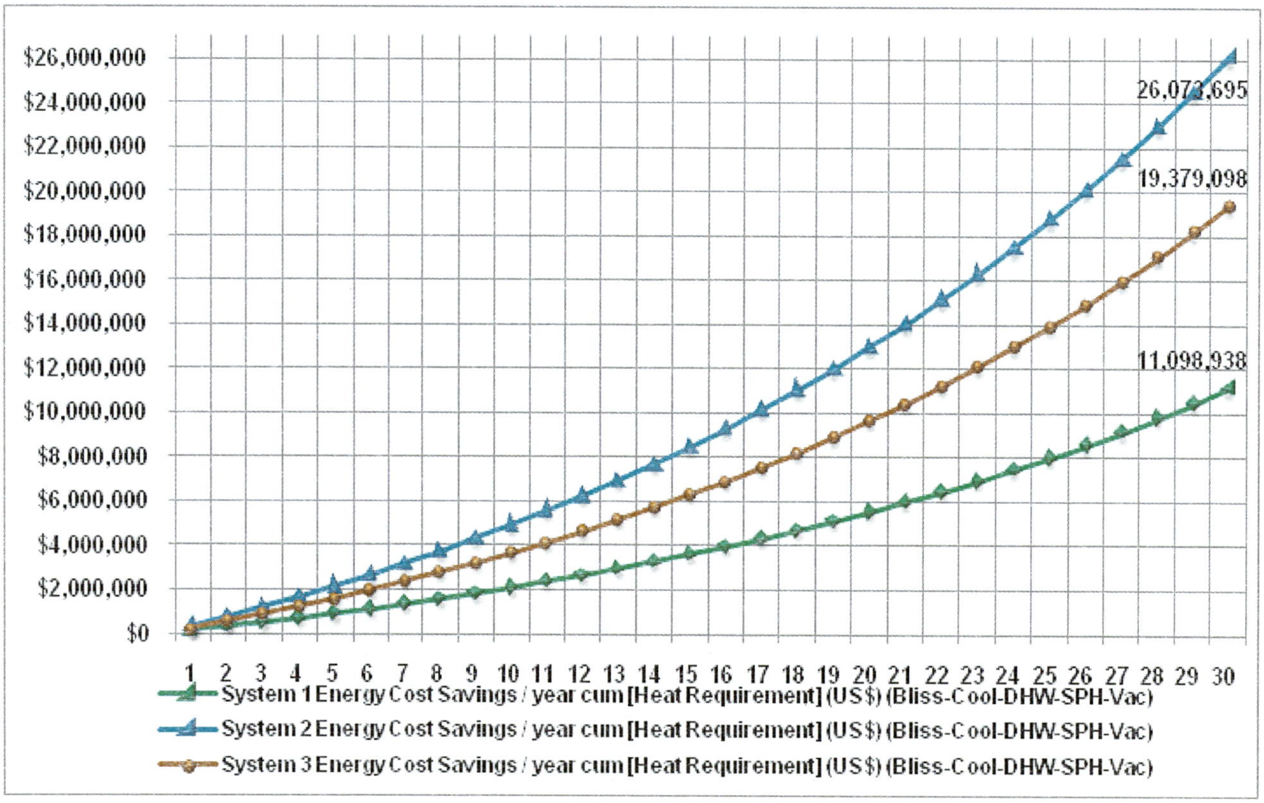

Figure B-27. Energy cost saving / year (cum) – Fort Bliss – DHW / SPH / cooling – vacuum tube.

Solar energy yield and simple payback time of a solar system with flat plate collectors in a district heating net in Fort Bliss for DHW and space heating

Table B-46 lists the calculated specific solar energy yield (kWh/[yr·m^2]) for a district heating net that covers demand for space heating and DHW. The calculations were done at a yearly average net return temperature of 60 °C using flat plate collectors.

Table B-46. Specific solar yield in Btu/(yr • sq ft) [kWh/(yr • m2)] Fort Bliss, flat-plate-collectors; demand: DHW and space heating.

IP	21,520 sq ft	32,280 sq ft	43,040 sq ft	53,800 sq ft	64,560 sq ft
26,420 gal	218,861	176,358	135,123	105,307	83,104
39,630 gal	220,447	186,508	147,493	116,726	93,254
52,840 gal	220,447	192,534	156,692	126,559	102,452
66,050 gal	220,447	194,437	163,987	134,489	110,065
79,260 gal	218,861	195,706	168,745	141,150	115,457
SI	**2,000 m^2**	**3,000 m^2**	**4,000 m^2**	**5,000 m^2**	**6,000 m^2**
100 m^3	690	556	426	332	262
150 m^3	695	588	465	368	294
200 m^3	695	607	494	399	323
250 m^3	695	613	517	424	347
300 m^3	690	617	532	445	364

The specific solar yield lies in the range of about 690 – 262 kWh/(yr·m^2) (219–83 kBtu/[yr*sq ft]). The highest solar yield per 10.8 sq ft (m^2) collector area is achieved with a 21,520 sq ft (2,000 m^2)-collector-field. A storage volume of 26,420 gal (100 m^3) will be sufficient at this size. The solar fraction sized at 2,000 m^2/100 m^3 (0.815 sq ft/gal) is about 14% (see Figure B-19). For collector areas greater than 21,520 sq ft (2,000 m^2) with higher solar fractions, the specific solar energy yield drops significantly (Table B-47) and the influence of the storage volume becomes more important.

Simple payback time

Table B-47 lists the simple payback time. The most efficiently systems lie in the range of 2,000 to 3,000 m^2 (21,520 to 32,280 sq ft) and have a simple payback time of 10–12 years. A storage volume of 100 –150 m^3 (26,420–39,630 gal) is enough for this size.

Table B-47. Simple payback time; Fort Bliss, flat-plate-collectors; demand: DHW and space heating.

IP	21,520 sq ft	32,280 sq ft	43,040 sq ft	53,800 sq ft	64,560 sq ft
26,420 gal	10 yr	12 yr	15 yr	20 yr	25 yr
39,630 gal	10 yr	11 yr	14 yr	18 yr	23 yr
52,840 gal	10 yr	11 yr	14 yr	17 yr	21 yr
66,050 gal	10 yr	11 yr	13 yr	16 yr	20 yr
79,260 gal	10 yr	11 yr	13 yr	16 yr	19 yr
SI	**2,000 m^2**	**3,000 m^2**	**4,000 m^2**	**5,000 m^2**	**6,000 m^2**
100 m^3	10 yr	12 yr	15 yr	20 yr	25 yr
150 m^3	10 yr	11 yr	14 yr	18 yr	23 yr
200 m^3	10 yr	11 yr	14 yr	17 yr	21 yr
250 m^3	10 yr	11 yr	13 yr	16 yr	20 yr
300 m^3	10 yr	11 yr	13 yr	16 yr	19 yr

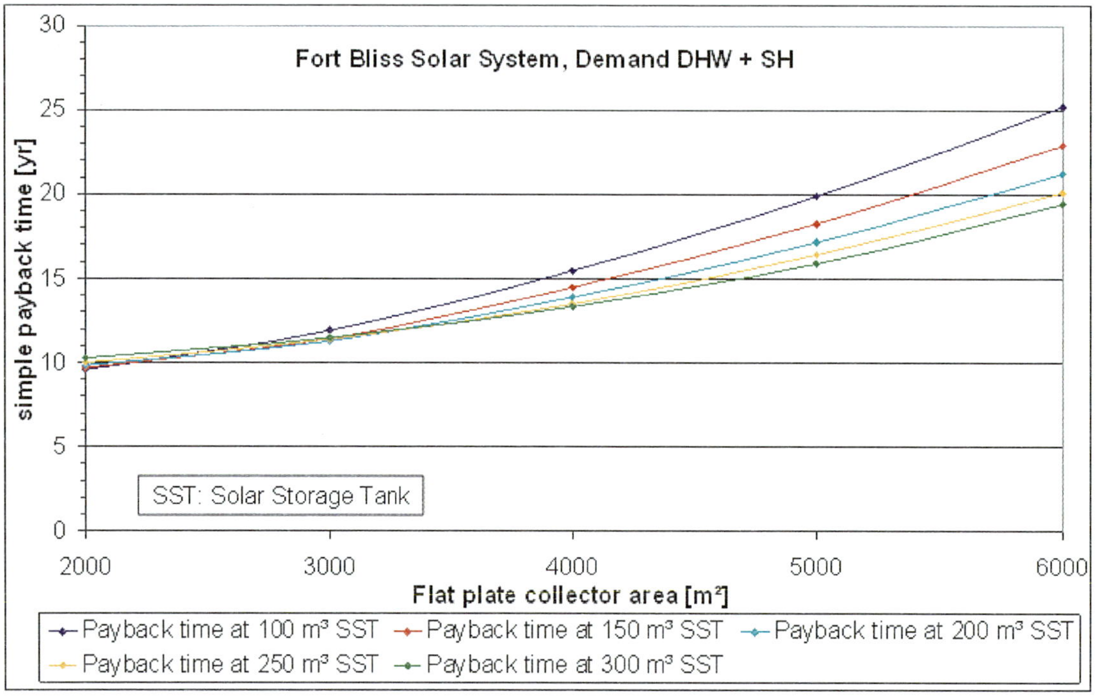

Figure B-28. Simple payback time depending on flat plate area, Fort Bliss, DHW and space heating.

Payback time — annuity method / economic results – Fort Bliss – DHW / SPH – flat plate collectors

Table B-48 lists the basic data used to calculate the payback time with the annuity method, the energy prices in the 30-year period, and the energy savings in the 30-year period.

Three dedicated solar systems were chosen for each simulation case. Table B-48 lists the system data for each solar system.

Table B-48. Basic data - Fort Bliss - DHW / SPH - flat plate collectors.

Basic Data	$ / MBTU	€ / kWh
Life time (years)		20
Interest Rate (%)		6,00%
Planning cost factor (% of Investment)		115,00%
Price increase Fuel / year (%)		5,00%
Price increase Maintenance / year (%)		2,00%
Fuel price (€/kWh) [$/kWh]	0,0390	0,0300
Boiler Efficiency (%)		70,00%
Assumed full load hrs / year		3.000
Maintenance Boiler / MWh (€) [$]	26.000	20.000
Heat Demand (kWh) [MBTU]	32.697.196	9.583.000
Boiler capacity (MW) [MBTUh]	10,899	3,19
Maintenance Cost Boiler (€) [$]	83.052,67	63.886,67
Maintenance Cost Solar (% System Invest)		2,00%
Degradation Solar (%/year)		0,00%

Calculation parameters:

• Heating Cost Fuel:	Cost of Conventional Energy – US$ / kWh
• Cost Heat Production Fuel:	Heating Cost Fuel (Energy Requirement) + Maintenance Cost Fuel
• Cost Heat Production Fuel + Solar:	Heating Cost Fuel + Solar (Energy Requirement) + Maintenance Cost (Fuel + Solar) + Interest Solar System
• Energy Cost Savings:	Cost Heat Production Fuel + Solar - Cost Heat Production Fuel
• Heat Production Price / kWh:	Cost Heat Production / Heating Demand
• Payback Time:	Total Investment – Energy Cost Savings

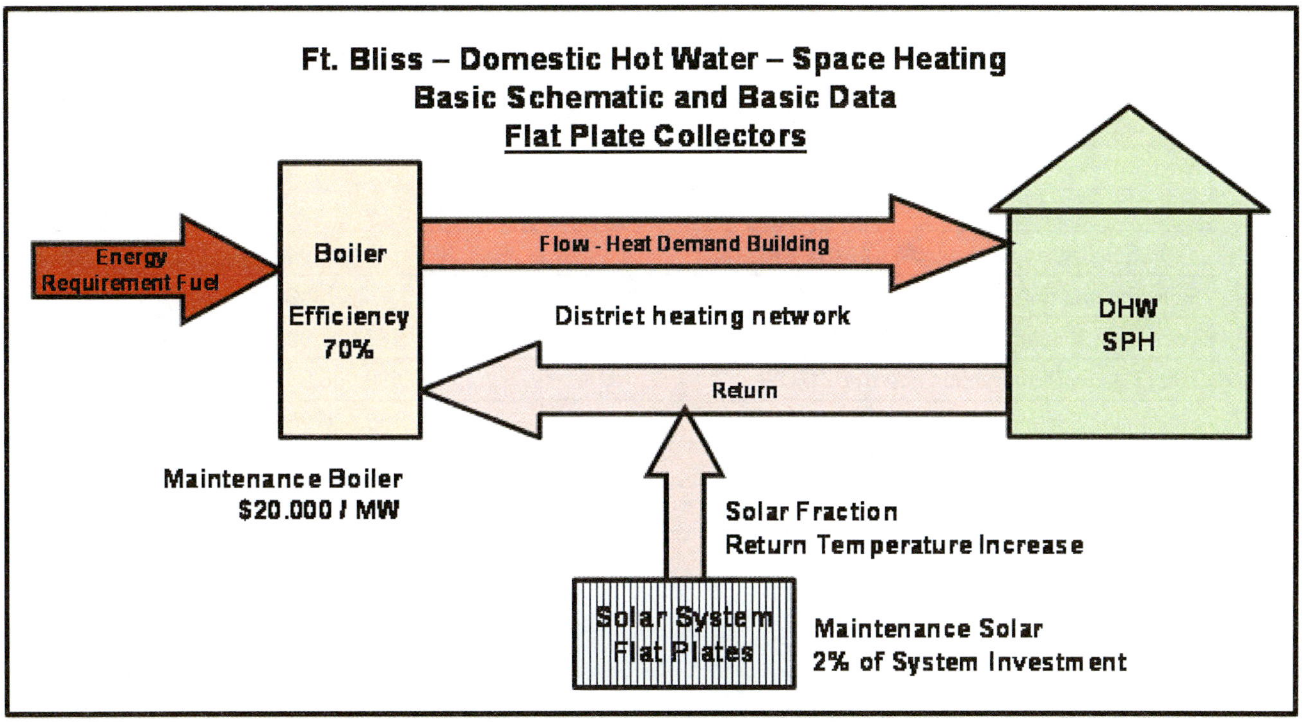

Figure B-29. Basic schematic – Fort Bliss – DHW / SPH – flat plate collectors.

Table B-49. Solar system data – Fort Bliss – DHW / SPH – flat plate collectors.

System 1		
Area Solar system (m^2)[sq ft]	2.000,00	21.528,00
Volume storage tank (m^3) [ft^3]	100,00	3.531.500,00
Yield Solar System / year (kWh)[MBTU]	1.379.952,00	4.708.396,22
Solar fraction (%)	**14,40%**	
Total Heat Demand (kWh) [MBTU]	9.583.000,00	32.697.196,00
Total Energy Requirement (kWh) [MBTU]	13.690.000,00	46.710.280,00
Investment System (€) [$]	530.114,78	689.149,22
Planning Cost (€) [$]	79.517,22	103.372,38
Total Investment (€) [$]	609.632,00	792.521,60

System 2		
Area Solar system (m^2)[sq ft]	6.000,00	64.584,00
Volume storage tank (m^3) [ft^3]	300,00	10.594.500,00
Yield Solar System / year (kWh)[MBTU]	2.184.924,00	7.454.960,69
Solar fraction (%)	**22,80%**	
Total Heat Demand (kWh) [MBTU]	9.583.000,00	32.697.196,00
Total Energy Requirement (kWh) [MBTU]	13.690.000,00	46.710.280,00
Investment System (€) [$]	1.703.290,43	2.214.277,57
Planning Cost (€) [$]	255.493,57	332.141,63
Total Investment (€) [$]	1.958.784,00	2.546.419,20

System 3		
Area Solar system (m^2)[sq ft]	4.000,00	43.056,00
Volume storage tank (m^3) [ft^3]	300,00	10.594.500,00
Yield Solar System / year (kWh)[MBTU]	2.127.426,00	7.258.777,51
Solar fraction (%)	**22,20%**	
Total Heat Demand (kWh) [MBTU]	9.583.000,00	32.697.196,00
Total Energy Requirement (kWh) [MBTU]	13.690.000,00	46.710.280,00
Investment System (€) [$]	1.135.526,96	1.476.185,04
Planning Cost (€) [$]	170.329,04	221.427,76
Total Investment (€) [$]	1.305.856,00	1.697.612,80

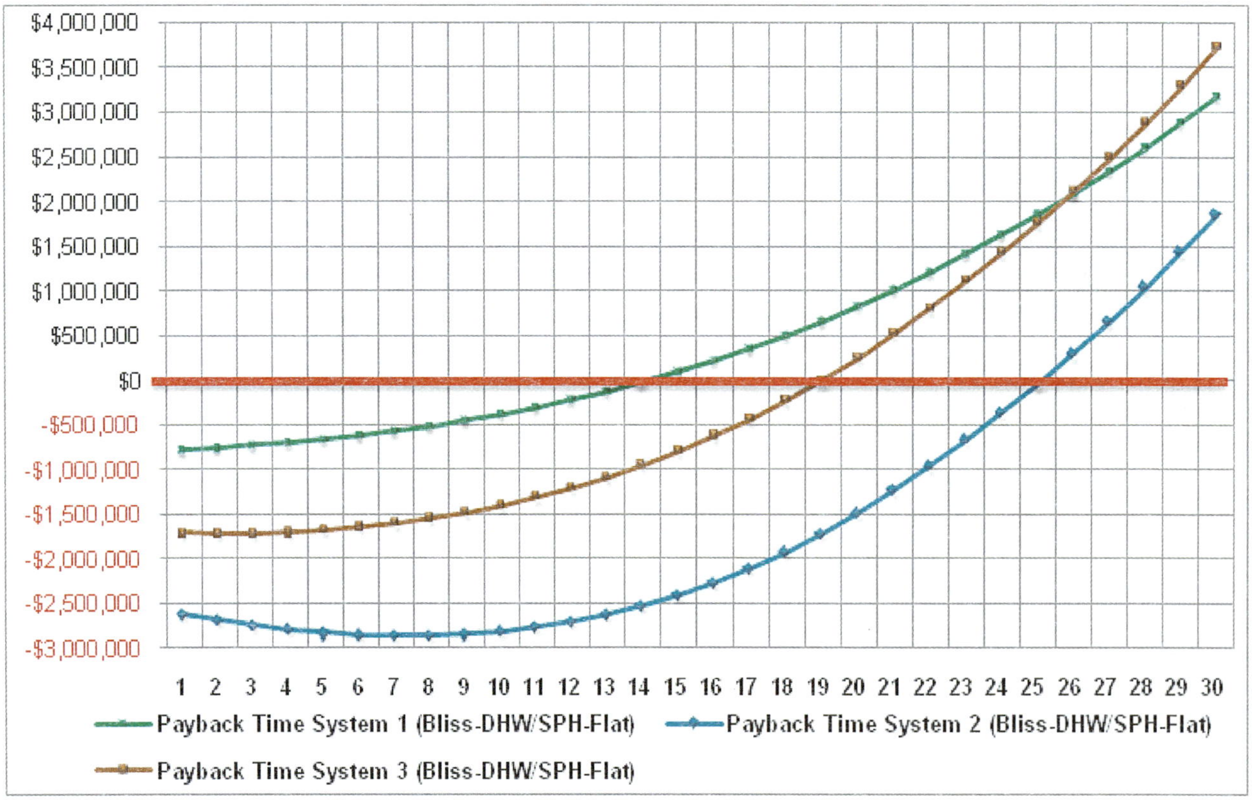

Figure B-30. Payback Time - Fort Bliss – DHW / SPH – flat plate collectors.

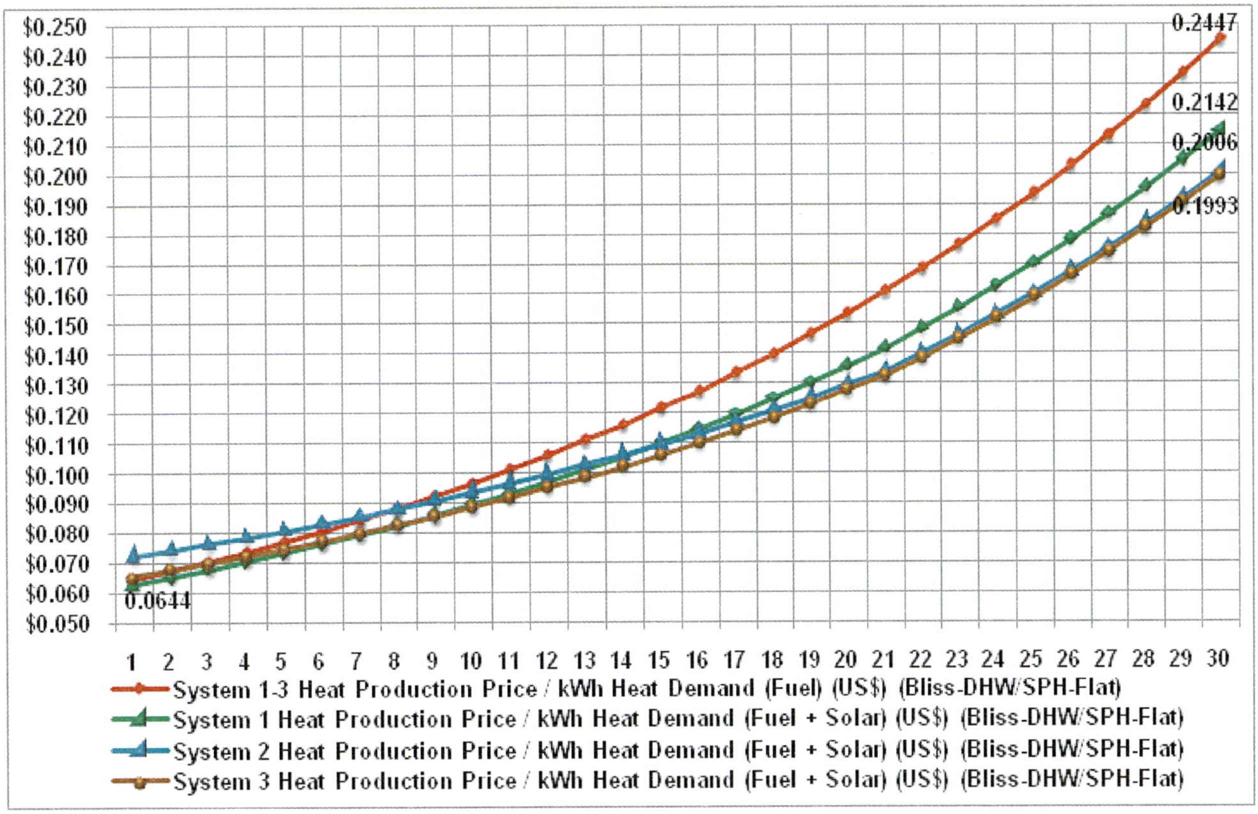

Figure B-31. Energy cost - Fort Bliss – DHW / SPH – flat plate collectors.

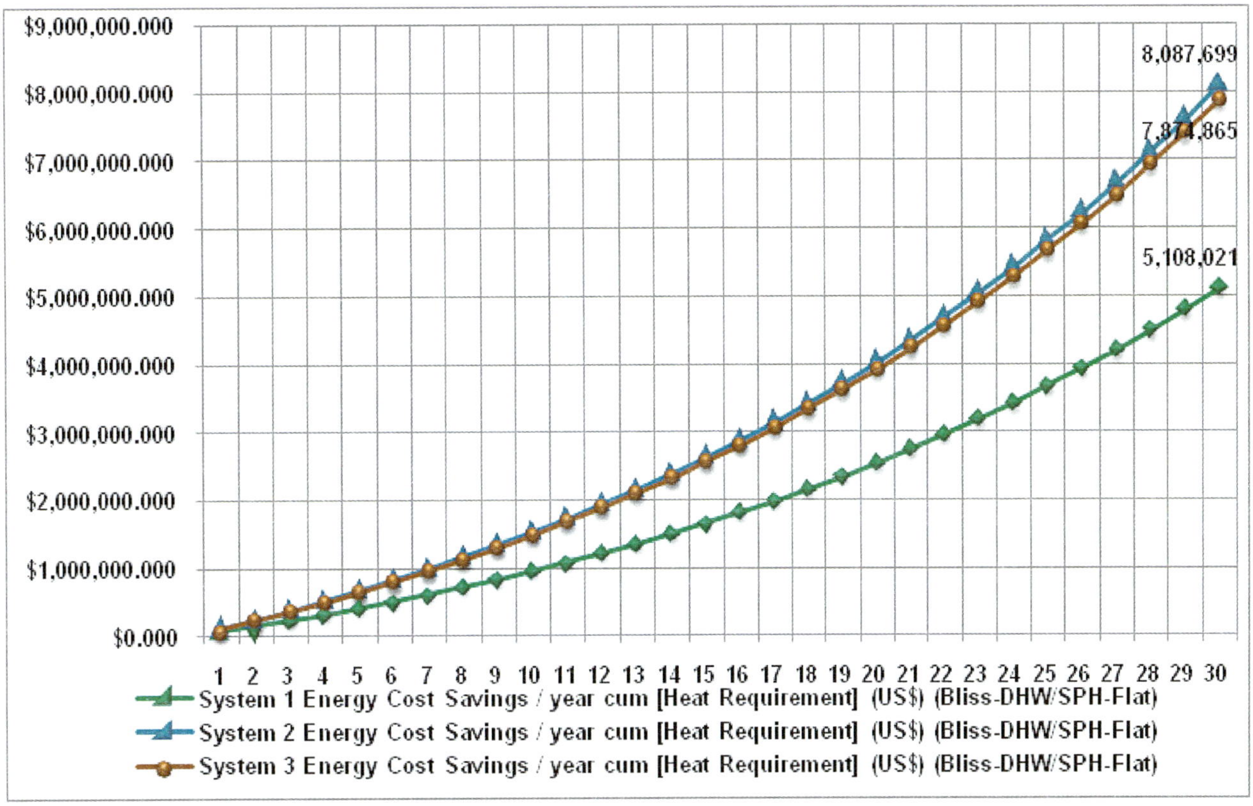

Figure B-32. Energy cost saving / year (cum) – Fort Bliss – DHW / SPH – flat plate collectors.

Overview of calculated payback-times for solar systems connected to district heating in Fort Bliss (simple payback)

Figure B-33 shows the payback times on the calculated systems in Fort Bliss, depending on the solar fraction. The legend describes the chosen system and solar storage size. For the solar system with flat plate collectors for DHW and space heating only, the values near optimum line of solar fraction depend on solar storage.

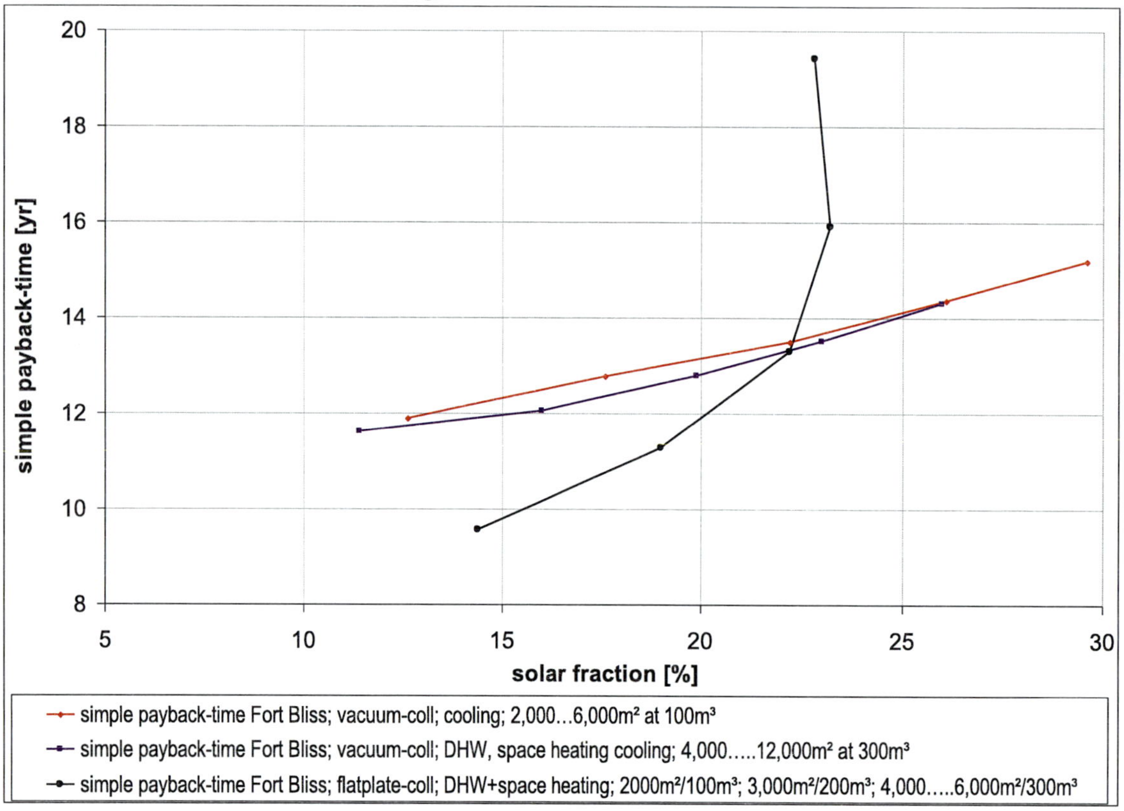

Figure B-33. Simple payback time as a function of solar fraction Fort Bliss

It is clearly shown that the payback-time of the two systems, which cover cooling, rises nearly linearly with the solar fraction. Even at a solar fraction of 25%, the payback time reaches an acceptable value of 14–15 years.

For the system with DHW and space heating with flat plate collectors, it is clearly shown that solar fractions greater than 20 lead to a significant increase in payback time. At low solar fractions (~15%), simple payback times below 10 years are possible.

Overview of calculated economic results for solar systems connected to district heating in Fort Bliss (payback – annuity method)

The applications of the two systems technologies are not directly comparable. Furthermore these results are only valid for the basic data used. A change of the basic data (interest rate, fuel price, specific cost collectors, etc.) will lead to different results.

Table B-50. Overview economic results (annuity method) solar systems for Fort Bliss.

	Cooling (Vac)	Cool/DHW/SPH (Vac)	DHW / SPH (Flat)
Heating Demand Network	4,162,152 Btu/a (14,215 MWh/a)	6,967,761 Btu/a (23,797 MWh/a)	2,805,902 Btu/a (9,583 MWh/a)
System 1			
Area Solar system	21,520 sq ft (2,000 m^2)	43,040 sq ft (4,000 m^2)	21,520 sq ft (2,000 m^2)
Volume storage tank	26,420 gal (100 m^3)	52,840 gal (200 m^3)	26,420 gal (100 m^3)
Solar fraction (%)	11.40%	12.60%	14.40%
Total Investment [$]	1,031,472	2,102,256	792,522
Payback Time	15	17	14
Energy Price (Fuel) 1. year ($)	0.0622	0.0609	0.0644
Energy Price (Fuel+Solar) 1. year ($)	0.0615	0.0607	0.0628
Energy Price (Fuel) 30. year ($)	0.2409	0.2386	0.2447
Energy Price (Fuel+Solar) 30. year ($)	0.2170	0.2124	0.2142
Energy savings (M$ / 30 years)	5,998	11,099	5,108
System 2			
Area Solar system	64,560 sq ft (6,000 m^2)	129,120 sq ft (12,000 m^2)	64,560 sq ft (6,000 m^2)
Volume storage tank	39,630 gal (150 m^3)	79,260 gal (300 m^3)	79,260 gal (300 m^3)
Solar fraction (%)	26,10%	29,60%	22,80%
Total Investment [$]	3,123,900	6,419,400	2,546,419
Payback Time	19	21	25
Energy Price (Fuel) 1. year ($)	0.0622	0.0609	0.0644
Energy Price (Fuel+Solar) 1. year ($)	0.0647	0.0653	0.0722
Energy Price (Fuel) 30. year ($)	0.2409	0.2386	0.2447
Energy Price (Fuel+Solar) 30. year ($)	0.1878	0.1790	0.2006
Energy savings (M$ / 30 years)	13,733	26,074	8,088
System 3			
Area Solar system	43,040 sq ft (4,000 m^2)	86,080 sq ft (8,000 m^2)	43,040 sq ft (4,000 m^2)
Volume storage tank	26,420 gal (100 m^3)	52,840 gal (200 m^3)	79,260 gal (300 m^3)
Solar fraction (%)	19.90%	22.00%	22.20%
Total Investment [$]	2,063,100	4,204,200	1,697,613
Payback Time	17	19	19
Energy Price (Fuel) 1. year ($)	0.0622	0.0609	0.0644
Energy Price (Fuel+Solar) 1. year ($)	0.0624	0.0623	0.0657
Energy Price (Fuel) 30. year ($)	0.2409	0.2386	0.2447
Energy Price (Fuel+Solar) 30. year ($)	0.1997	0.1936	0.1993
Energy savings (M$ / 30 years)	10,471	19,379	7,875

Comparison of the energy yield and simple payback time of solar systems with flat plate and vacuum pipe collectors in a district net in Fort Bragg

Tables B-51 and B-52. list the calculated specific solar energy yield (kWh/[yr·m^2]) for solar systems with flat plate and vacuum pipe collectors in a district net in Fort Bragg (Table B-51: flat plate, Table B-52: vacuum pipe). The calculations were done at a yearly average net return temperature of 140 °F (60 °C).

For flat plate collectors (Table B-51), the specific solar yield per m^2 collector area with reasonable storage volume > 50 m^3 (13,210.0 gal) lies in the range of about 430 - 220 kWh/(yr*m^2) (136–69.7kBtu/[yr*sq ft]). The highest solar yield per m^2 collector area is achieved with a 1,000 m^2 (10,760 sq ft) collector-field. A storage volume of 100 m^3 (26,420 gal) will be sufficient. The solar fraction at this size is about 11% (see Figure B-19).

Table B-51. Specific solar yield in kWh/(yr • m²) Fort Bragg, flat plate-collectors, demand: DHW and space heating.

IP	**10,760 sq ft**	**16,140 sq ft**	**21,520 sq ft**	**26,900 sq ft**	**32,280 sq ft**
13,210 gal	131,951	104,990	95,157	77,394	62,169
19,815 gal	136,709	109,748	101,184	84,373	69,147
26,420 gal	137,978	113,871	104,673	88,496	72,954
33,025 gal	137,978	114,506	107,210	91,985	77,077
39,630 gal	137,978	114,506	108,479	93,571	79,615
SI	**1,000 m^2**	**1,500 m^2**	**2,000 m^2**	**2,500 m^2**	**3,000 m^2**
50 m^3	416	331	300	244	196
75 m^3	431	346	319	266	218
100 m^3	435	359	330	279	230
125 m^3	435	361	338	290	243
150 m^3	435	361	342	295	251

Table B-52 lists the results of calculations done for systems using vacuum pipe collectors and with collector areas of 500–2,500 m^2 (5,380– 26,900 sq ft). The specific yield lies in the range of about 670–400 kWh/(yr*m^2) (212.5–126.9 kBtu/[yr*sq ft])with a storage > 25 m^3 (6,605 gal). The highest solar yield per m^2 (10.8 sq ft) collector area is achieved with a 500 m^2 (5,380 sq ft) collector-field. The specific energy yield of vacuum pipe collectors is in general higher than with flat plate collectors.

Table B-52. Specific solar yield in kWh/(yr • m²) Fort Bragg, vacuum pipe-collectors; demand: DHW and space heating.

IP	**5,380 sq ft**	**10,760 sq ft**	**16,140 sq ft**	**21,520 sq ft**	**26,900 sq ft**
6,605 gal	213,152	173,186	149,396	134,489	118,946
13,210 gal	213,152	182,701	158,912	143,370	126,242
19,815 gal	215,372	185,239	162,084	148,128	131,634
26,420 gal	215,372	186,508	164,622	151,300	135,440
33,025 gal	213,152	186,508	166,208	153,837	137,978
SI	**500 m^2**	**1,000 m^2**	**1,500 m^2**	**2,000 m^2**	**2,500 m^2**
25 m^3	672	546	471	424	375
50 m^3	672	576	501	452	398
75 m^3	679	584	511	467	415
100 m^3	679	588	519	477	427
125 m^3	672	588	524	485	435

Simple payback time

Figure B-34 shows an overview of the payback times on the calculated systems in Fort Bragg for DHW and space heating, as a function of the solar fraction, based on the values with the same solar fraction in Table B-53. The legend describes the chosen system and solar storage size.

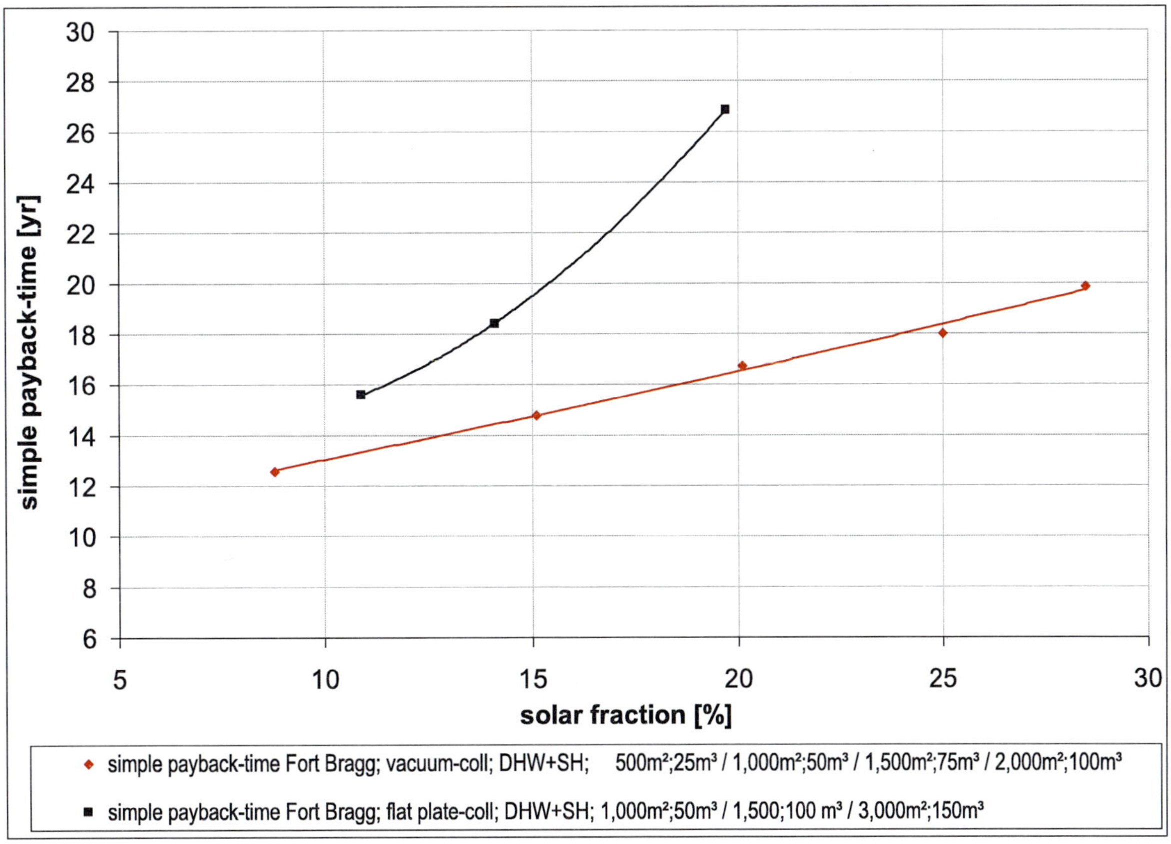

Figure B-34. Simple payback time as a function of solar fraction Fort Bragg, DHW and space heating.

One can see that the pay back times of the vacuum pipe collectors are lower than with flat plate collectors. The higher specific cost of the vacuum pipe collectors will be overcompensated by a higher efficiency.

Payback time – annuity method / economic results – Fort Bragg – DHW / SPH – vacuum tube collectors

Table B-53 lists the Basic Data used to calculate the payback time with the annuity method, the energy prices in the 30-year period, and the energy savings in the 30-year period.

Three dedicated solar systems are chosen for each simulation case. Table B-54 lists the system data for each solar system.

Table B-53. Basic data - Fort Bragg - DHW / SPH - vacuum tube collectors.

Basic Data	**$ / MBTU**	**€ / kWh**
Life time (years)		20
Interest Rate (%)		6,00%
Planning cost factor (% of Investment)		115,00%
Price increase Fuel / year (%)		5,00%
Price increase Maintenance / year (%)		2,00%
Fuel price (€/kWh) [$/kWh]	0,0390	0,0300
Boiler Efficiency (%)		70,00%

Basic Data	$ / MBTU	€ / kWh
Assumed full load hrs / year		3.000
Maintenance Boiler / MWh (€) [$]	26.000	20.000
Heat Demand (kWh) [MBTU]	13.020.192	3.816.000
Boiler capacity (MW) [MBTUh]	4,340	1,272
Maintenance Cost Boiler (€) [$]	33.072,00	25.440,00
Maintenance Cost Solar (% System Invest)		2,00%
Degradation Solar (%/year)		0,20%

Calculation parameters:

- Heating Cost Fuel: Cost of Conventional Energy – US$ / kWh
- Cost Heat Production Fuel: Heating Cost Fuel (Energy Requirement) + Maintenance Cost Fuel
- Cost Heat Production Fuel + Solar: Heating Cost Fuel + Solar (Energy Requirement) + Maintenance Cost (Fuel + Solar) + Interest Solar System
- Energy Cost Savings: Cost Heat Production Fuel + Solar - Cost Heat Production Fuel
- Heat Production Price / kWh: Cost Heat Production / Heating Demand
- Payback Time: Total Investment – Energy Cost Savings

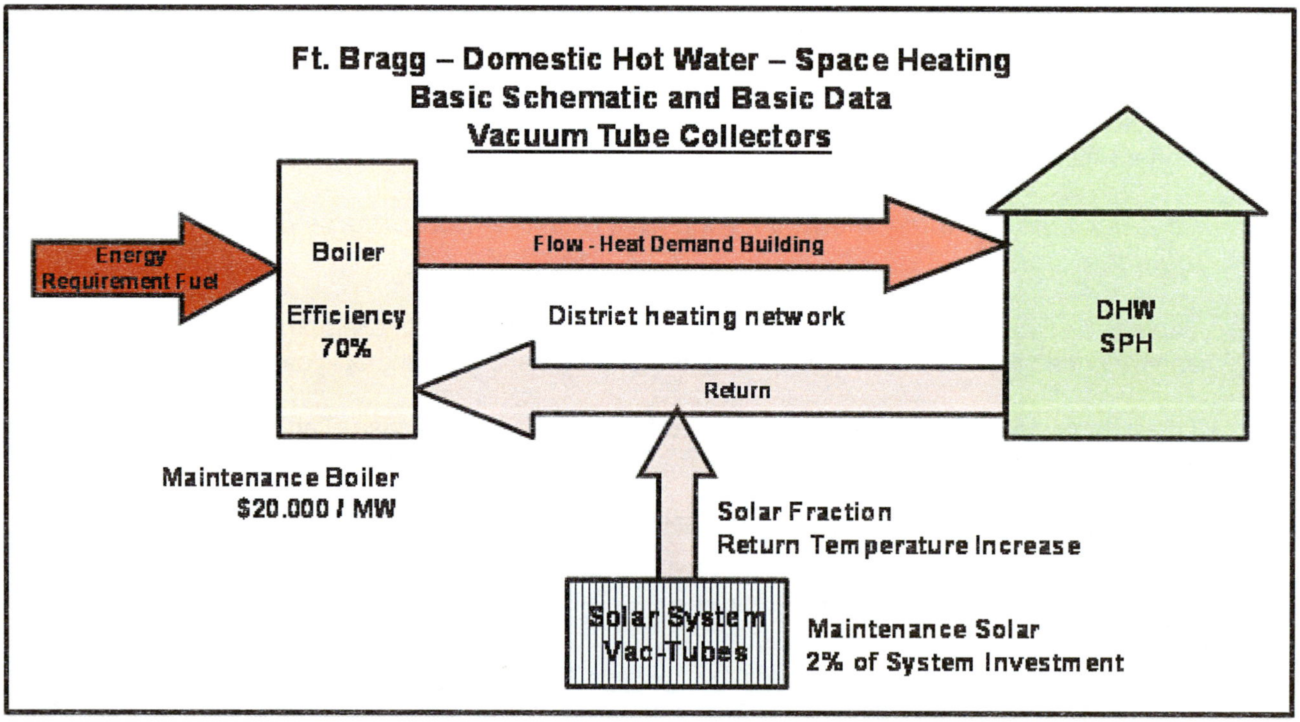

Figure B-35. Basic schematic – Fort Bragg – DHW / SPH – vacuum tube collectors.

Table B-54. Solar system data – Fort Bragg – DHW / SPH – vacuum collectors.

System 1		
Area Solar system (m²)[sq ft]	500.00	5.382.00
Volume storage tank (m³) [ft³]	25.00	882.875.00
Yield Solar System / year (kWh)[MBTU]	335,808.00	1.145.776,90
Solar fraction (%)	**8.80%**	
Total Heat Demand (kWh) [MBTU]	3,816,000.00	13.020.192,00
Total Energy Requirement (kWh) [MBTU]	5,451,428.57	18.600.274,29
Investment System (€) [$]	169,919.02	220.894,73
Planning Cost (€) [$]	25,487.85	33.134,21
Total Investment (€) [$]	195,406.88	254.028,94

System 2		
Area Solar system (m²)[sq ft]	2.500,00	26.910,00
Volume storage tank (m³) [ft³]	125,00	4.414.375,00
Yield Solar System / year (kWh)[MBTU]	1.087.560,00	3.710.754,72
Solar fraction (%)	**28,50%**	
Total Heat Demand (kWh) [MBTU]	3.816.000,00	13.020.192,00
Total Energy Requirement (kWh) [MBTU]	5.451.428,57	18.600.274,29
Investment System (€) [$]	866.956,52	1.127.043,48
Planning Cost (€) [$]	130.043,48	169.056,52
Total Investment (€) [$]	997.000,00	1.296.100,00

System 3		
Area Solar system (m²)[sq ft]	1.500,00	16.146,00
Volume storage tank (m³) [ft³]	75,00	2.648.625,00
Yield Solar System / year (kWh)[MBTU]	767.016,00	2.617.058,59
Solar fraction (%)	**20,10%**	
Total Heat Demand (kWh) [MBTU]	3.816.000,00	13.020.192,00
Total Energy Requirement (kWh) [MBTU]	5.451.428,57	18.600.274,29
Investment System (€) [$]	514.782,61	669.217,39
Planning Cost (€) [$]	77.217,39	100.382,61
Total Investment (€) [$]	592.000,00	769.600,00

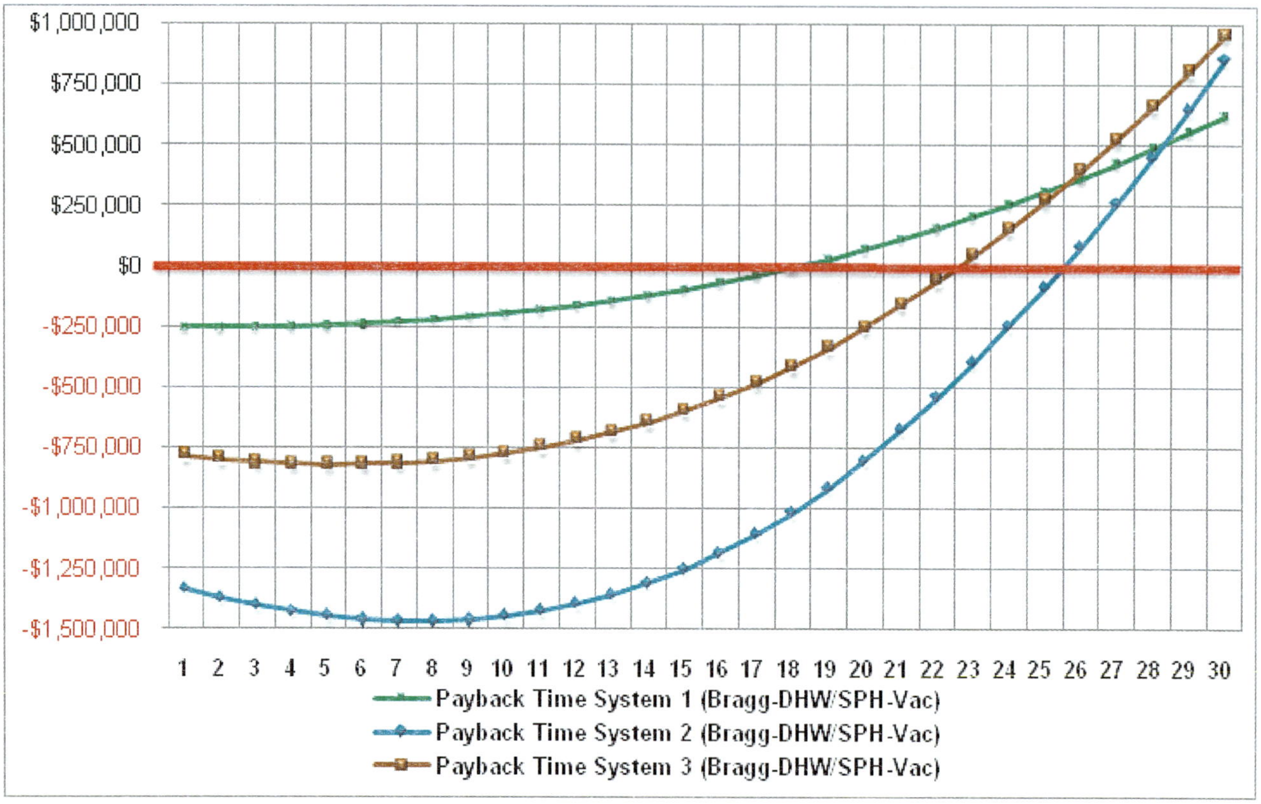

Figure B-36. Payback time - Fort Bragg – DHW / SPH – vacuum tube collectors.

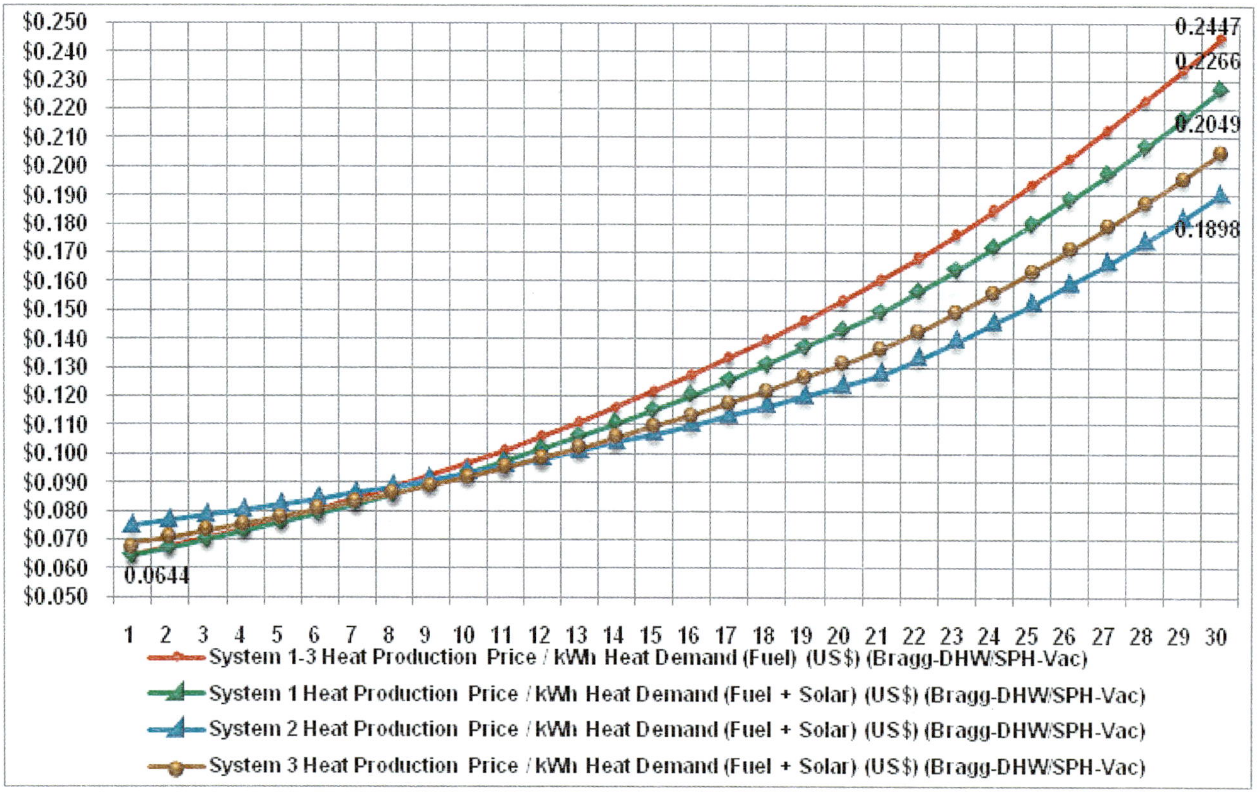

Figure B-37. Energy cost - Fort Bragg – DHW / SPH – vacuum tube collectors.

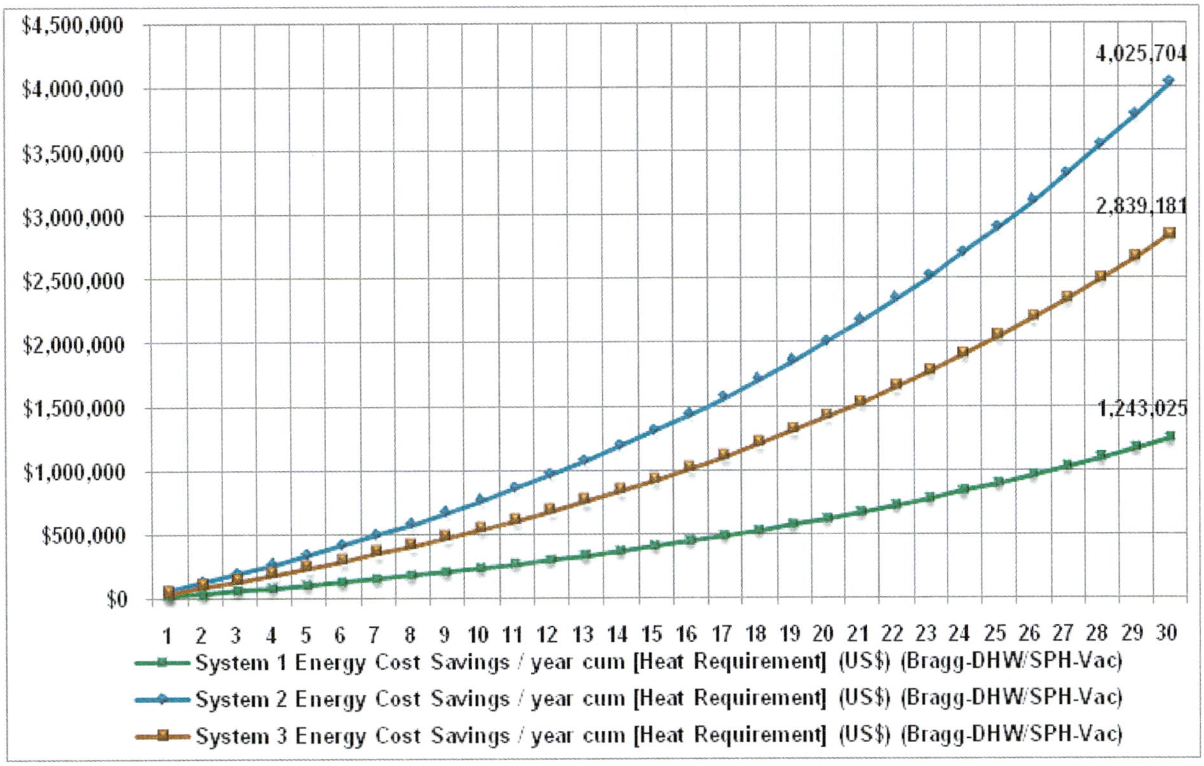

Figure B-38. Energy cost saving / year (cum) – Fort Bragg – DHW / SPH – vacuum tube collectors.

Payback time – annuity method / economic results – Fort Bragg – DHW / SPH – flat plate collectors

Table B-55 lists the basic data used to calculate the payback time with the annuity method, the energy prices in the 30-year period, and the energy savings in the 30-year period. Three dedicated solar systems are chosen for each simulation case. Table B-56 lists the system data.

Table B-55. Basic data - Fort Bragg - DHW / SPH - vacuum tube collectors.

Basic Data	$ / MBTU	€ / kWh
Life time (years)		20
Interest Rate (%)		6,00%
Planning cost factor (% of Investment)		115,00%
Price increase Fuel / year (%)		5,00%
Price increase Maintenance / year (%)		2,00%
Fuel price (€/kWh) [$/kWh]	0,0390	0,0300
Boiler Efficiency (%)		70,00%
Assumed full load hrs / year		4.000
Maintenance Boiler / MWh (€) [$]	26.000	20.000
Heat Demand (kWh) [MBTU]	13.020.192	3.816.000
Boiler capacity (MW) [MBTUh]	3,255	0,95
Maintenance Cost Boiler (€) [$]	24.804,00	19.080,00
Maintenance Cost Solar (% System Invest)		2,00%
Degradation Solar (%/year)		0,00%

Calculation parameters

- Heating Cost Fuel: Cost of Conventional Energy – US$ / kWh
- Cost Heat Production Fuel: Heating Cost Fuel (Energy Requirement) + Maintenance Cost Fuel
- Cost Heat Production Fuel + Solar: Heating Cost Fuel + Solar (Energy Requirement) + Maintenance Cost (Fuel + Solar) + Interest Solar System
- Energy Cost Savings: Cost Heat Production Fuel + Solar - Cost Heat Production Fuel
- Heat Production Price / kWh: Cost Heat Production / Heating Demand
- Payback Time: Total Investment – Energy Cost Savings

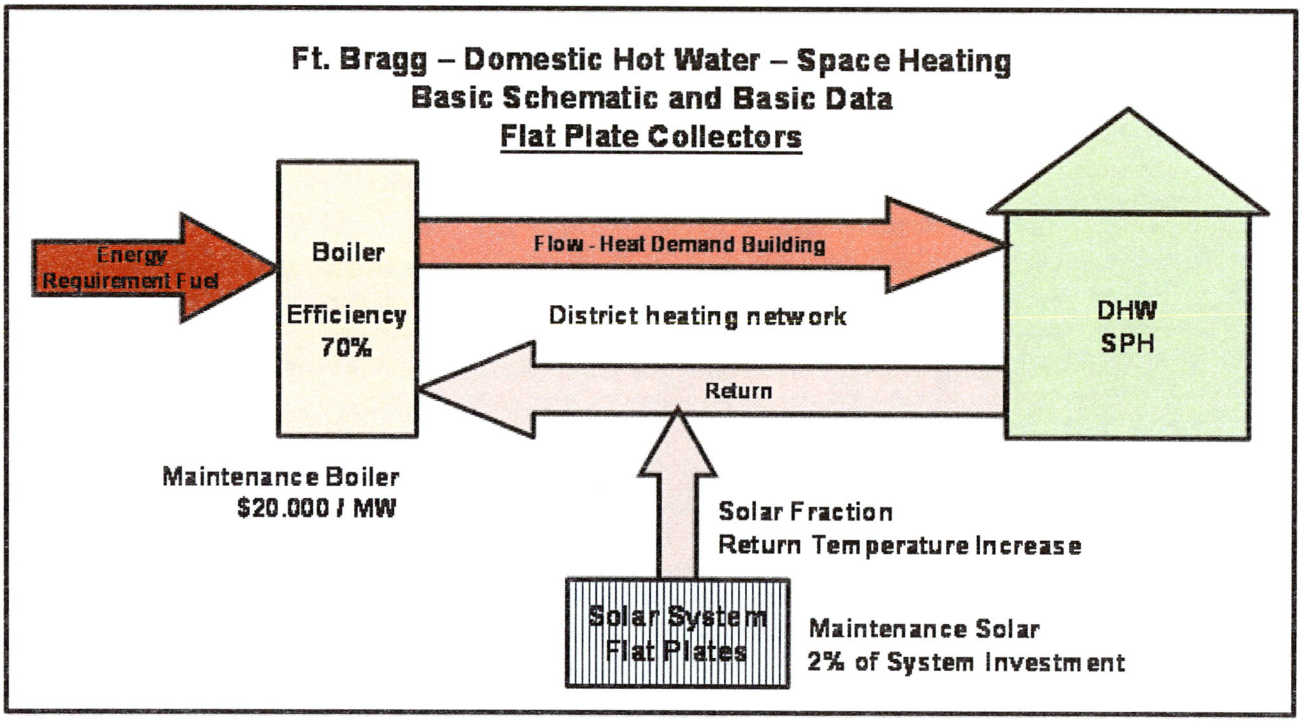

Figure B-39. Basic schematic – Fort Bragg – DHW / SPH – flat plate collectors.

Table B-56. Solar system data - Fort Bragg - DHW / SPH - flat plate collectors.

System 1		
Area Solar system (m^2)[sq ft]	1.000,00	10.764,00
Volume storage tank (m^3) [ft^3]	75,00	2.648.625,00
Yield Solar System / year (kWh)[MBTU]	431.208,00	1.471.281,70
Solar fraction (%)	**11,30%**	
Total Heat Demand (kWh) [MBTU]	3.816.000,00	13.020.192,00
Total Energy Requirement (kWh) [MBTU]	5.451.428,57	18.600.274,29
Investment System (€) [$]	262.575,22	341.347,78
Planning Cost (€) [$]	39.386,28	51.202,17
Total Investment (€) [$]	301.961,50	392.549,95

System 2		
Area Solar system (m^2)[sq ft]	3.000,00	32.292,00
Volume storage tank (m^3) [ft^3]	150,00	5.297.250,00
Yield Solar System / year (kWh)[MBTU]	751.752,00	2.564.977,82
Solar fraction (%)	**19,70%**	
Total Heat Demand (kWh) [MBTU]	3.816.000,00	13.020.192,00
Total Energy Requirement (kWh) [MBTU]	5.451.428,57	18.600.274,29
Investment System (€) [$]	809.806,96	1.052.749,04
Planning Cost (€) [$]	121.471,04	157.912,36
Total Investment (€) [$]	931.278,00	1.210.661,40

System 3		
Area Solar system (m^2)[sq ft]	2.000,00	21.528,00
Volume storage tank (m^3) [ft^3]	100,00	3.531.500,00
Yield Solar System / year (kWh)[MBTU]	660.168,00	2.252.493,22
Solar fraction (%)	**17,30%**	
Total Heat Demand (kWh) [MBTU]	3.816.000,00	13.020.192,00
Total Energy Requirement (kWh) [MBTU]	5.451.428,57	18.600.274,29
Investment System (€) [$]	530.114,78	689.149,22
Planning Cost (€) [$]	79.517,22	103.372,38
Total Investment (€) [$]	609.632,00	792.521,60

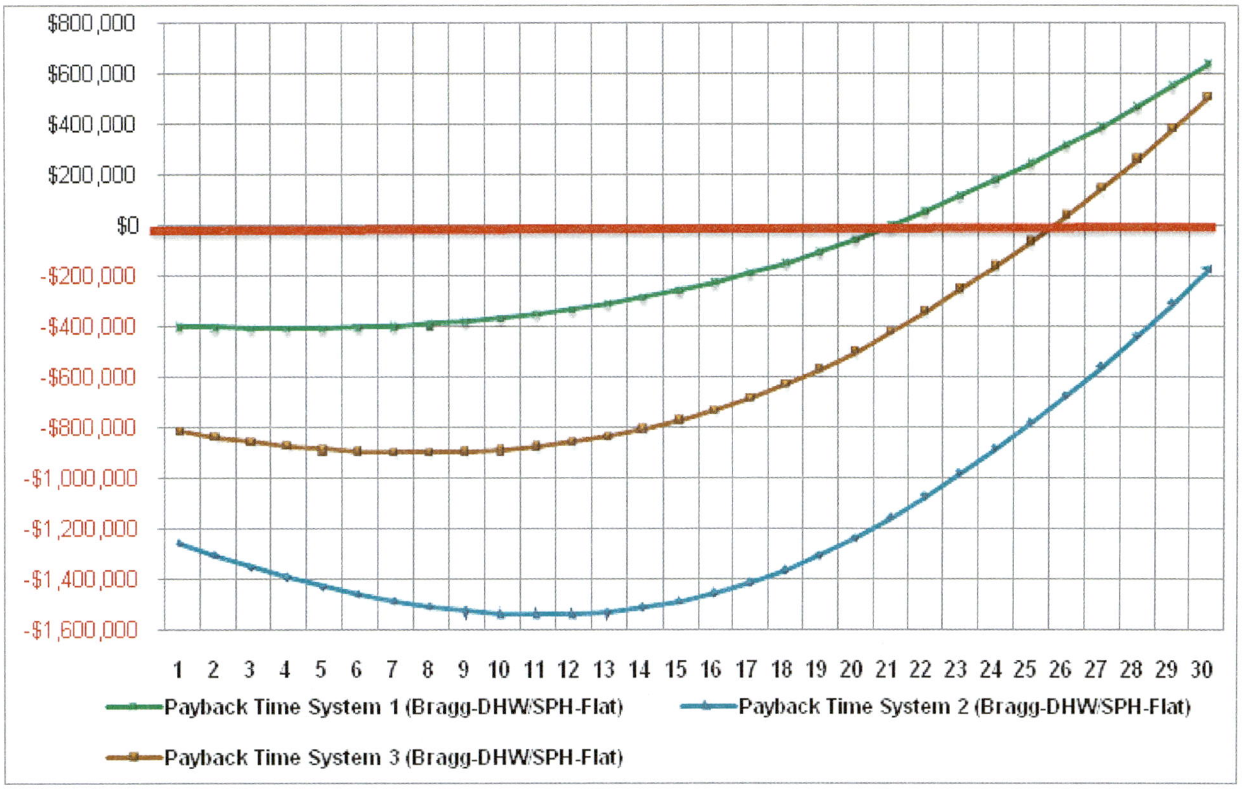

Figure B-40. Payback time - Fort Bragg – DHW / SPH – flat plate collectors.

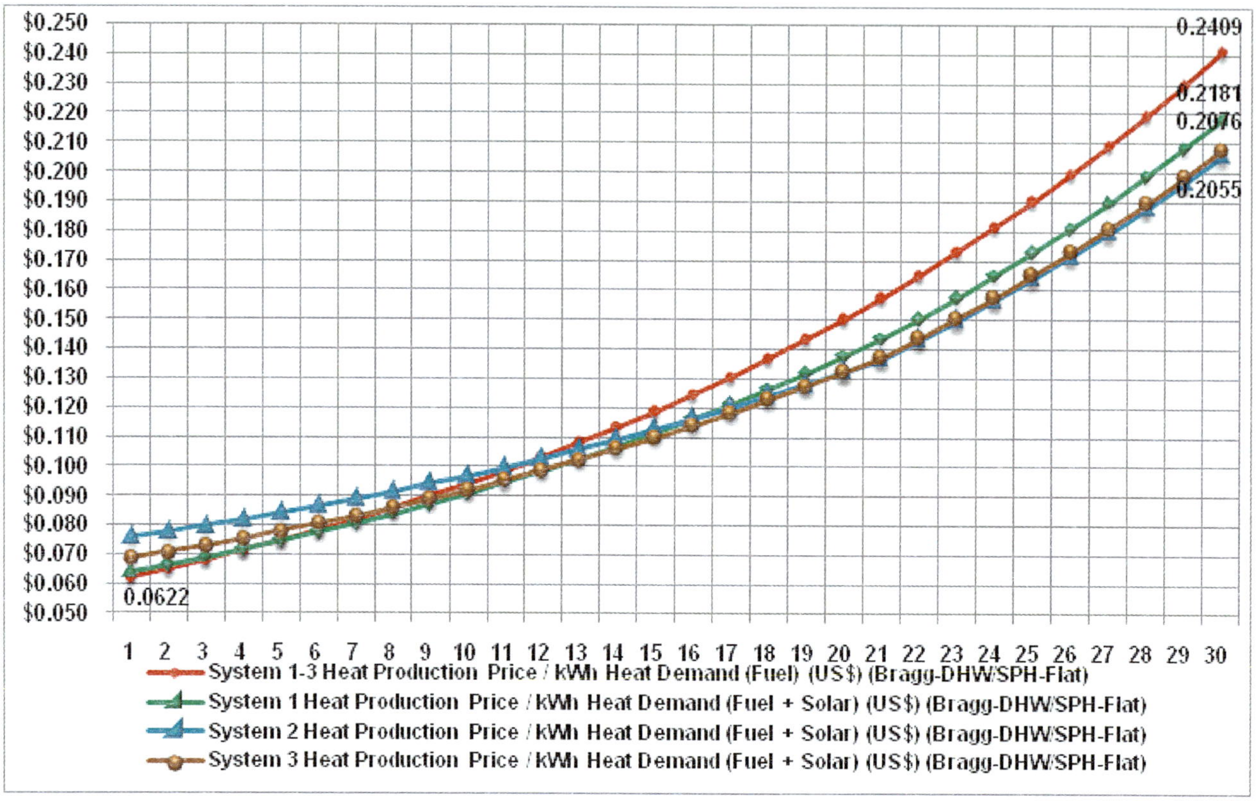

Figure B-41. Energy cost - Fort Bragg – DHW / SPH – flat plate collectors.

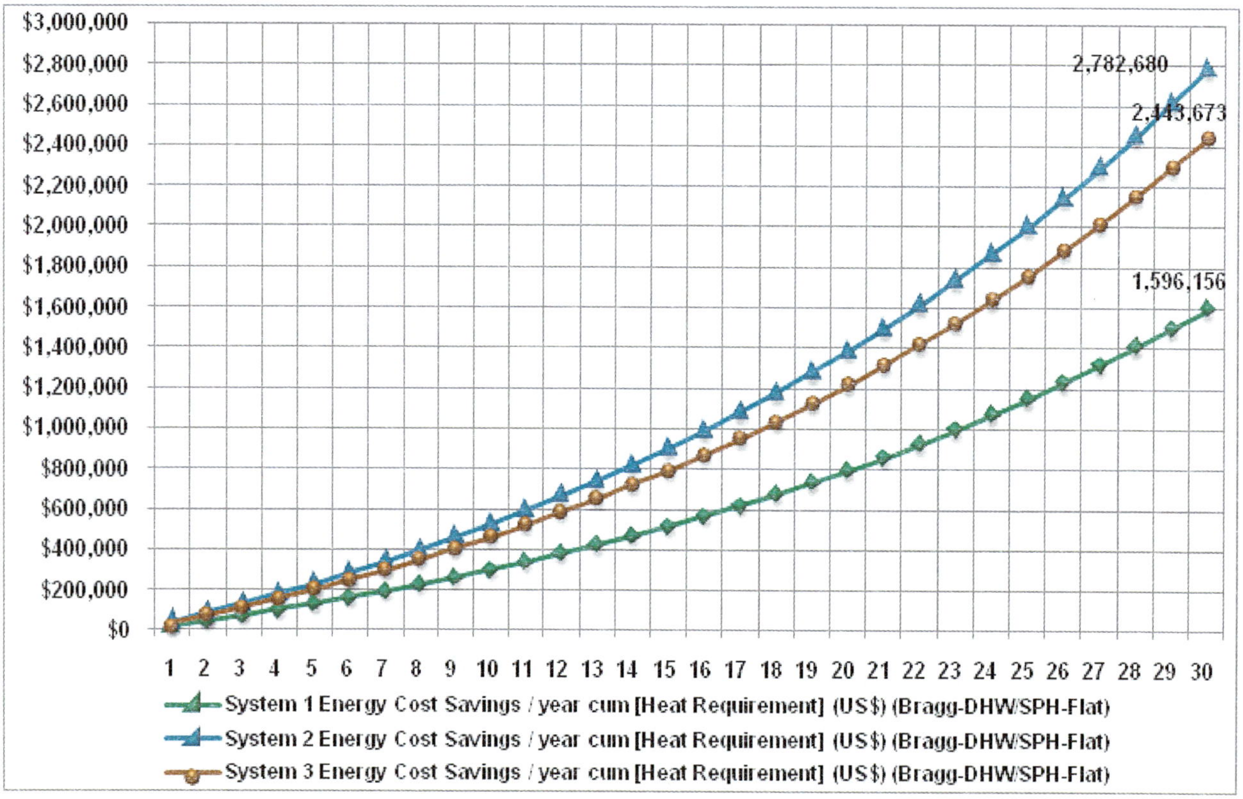

Figure B-42. Energy cost saving / year (cum) – Fort Bragg – DHW / SPH – flat plate collectors.

Overview of calculated economic results for solar systems connected to district heating in Fort Bragg (payback — annuity method)

The applications of the two systems technologies are directly comparable. Furthermore these results are only valid for the basic data used. A change of the basic data (interest rate, fuel price, specific cost collectors, etc.) will lead to different results.

Table B-57. Overview economic results (annuity method) solar systems for Fort Bliss.

	DHW / SPH (Vac)	DHW / SPH (Flat)
Heating Demand Network	1,117,325 Btu/a (3,816 MWh/a)	1,117,325 Btu/a (3,816 MWh/a)
System 1		
Area Solar system	5,380 sq ft (500 m^2)	10,760 sq ft (1,000 m^2)
Volume storage tank	6,605 gal (25 m^3)	19,815 gal (75 m^3)
Solar fraction (%)	8.80%	11.30%
Total Investment [$]	254.029	392.550
Payback Time	18	21
Energy Price (Fuel) 1. year ($)	0.0644	0.0622
Energy Price (Fuel+Solar) 1. year ($)	0.0646	0.0639
Energy Price (Fuel) 30. year ($)	0.2447	0.2409
Energy Price (Fuel+Solar) 30. year ($)	0.2266	0.2181
Energy savings (M$ / 30 years)	1.243	1.596
System 2		
Area Solar system	26,900 sq ft (2,500 m^2)	32,280 sq ft (3,000 m^2)
Volume storage tank	33,025 gal (125 m^3)	39,630 gal (150 m^3)
Solar fraction (%)	28.50%	19.70%
Total Investment [$]	1,296,100	1,210,661
Payback Time	25	30
Energy Price (Fuel) 1. year ($)	0.0644	0.0622
Energy Price (Fuel+Solar) 1. year ($)	0.0748	0.0758
Energy Price (Fuel) 30. year ($)	0.2447	0.2409
Energy Price (Fuel+Solar) 30. year ($)	0.1898	0.2055
Energy savings (M$ / 30 years)	4.026	2.783
System 3		
Area Solar system	16,140 sq ft (1,500 m^2)	21,520 sq ft (2,000 m^2)
Volume storage tank	19,815 gal (75 m^3)	26,420 gal (100 m^3)
Solar fraction (%)	20.10%	17.30%
Total Investment [$]	769,600	792,522
Payback Time	22	25
Energy Price (Fuel) 1. year ($)	0.0644	0.0622
Energy Price (Fuel+Solar) 1. year ($)	0.0688	0.0686
Energy Price (Fuel) 30. year ($)	0.2447	0.2409
Energy Price (Fuel+Solar) 30. year ($)	0.2049	0.2076
Energy savings (M$ / 30 years)	2.839	2.444

Operational conditions of flat plate and vacuum pipe collectors

The decision to use a flat plate or a vacuum pipe collector will depend mainly of the return temperature of the user and the desired solar fraction at the existing weather conditions. Weather conditions and solar fraction alone not a sufficient criterion; the return temperature must always be considered. Therefore, it is difficult to recommend either vacuum pipes or flat plate collectors based solely on weather conditions and solar fractions.

Vacuum pipe collectors can more easily achieve a high solar fraction at reasonable storage volume for users with high return temperatures. The calculations in Fort Bliss and Fort Bragg show that the use of flat plate collectors at a return temperature of 140 °F (60 °C) leads to a maximum solar fraction of 24% (Bliss) and 19% (Bragg).

At return temperatures higher than 140 °F (60 °C), the efficiency of flat plate collectors decreases more dramatically than with vacuum pipes. Therefore, all applications with cooling (Fort Bliss) were calculated only with vacuum pipe collectors.

The decision to use flat plate or vacuum collectors in district heating systems just for DHW and SH-demand (140.0 °F [60 °C] return temperature), should be made based on both economic and technical aspects of the solar system. The different construction of flat plate systems and the Paradigma-pipe collector system (flat plate: with collector-heat exchanger; Paradigma: no collector-heat exchanger) justifies the demonstration of both systems – even if one system type has slightly economical disadvantages in the specific application.

Recommendation – solar technology versus application area

The calculations show that both types of collectors, FPCs and ETCs, have their specific applications.

Especially at higher temperature levels, flat plat collectors earn only a small amount of energy; the economic efforts to reach the target figures are very high. This is valid e.g., with cooling and heating at a higher temperature level. On the other hand, if a low temperature level is sufficient, e.g., with DHW or space heating at moderate net temperatures, flat plate collectors are the better choice due to their lower optical losses.

ETCs have a higher energy gain than FPCs with same gross area so ETCs' solar fraction of energy demand can also be higher. Up to a certain point, this can be compensated by the use of a storage tank. Under different conditions (i.e., at moderate net temperatures, and a higher level of solar fraction), the two types will be in strong competition. The decision to select on over the other must be based on simulation and economic analyses.

Figure B-43 illustrates the characteristic advantages of the two technical options.

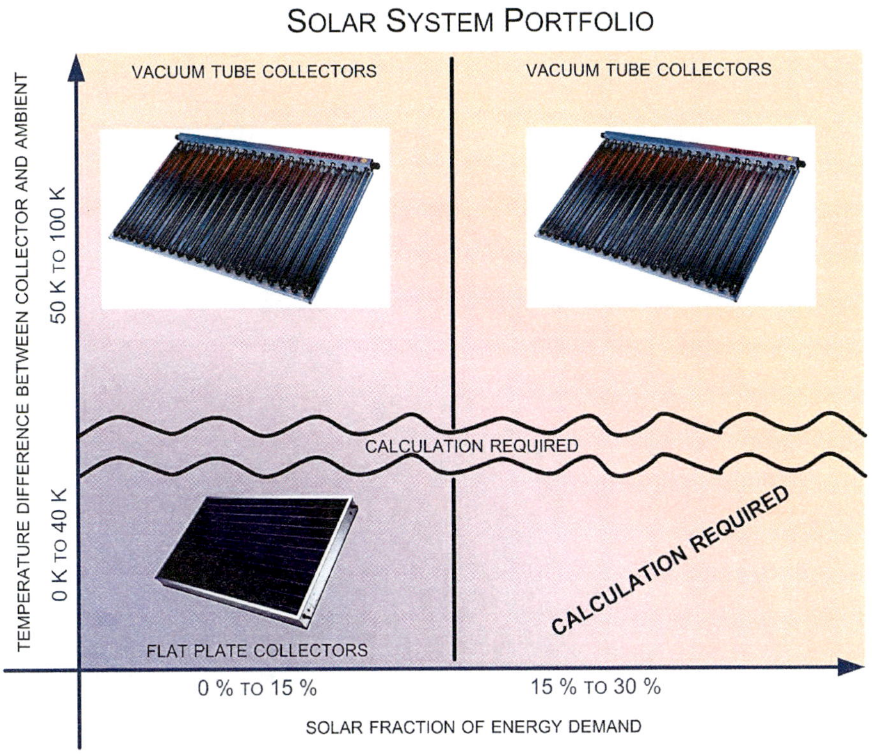

Figure B-43. Solar fraction of energy demand.

Conclusion of the simulation

The analysis regarding the different temperatures highlights the impacts of different operation modes with regards to the uses of the heat from solar fields. For example, heat can be used for space heating in the winter and DHW throughout the year. Since the space heating demand in arid climates occurs in the summer, at its minimum, the heat can be used to operate an absorption chiller. In this case, the question is not only of the energy balance, but also — and even more importantly — a question of the impacts on the system's temperature. Since an absorption chiller requires supply temperatures of about 195 to 205 °F (91 to 96 °C), the temperature of the solar field, i.e., "heat generation," must be high enough to maintain supply temperatures sufficient for DHW (150 °F [66 °C]). Moreover absorption chillers reinject temperatures of 160 to 185 °F (71 to 85 °C) into the system's return, decreasing the temperature difference of the system.

This short discussion should help to explain the difficulties inherent in optimizing the solar harvest factor. This optimization of the solar harvest factor does not necessarily lead to an optimization of the entire system. System optimization requires a detailed analysis of summer heat use and an analysis of its impacts on the entire system.

An additional option to consider may be adsorption or sorption chillers, which might provide an alternative that would reduce the temperature issues since both adsorption or sorption chillers require lower supply temperatures. "Solar cooling via absorption" can reduce stagnation of a solar filed. However, absorption systems will cause higher first costs since the solar field needs to meet the summer load as well, because such a system will probably require a larger square footage, and also because O&M efforts might possibly increase. By their nature, absorption systems are most likely not suited to humid climates. But on the other hand both, adsorption and sorption chillers are relatively recent innovations; they are not "state of the art" technologies, and thus may not be well suited to ensure a secure energy supply.

Appendix C: Market Price Scenario Europe – Climate related Economic Comparisons of Solar Systems

Market price scenarios of solar systems in Europe

Table 4.4 (p 67) defines large scale solar systems as those with areas greater than 100 m^2 requiring an investment per square meter of 500–750 €/m^2 (~\$650 – \$975). Small scale solar systems are those with areas less than 100 m^2 requiring an investment per square meter of 800–1,000 €/m^2 (~\$1,040 – \$1,300/sq ft).

Table C-1 lists two (Paradigm and Arcon) quotations for a 5,000 m^2 (53,800 sq ft) solar system. Since the basic requirements for both solar systems are comparable, the pricing of flat collectors and vacuum tube collectors in these two quotations are made comparable for this particular project. (Note that quotations for other project scenarios may yield different results.)

Table C-1. Example - 5000 m^2 Solar System Investment – Paradigma and Arcon.

	Paradigma	Arcon
Solar System Area (m^2)	5,000	5,000
Collector, Connections, Piping (\$)	1,417,000	1,144,000
Mounting Kits (\$)	325,000	325,000*
Heat Exchanger, Pumps, Filter, Monitoring, Expansion Devices (\$)	135,200	182,000
Storage Tank (\$)	195,000	286,000
Shipping (\$)	91,000	123,500
Commissioning (\$)	65,000	58,500
Labor on site (\$)	130,000*	110,500*
Total Investment (\$)	2,358,200	2,229,500
Total Investment (\$ /m^2)	471.64	445.90

Also, the investment prices for a solar system in Austria (Graz-Liebenau) based on flat plate collectors amounts to a total investment per square meter of € 390 or \$ 507. Figure C-1 shows the price ranges of Solar Systems (dependent on system size) in Europe.

The next section compares identical solar systems installed in different climate zones in the United States, based on this price information.

Climate related comparison of solar systems

The solar system used for these comparisons is designed for a DHW application only based on a flat plate collector system. Table C-2 lists the design criteria. Figure C-2 shows the solar system schematic.

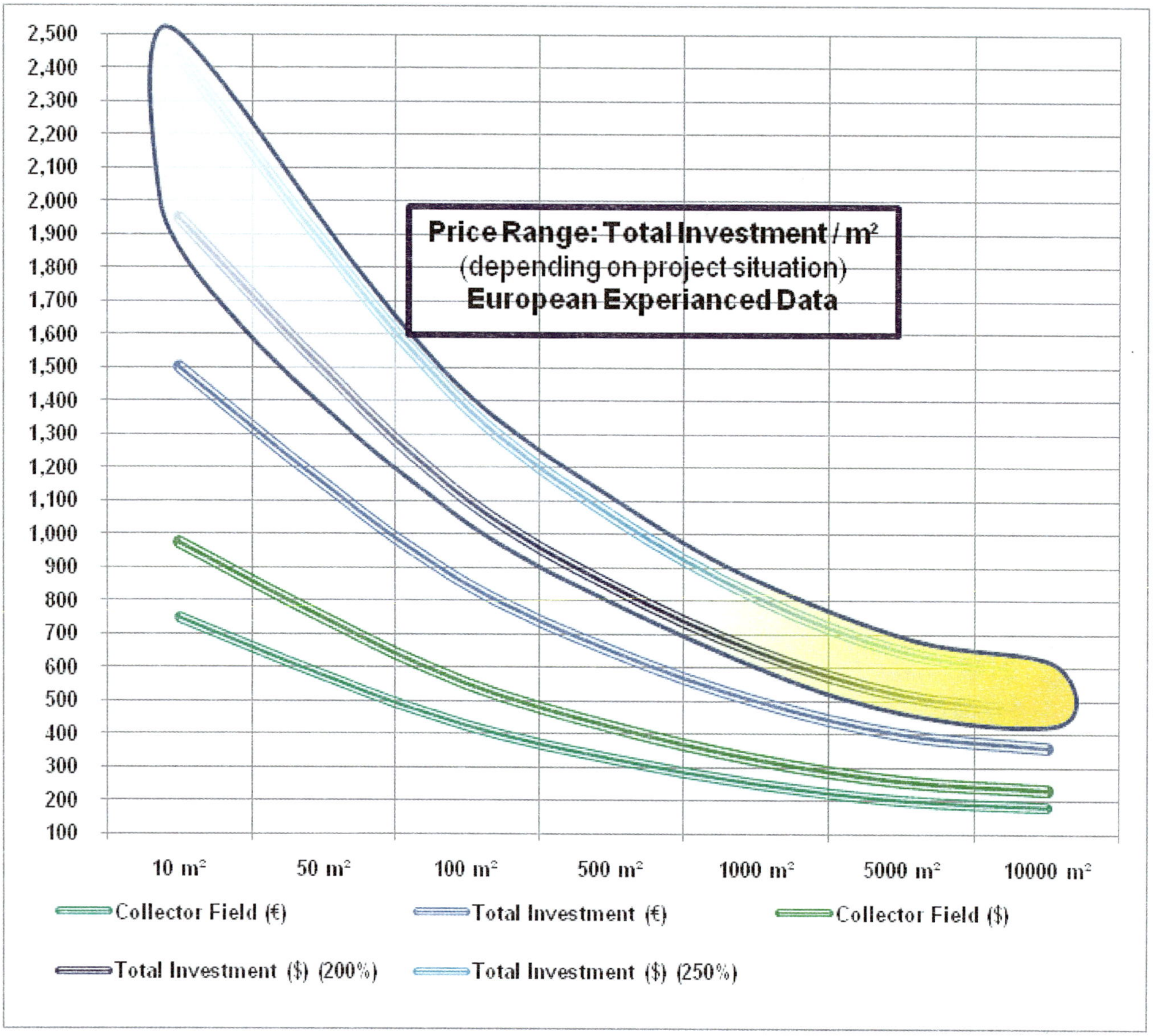

Figure C-1. Price range total investment solar system depending on solar field area.

Table C-2. Design criteria for solar system comparison.

Site of Installation		Chicago, IL	Boston, MA	New York, NY	Orlando, FL	Kansas City, KS	Seattle, WA	Fort Irwin, CA
Climate (from T*SOL library)		Chicago, IL	Boston, MA	New York, NY	Tampa, FL	Kansas City, KS	Seattle, WA	Las Vegas, NV
Global irradiation (horizontal)	Btu/sq ft (kWh/[m^2*yr])	448,507 (1,414)	451,996 (1,425)	451,679 (1,424)	561,426 (1,770)	494,816 (1,560)	387,289 (1,221)	652,460 (2,057)
Share of diffuse irradiation	%	50.3	50.9	50.8	47.2	48.2	53.5	37.2
Lowest ambient temperature	°F (°C)	-11 (-23.7)	-3 (-19.3)	-1 (-18.1)	29 (-1.6)	-5 (-20.7)	23 (-5.2)	15 (-9.4)
Year average temperature	°F (°C)	49 (9.5)	51 (10.8)	55 (12.5)	73 (22.5)	54 (12.1)	53 (11.6)	67 (19.5)
Cold water temperature summer	°F (°C)	57 (14)	57 (14)	57 (14)	57 (14)	57 (14)	57 (14)	57 (14)
Cold water temperature winter	°F (°C)	46 (8)	46 (8)	46 (8)	46 (8)	46 (8)	46 (8)	46 (8)
DHW storage tank outlet temperature	°F (°C)	140 (60)	140 (60)	140 (60)	140 (60)	140 (60)	140 (60)	140 (60)
DHW demand	gal (m^3/day)	2,642 (10)	2,642 (10)	2,642 (10)	2,642 (10)	2,642 (10)	2,642 (10)	2,642 (10)

Site of Installation		Chicago, IL	Boston, MA	New York, NY	Orlando, FL	Kansas City, KS	Seattle, WA	Fort Irwin, CA
DHW demand	gal (m^3/yr)	964,330 (3,650)	964,330 (3,650)	964,330 (3,650)	964,330 (3,650)	964,330 (3,650)	964,330 (3,650)	964,330 (3,650)
DHW demand profile (from T*SOL library)		Multi. dwelling	Multi. dwelling	Multi. dwelling	Multi. dwelling	Multi. dwelling	Multi. dwelling	Multi. dwelling
DHW energy demand	Btu (MWh/yr)	60,794 (207.63)	60,794 (207.63)	60,794 (207.63)	60,794 (207.63)	60,794 (207.63)	60,791 (207.62)	60,794 (207.63)
DHW circulation energy demand	Btu (MWh/yr)	33,599 (114.75)	33,599 (114.75)	33,596 (114.74)	33,587 (114.71)	33,596 (114.74)	33,596 (114.74)	33,593 (114.73)
DHW storage tank energy losses	Btu (MWh/yr)	735 (2.51)	738 (2.52)	738 (2.52)	735 (2.51)	738 (2.52)	738 (2.52)	738 (2.52)
DHW total energy demand	***Btu (MWh/yr)***	***95,131 (324.90)***	***95,131 (324.90)***	***95,128 (324.89)***	***95,119 (324.86)***	***95,128 (324.89)***	***95,128 (324.89)***	***95,125 (324.88)***
Collector area (aperture area)	Sq ft ($m^{2)}$	1,356 (126)	1,248 (116)	1,270 (118)	861 (80)	1,076 (100)	1,765 (164)	689 (64)
Type of collector		Flat plate	Flat plate	Flat plate	Flat plate	Flat plate	Flat plate	Flat plate
Orientation		South	South	South	South	South	South	South
Tilt angle	°	35	35	35	35	35	35	35
Shading (objects above horizon)	°	10	10	10	10	10	10	10
Length of collector area piping, outdoor	ft (m)	207 (63)	190 (58)	194 (59)	131 (40)	164 (50)	269 (82)	105 (32)
Length of collector area piping, indoor	ft (m)	21 (6.3)	19 (5.8)	19 (5.9)	13 (4.0)	16 (5.0)	27 (8.2)	10 (3.2)
Flow velocity in collector area piping	m/s	0.5	0.5	0.5	0.5	0.5	0.5	0.5
Freeze protection, glycol content	%	45	40	40	15	40	25	30
Irradiation in collector area	MWh/yr	190.77	179.59	178.79	143.52	167.49	213.70	143.51
Irradiation in collector area	kWh/(m^2*yr)	1,514	1,549	1,516	1,794	1,675	1,303	2,242
Capacity solar storage tank	m^3	6.3	5.8	5.9	4.0	5.0	8.2	3.2
Capacity DHW storage tank	m^3	3.0	3.0	3.0	3.0	3.0	3.0	3.0
Energy from solar system	MWh/yr	97.52	97.44	97.44	98.12	96.83	97,71	97.94
Solar efficiency	%	51.1	54.3	54.5	68.4	57.8	45.7	68.2
Solar fraction at DHW total energy demand	*%*	***30.0***	***30.0***	***30.0***	***30.2***	***29.8***	***30.1***	***30.1***

The solar fractions of all systems are identical. The solar system areas are adapted according to the solar irradiation in the different climate zones.

In all cases, the solar system increases the return of the total DHW-heating system. In this case, the return is on a low temperature level. In such an application flat plate collectors show a higher economic efficiency than do the vacuum tube collectors.

Vacuum tube collectors have an advantage at increased temperature levels. Different schematics for these solar applications must be considered.

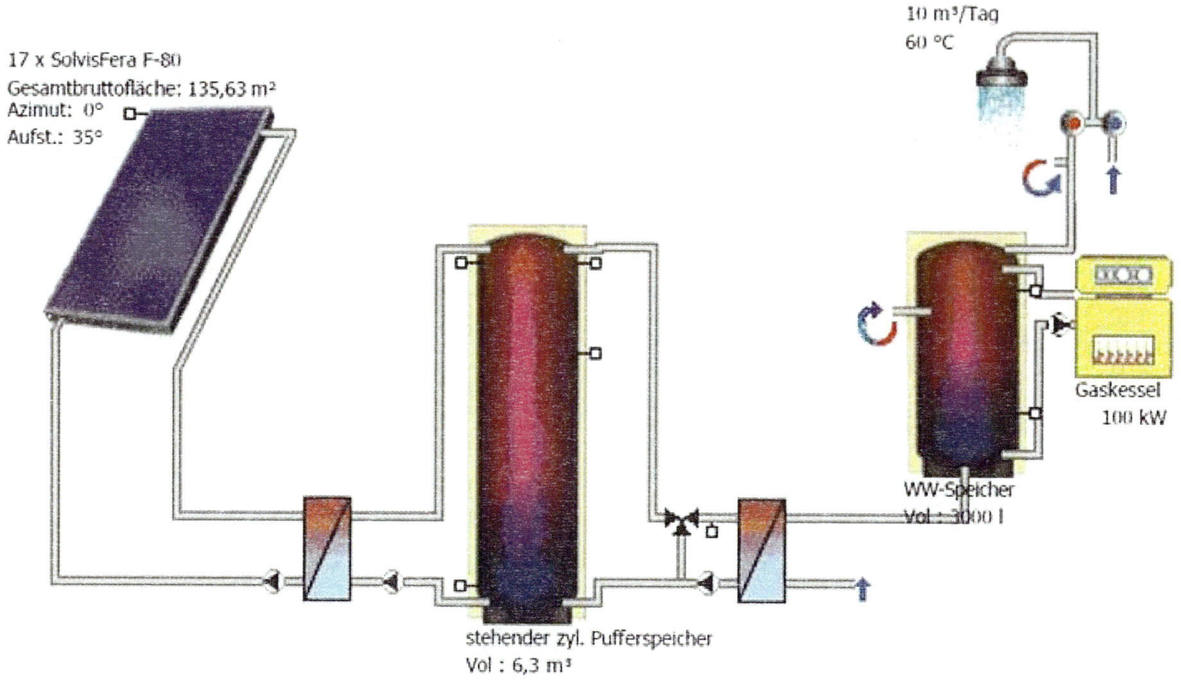

Figure C-2. Schematics of solar system – comparison.

Table C-3 lists the data used for investment calculations. Figure C-3 shows the investment calculation results based on the experienced data.

Investment levels and economic efficiencies will differ due to the variations in solar irradiation in the different climate zones.

Table C-3. Overview solar system comparison in different climate zones.

Investment	Factors	Chicago, IL	Boston, MA	New York, NY	Orlando, FL	Kansas City, KS	Seattle, WA	Fort Irwin, CA
Collector Type		Flat	Flat	Flat	Flat	Flat	Flat	Flat
Solarfield sq ft (m^2)		1,313 (122)	1,248 (116)	1,270 (118)	861 (80)	1,076 (100)	1,765 (164)	689 (64)
Collector Price \$/sq ft (\$/m^2)		46.47 (500)	46.47 (500)	46.47 (500)	46.47 (500)	46.47 (500)	46.47 (500)	46.47 (500)
Collector Invest (\$)		61,000	58,000	59,000	40,000	50,000	82,000	32,000
Mounting Collectors (% of Collector Invest) (\$)	20%	12,200	11,600	11,800	8,000	10,000	16,400	6,400
Piping incl. Labor (% of Collector Invest) (\$)	20%	12,200	11,600	11,800	8,000	10,000	16,400	6,400
Storage Volume gal/sq ft (l/m^2)	1.23 (50)	149.76 (6,100)	142.39 (5,800)	144.85 (5,900)	98.20 (4,000)	122.75 (5,000)	201.31 (8,200)	78.56 (3,200)
Storage incl. Labor \$/gal (\$/m^3)	2.83 (750)	17.31 (4,575)	16.46 (4,350)	16.74 (4,425)	11.36 (3,000)	14.19 (3,750)	23.28 (6,150)	9.08 (2,400)
Control- / Monitoring Systems (\$ of Collector Invest) (\$)	5%	3,050	2,900	2,950	2,000	2,500	4,100	1,600
System Invest		**93,025**	**88,450**	**89,975**	**61,000**	**76,250**	**125,050**	**48,800**
Commissioning (% of System Invest)	**10%**	**9,303**	**8,845**	**8,998**	**6,100**	**7,625**	**12,505**	**4,880**
Planning Cost (% of System Invest) (\$)	**15%**	**13,954**	**13,268**	**13,496**	**9,150**	**11,438**	**18,758**	**7,320**
Total Investment (\$)		116,281	110,563	112,469	76,250	95,313	156,313	61,000
Total Investment / m^2		953 m^2	953 m^2	953 m^2	953 m^2	953 m^2	953 m^2	953 m^2
Global Irradiation (horizontal)		1414	1425	1424	1770	1560	1221	2057
Solar Fraction (%)		30%	30%	30%	30%	30%	30%	30%
Yield Solar System Btu (kWh)		332,835.76 (97.520)	332,562.72 (97.440)	332,562.72 (97.440)	334,883.56 (98.120)	330,480.79 (96.830)	333,484.23 (97.710)	334,269.22 (97.940)

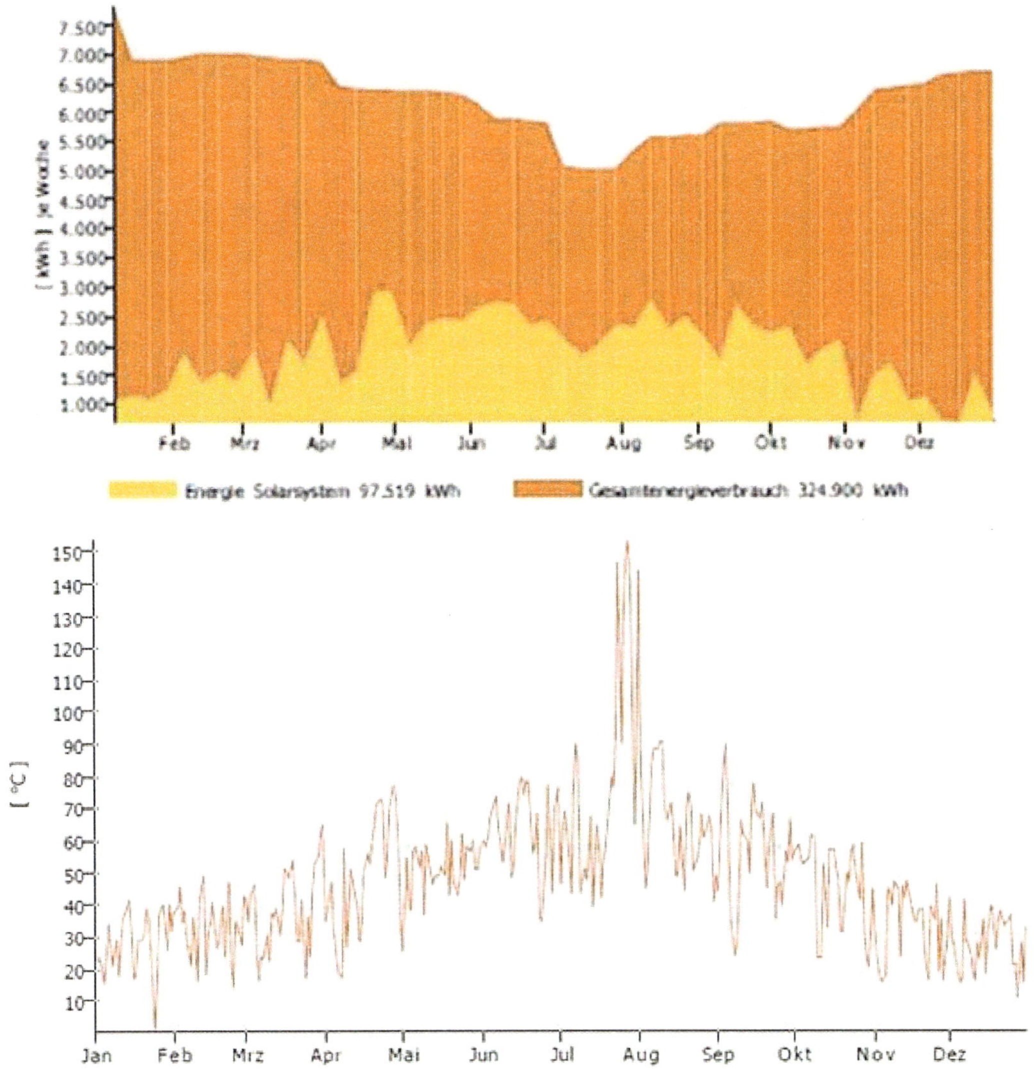

Figure C-3. Solar contribution to energy demand / daily maximum temperature at collector.

Comparison of payback time

This section summarizes two payback time scenarios. In both cases, the energy savings between a conventional DHW supply with a fuel based heating and a combined DHW supply with a fuel based heating and a solar system are the factors for the payback calculation.

In Scenario 1, payback time is based on energy savings with the annuity method (interest included). The range of the payback times in Scenario 1 is between 15 years (California – Fort Irwin) and more than 20 years (Seattle – Washington) (Figure C-4).

In Scenario 2, the range of the payback times is between 10 years (California – Fort Irwin) and 23 years (Seattle – Washington) (Figure C-5).

Heating price development

The comparison of the heat production price of a conventional fuel based DHW supply system and a combined fuel and solar based DHW supply system leads to the results shown in Figure C-6.

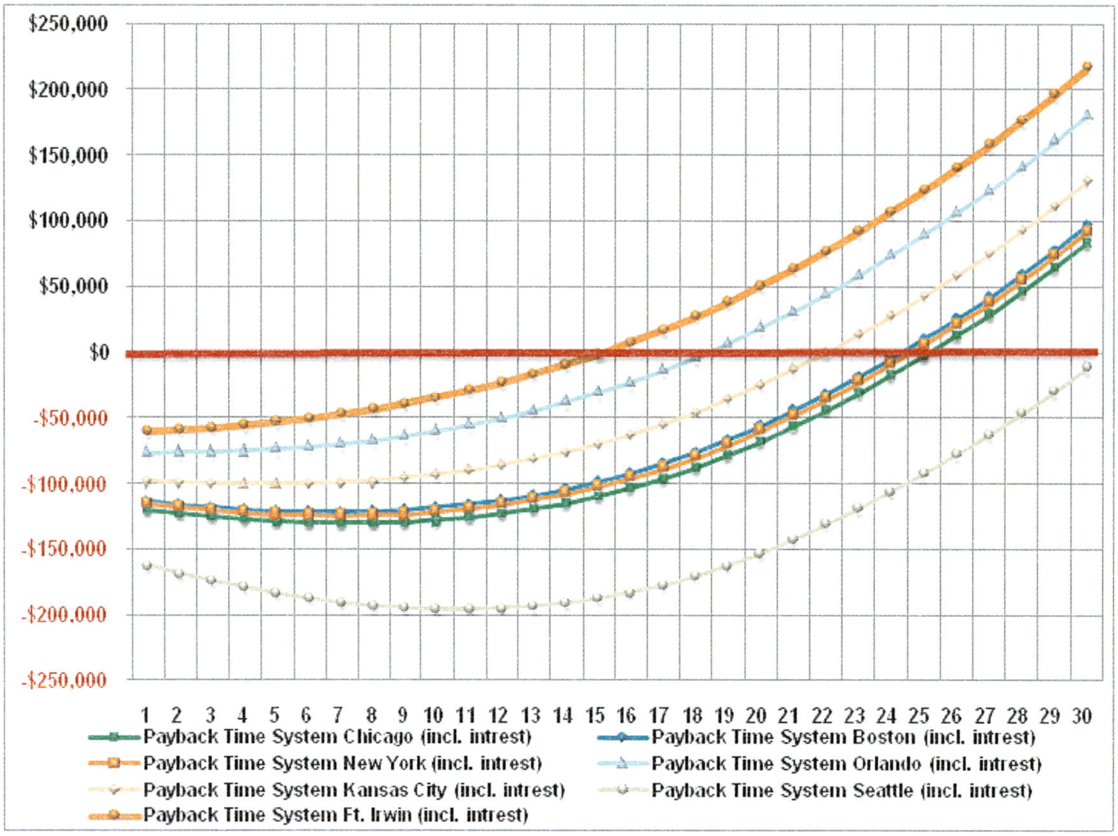

Figure C-4. Comparison of payback time scenario 1 – climate zones.

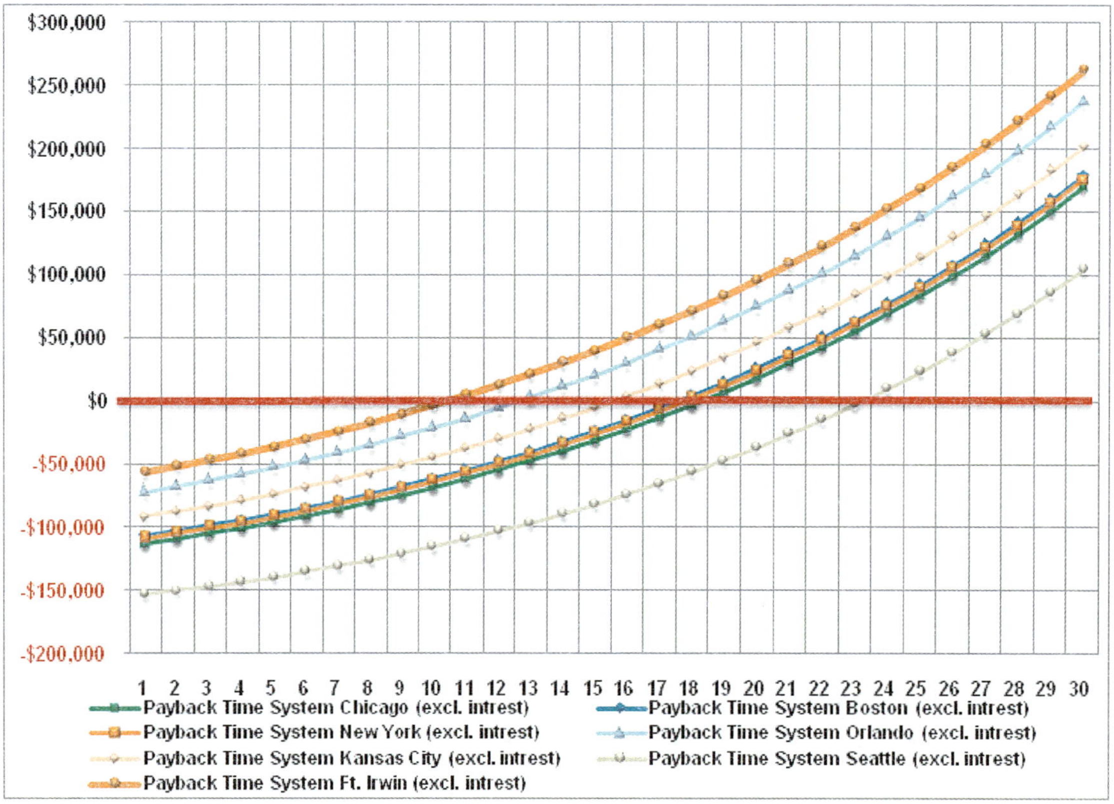

Figure C-5. Comparison of payback time scenario 2 – climate zones.

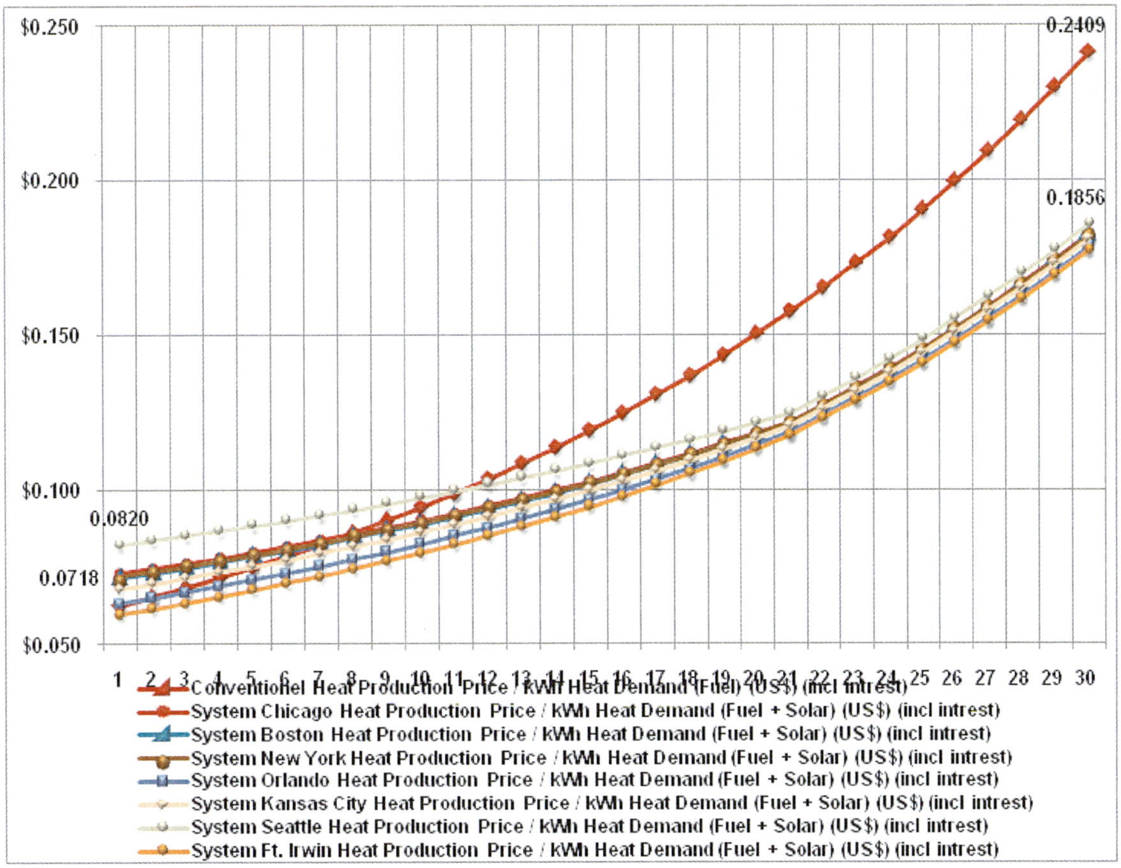

Figure C-6. Comparison of heat production price.

The heat production price includes the maintenance cost of the heating and solar systems and the interest of the investment. A yearly fuel price increase of 5% leads to total price increases as shown in Table C-4.

Without a solar system, the calculated price increase for the heat production cost in the 30-year period may be ~380% compared to the current situation with a fuel price of $0.039 per kwh.

This price increase may only be ~280% – 300% depending on the climate zones if a solar system is used as described before.

Table C-4. Comparison price Increase – heat production cost.

Heat Production Price	Chicago, IL	Boston, MA	New York, NY	Orlando, FL	Kansas City, KS	Seattle, WA	Fort Irwin, CA
Conventional System 1. Year	0.062	0.062	0.062	0.062	0.062	0.062	0.062
Conventional System 30. Year	0.241	0.241	0.241	0.241	0.241	0.241	0.241
Total Price Increase without Solar System	387.2%	387.2%	387.2%	387.2%	387.2%	387.2%	387.2%
Combined System 1. Year	0.051	0.051	0.051	0.049	0.050	0.053	0.048
Combined System 30. Year	0.182	0.182	0.182	0.178	0.181	0.186	0.177
Total Price Increase with Solar System	292.9%	292.2%	292.4%	286.6%	290.7%	298.3%	284.6%

Energy savings – with solar system

The energy savings at each location is identical in this case because of the identical solar fraction.

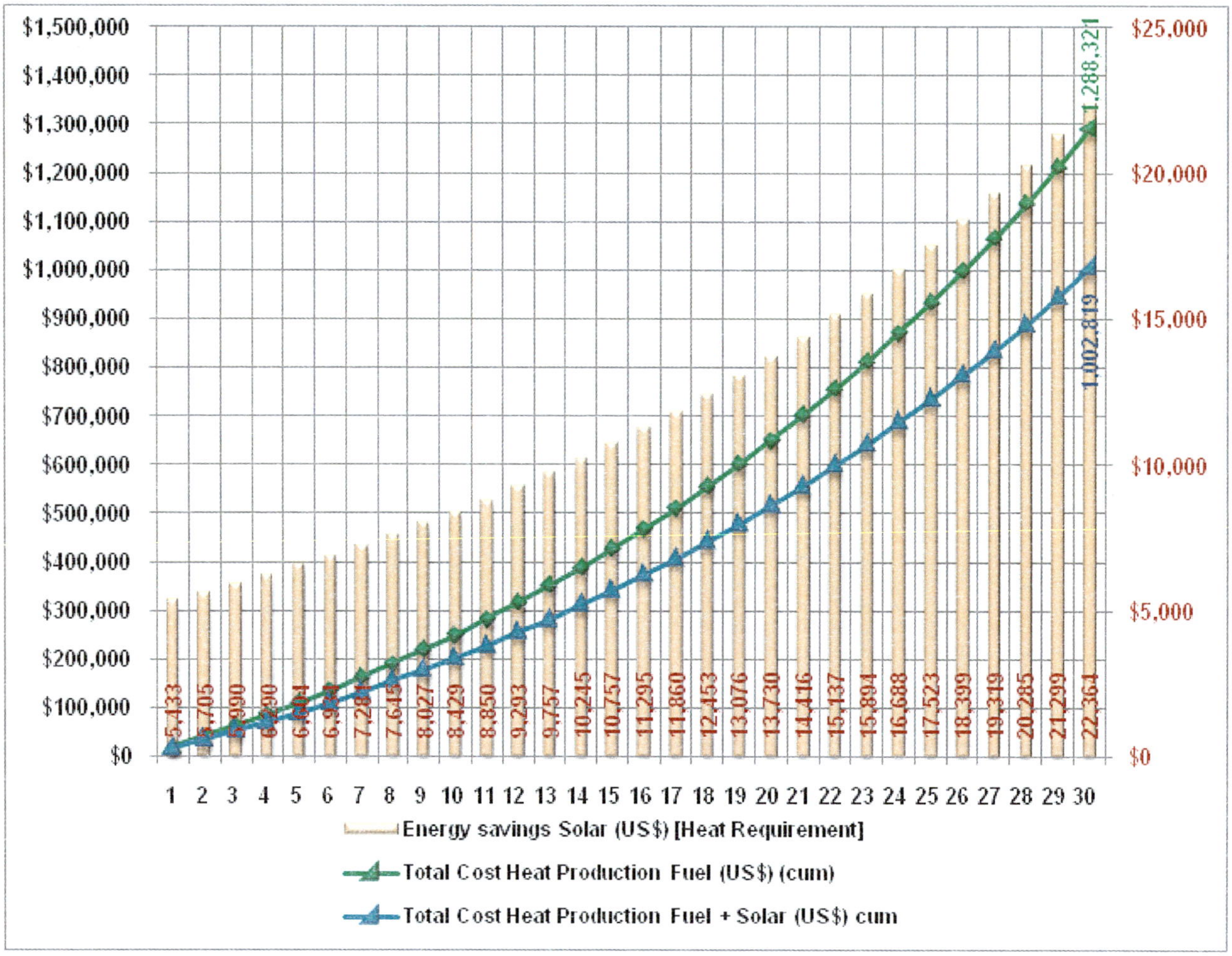

Figure C-7. Energy savings within 30 years.

The energy savings in this case study is ~$280,000 independent of the climate zone.

The investment prices in the different climate zones are in a range of $45,000 (Fort Irwin) and $125,000 (Seattle), see also Table C-3.

Appendix D: Sample SRCC Rating Page (Flat-Plate Collector)

SOLAR COLLECTOR
CERTIFICATION AND RATING

SRCC OG-100

CERTIFIED SOLAR COLLECTOR

SUPPLIER:	**Fafco, Inc.** 435 Otterson Dr. Chico, CA 95928 USA
MODEL:	Sunsaver
COLLECTOR TYPE:	Unglazed Flat-Plate
CERTIFICATION#:	2007051A

ALL SIZES OF THIS COLLECTOR MODEL ARE CERTIFIED

COLLECTOR THERMAL PERFORMANCE RATING

Megajoules Per Square Meter Per Day			
CATEGORY (Ti-Ta)	CLEAR DAY	MILDLY CLOUDY	CLOUDY DAY
A (-5 °C)	21.2	16.6	12.1
B (5 °C)	14.9	10.4	6.0
C (20 °C)	6.9	3.0	0.2
D (50 °C)	0.0	0.0	0.0
E (80 °C)	0.0	0.0	0.0

Thousands of BTU Per Square Foot Per Day			
CATEGORY (Ti-Ta)	CLEAR DAY	MILDLY CLOUDY	CLOUDY DAY
A (-9 °F)	1.9	1.5	1.1
B (9 °F)	1.3	0.9	0.5
C (36 °F)	0.6	0.3	0.0
D (90 °F)	0.0	0.0	0
E (144 °F)	0.0	0.0	0.0

A- Pool Heating (Warm Climate) **B**- Pool Heating (Cool Climate) **C**- Water Heating (Warm Climate) **D**- Water Heating (Cool Climate) **E**- Air-conditioning.

Original Certification Date: 27-DEC-07

COLLECTOR SPECIFICATIONS

Gross Area:	3.631 m^2	39.08 sq ft
Dry Weight:	5.5 kg	12. lb
Test Pressure:	240. KPa	35. psg

Net Aperature Area: 3.63 m^2 39.08 sq ft
Fluid Capacity: 13.6 liter 3.6 gal

COLLECTOR MATERIALS

Frame:	None
Cover (Outer):	None
Cover (Inner):	

Pressure Drop

Flow		ΔP	
ml/s	**gpm**	**Pa**	**in H_2O**

Absorber Material:	Tube - □V Sta□il□zed Plastic Polymer / Plate □ None	**Insulation Side:**	None
Absorber Coating:	None	**Insulation Back:**	

TECHNICAL INFORMATION

Efficiency Equation [NOTE: Based on gross area and (P)=Ti-Taj]				**Y INTERCEPT**	**SLOPE**
S I UNITS:	**η= 0.834**	**-15.92210 (P)/I**	**-0.04951 (P)²/I**	**0.838**	**-17.249 W/m².°C**
I P UNITS:	**η= 0.834**	**-2.80468 (P)/I**	**-0.00485 (P)²/I**	**0.838**	**-3.038 Btu/hr.sq ft.°F**

Incident Angle Modifier [(S)=1/cosθ - 1, 0°<θ<=60°]		
Kα = 1	-0.030 (S)	0.028 (S)²
Kα□=□1	0.00 (S	Line□r □it

Model Tested:	922
Test Fluid:	Water
Test Flow Rate:	146.7 □l/s 2.33 gpm

REMARKS: Thermal performance tests were done indoors with a so□□r irradiance simulator.